Bernd Möller · Uwe Reuter

Uncertainty Forecasting in Engineering

Bernd Möller · Uwe Reuter

Uncertainty Forecasting in Engineering

With 101 Figures and 15 Tables

Springer

Universitätsprofessor Dr.-Ing. habil. Bernd Möller

Technische Universität Dresden
Fakultät Bauingenieurwesen
Institut für Statik und Dynamik der Tragwerke

01062 Dresden
Germany

bernd.moeller@tu-dresden.de

Dr.-Ing. Uwe Reuter

Technische Universität Dresden
Fakultät Bauingenieurwesen
Institut für Statik und Dynamik der Tragwerke

01062 Dresden
Germany

uwe.reuter@tu-dresden.de

Library of Congress Control Number: 2007930551

ISBN 978-3-540-37173-1 Springer Berlin Heidelberg New York

Springer is a part of Springer Science+Business Media

springer.com

Typesetting: Digital data supplied by the author
Production: LE-TeX Jelonek, Schmidt & Vöckler GbR, Leipzig
Cover: Frido Steinen-Broo, eStudio Calamar, Spain

Printed on acid-free paper 68/3180/YL - 5 4 3 2 1 0

Preface

Forecasting is fascinating. Who wouldn't like to cast a glimpse into the future? Far removed from metaphysics, mathematical methods such as time-lapse techniques, time series or artificial neural netwoks offer a rational means of achieving this. A precondition for the latter is the availability of a sequence of observed values from the past whose temporal classification permits the deduction of attributes necessary for forecasting purposes.

The subject matter of this book is uncertain forecasting using time series and neural networks based on uncertain observed data. 'Uncertain' data implies information exhibiting inaccuracy, uncertainty and questionability. The uncertainty of individual observations is modeled in this book by fuzziness. Sequences of uncertain observations hence constitute fuzzy time series. By means of new discretization techniques for uncertain data it is now possible to correctly and completely retain data uncertainty in forecasting work. The book presents numerical methods which permit successful forecasting not only in engineering but also in many other fields such as environmental science or economics, assuming of course that a suitable sequence of observed data is available. By taking account of data uncertainty, the indiscriminate reduction of uncertain observations to real numbers is avoided. The larger information content described by uncertainty is retained, and compared with real data, provides a deeper insight into causal relationships. This in turn has practical consequences as far as the fullfilment of technical requirements in engineering applications is concerned.

The book is aimed at engineers as well as professionals working in related fields. For readers with a basic engineering training, a knowledge of classical time series analysis and random processes would be helpful. The book is structured in such a way, however, that the reader will find no difficulty in working through the material without any special prior knowledge.

The book is mainly based on research work funded by the German Research Foundation (DFG), whose financial support the authors gratefully acknowledge. We also express our thanks to Dr Ian Westwood (PhD, Civil Engineering) for translating Sect. 1 and Sects. 3 to 5. Finally, we wish to thank the publishers 'Springer-Verlag' for their receptiveness regarding the subject matter, their valuable support during the development of the manuscript, and the final printing of the book.

Dresden, April 2007 Bernd Möller Uwe Reuter

Contents

Abbreviations

Arithmetics

$a, ..., z$	variables, $a, ..., z \in \mathbb{R}$
$\underline{a}, ..., \underline{z}$	vectors
$\underline{A}, ..., \underline{Z}$	matrices
$\sum$	addition operator of real-valued variables
$\prod$	multiplication operator of real-valued variables
$\oplus$	$l_\alpha r_\alpha$-addition operator of fuzzy variables
ln	natural logarithm

Analysis

$(... ; ...)^T$	elements of a column matrix
$(... ; ...)$	open interval
$[... ; ...]$	closed interval
$\vert ... \vert$	absolute value
lim	limes, limit
∞	infinity
d	differentiation
∂	partial differentiation
$\int$	integration
$\rightarrow$	mapping

Set Theory

$\mathbf{A}$, ..., $\mathbf{Z}$	fundamental sets
A, ..., Z	sets
$\mathbb{N}$	natural numbers
$\mathbb{R}$	real numbers
$\mathbb{R}^n$	n-dimensional Euclidean space
$\{...;...\}$	set of ..., elements of a set
$\in$	element of
$\subseteq$	subset of
$\cap$	intersection
$\cup$	union
$\emptyset$	empty set

Logic

$\wedge$	conjunction, logical *and*
$\vee$	alternative, logical *or*
$\mid$	for which the following holds
$\forall$	universal quantifier, for all
$\exists$	existential quantifier, there exists

Fuzzy Set Theory

$\sim$	fuzziness
$\tilde{a}$, ..., $\tilde{z}$	fuzzy variables
x_l	peak point of $\tilde{x}$
$S_{\tilde{x}}$	support of $\tilde{x}$
$\mu_{\tilde{x}}(x)$	membership function of $\tilde{x}$
α_i	ith α-level
X_{α_i}	α-level set of $\tilde{x}$ for $\alpha = \alpha_i$
n	number of discrete α-levels
$\Delta x_{\alpha_i l}$	l_α-increment of $\tilde{x}$ for $\alpha = \alpha_i$
$\Delta x_{\alpha_i r}$	r_α-increment of $\tilde{x}$ for $\alpha = \alpha_i$
Δx_i	ith element of the $l_\alpha r_\alpha$-increment representation of $\tilde{x}$
$L_{\alpha_{i,i+1}}(\cdot)$	sub-function of $\mu_{\tilde{x}}(x)$ between $\alpha_i \leqslant \alpha < \alpha_{i+1}$ (left)
$R_{\alpha_{i,i+1}}(\cdot)$	sub-function of $\mu_{\tilde{x}}(x)$ between $\alpha_i \leqslant \alpha < \alpha_{i+1}$ (right)
$\hat{\tilde{z}}$	best possible approximation of $\tilde{z}$

$d_F(\tilde{a};\tilde{b})$	distance between two fuzzy variables $\tilde{a}$ and $\tilde{b}$
$d_H\,(A;B)$	Hausdorff distance between two sets A and B
$d_E(a;b)$	Euclidean distance between two real-valued variables a and b
sup	supremum operator
inf	infimum operator
max	maximum operator
min	minimum operator

Fuzzy Probabilistics

A, ..., Z	random variables
$\tilde{\text{A}}$, ..., $\tilde{\text{Z}}$	fuzzy random variables
ω	random elementary event
Ω	space of the random elementary events
$P(\square)$	probability of occurence of $\square$
$\tilde{P}(\square)$	fuzzy probability of occurence of $\square$
$F_{\text{X}}(x)$	probability distribution function of X
$f_{\text{X}}(x)$	probability density function of X
$\tilde{F}_{\tilde{\text{X}}}(x)$	fuzzy probability distribution function form I of $\tilde{\text{X}}$
$F_{\tilde{\text{X}}}(\tilde{x})$	fuzzy probability distribution function form II of $\tilde{\text{X}}$
${}_{lr}F_{\tilde{\text{X}}}(\tilde{x})$	fuzzy probability distribution function form II of $\tilde{\text{X}}$ using $l_\alpha r_\alpha$-discretization
${}_{lr}f_{\tilde{\text{X}}}(\tilde{x})$	fuzzy probability density function form II of $\tilde{\text{X}}$ using $l_\alpha r_\alpha$-discretization

Artificial Neural Networks for Fuzzy Variables

I	input layer
H_i	ith hidden layer
O	output layer
n_I	number of neurons in the input layer
n_{H_i}	number of neurons in the ith hidden layer
n_O	number of neurons in the output layer
$\tilde{o}_i^\square$	fuzzy output value of the ith neuron of the layer $\square$
$\underline{W}_{ij}^\square$	weighting matrix of the jth neuron of the layer $\square$ referring to the ith neuron of the preceding layer
$\tilde{net}_i^\square$	fuzzy netto input of the ith neuron of the layer $\square$
$\tilde{f}_A(\cdot)$	fuzzy activation function

$f_A(\cdot)$	deterministic activation function
$\tilde{f}_O(\cdot)$	fuzzy output function
$f_O(\cdot)$	deterministic output function
$\tilde{\theta}_i^{\square}$	fuzzy bias of the ith neuron of the layer $\square$
$\underline{\tilde{x}}_i$	ith fuzzy input vector
$\underline{\tilde{y}}_i$	ith fuzzy control vector

Analysis of Time Series comprised of Fuzzy Data

τ	time coordinate
$\mathbf{T}$	set of equidistant points in time
$(\tilde{x}_\tau)_{\tau\in\mathbf{T}}$	fuzzy time series
$\tilde{x}_\tau$	fuzzy variable at point in time τ
$\tilde{t}(\tau)$	fuzzy trend function
$\tilde{t}_\tau$	functional value of $\tilde{t}(\tau)$ at point in time τ
$\tilde{z}(\tau)$	fuzzy cycle function
$\tilde{z}_\tau$	functional value of $\tilde{z}(\tau)$ at point in time τ
$(\tilde{\mathrm{R}}_\tau)_{\tau\in\mathbf{T}}$	fuzzy random residual process
$\tilde{r}_\tau$	realization of $(\tilde{\mathrm{R}}_\tau)_{\tau\in\mathbf{T}}$ at point in time τ, fuzzy residual component
$t_j^*(\tau)$	trend function of the jth $l_\alpha r_\alpha$-increment
λ_j	frequency of the cyclic variation of the jth $l_\alpha r_\alpha$-increment
$z_j^*(\tau)$	cycle function of the jth $l_\alpha r_\alpha$-increment
$\tilde{\bar{x}}$	fuzzy mean value of $(\tilde{x}_\tau)_{\tau\in\mathbf{T}}$
${}_{lr}\underline{s}^2_{\tilde{x}_\tau}$	$l_\alpha r_\alpha$-variance of $(\tilde{x}_\tau)_{\tau\in\mathbf{T}}$
${}_{lr}\underline{s}_{\tilde{x}_\tau}$	$l_\alpha r_\alpha$-standard deviation of $(\tilde{x}_\tau)_{\tau\in\mathbf{T}}$
${}_{lr}\underline{\hat{K}}_{\tilde{x}_\tau}(\Delta\tau)$	$l_\alpha r_\alpha$-covariance function of $(\tilde{x}_\tau)_{\tau\in\mathbf{T}}$
${}_{lr}\underline{\hat{R}}_{\tilde{x}_\tau}(\Delta\tau)$	$l_\alpha r_\alpha$-correlation function of $(\tilde{x}_\tau)_{\tau\in\mathbf{T}}$
L	linear filter for fuzzy time series
c_i	ith filter coefficient
D^p	fuzzy difference filter of the order p
L_e	extended linear filter for fuzzy time series
$\underline{C}_i$	ith coefficient matrix
D_e^p	extended fuzzy difference filter of the order p
$(\tilde{\mathrm{X}}_\tau)_{\tau\in\mathbf{T}}$	fuzzy random process
$\tilde{\mathrm{X}}_\tau$	fuzzy random variable at point in time τ
$E[\tilde{\mathrm{X}}_\tau]$, $\tilde{m}_{\tilde{\mathrm{X}}_\tau}$	fuzzy expected value of $\tilde{\mathrm{X}}_\tau$
${}_{lr}Var[\tilde{\mathrm{X}}_\tau]$, ${}_{lr}\underline{\sigma}^2_{\tilde{\mathrm{X}}_\tau}$	$l_\alpha r_\alpha$-variance of $\tilde{\mathrm{X}}_\tau$

${}_{lr}\underline{\sigma}_{\tilde{X}_\tau}$	$l_\alpha r_\alpha$-standard deviation of of $\tilde{X}_\tau$
${}_{lr}\underline{K}_{\tilde{X}_\tau}(\tau_a, \tau_b)$	$l_\alpha r_\alpha$-covariance function of $(\tilde{X}_\tau)_{\tau \in \mathbf{T}}$
${}_{lr}\underline{R}_{\tilde{X}_\tau}(\tau_a, \tau_b)$	$l_\alpha r_\alpha$-correlation function of $(\tilde{X}_\tau)_{\tau \in \mathbf{T}}$
AR	autoregressive
MA	moving average
ARMA	autoregressive moving average
$(\tilde{\mathcal{E}}_\tau)_{\tau \in \mathbf{T}}$	fuzzy white-noise prozess
ε_τ	realization of $(\tilde{\mathcal{E}}_\tau)_{\tau \in \mathbf{T}}$ at point in time τ
p	order of fuzzy AR processes
q	order of fuzzy MA processes
$\underline{A}_i$	ith coefficient matrix of fuzzy AR processes
$\underline{B}_i$	ith coefficient matrix of fuzzy MA processes
$\underline{C}_i$	ith coefficient matrix of inverted fuzzy MA processes
$\underline{\Theta}_i$	ith parameter matrix of the WILSON-algorithm for fuzzy MA processes
$\underline{\Delta}_i$	ith correction matrix of the WILSON-algorithm for fuzzy MA processes

Forecasting of Time Series comprised of Fuzzy Data

N	number of observed fuzzy variables of a fuzzy time series
h	forecasting step size
$(\vec{\tilde{X}}_\tau)_{\tau \in \mathbf{T}}$	fuzzy random forecasting process
$\vec{\tilde{X}}_\tau$	conditional fuzzy random variable at point in time τ
$\vec{\tilde{x}}_{N+h}$	realization of $\vec{\tilde{X}}_\tau$ at point in time $\tau = N + h$
$\mathring{\tilde{x}}_{N+h}$	optimum forecast at point in time $\tau = N + h$
$\tilde{x}^\kappa_{N+h}$	fuzzy forecast interval with confidence level κ
$\tilde{x}^*_{N+h}$	conditional fuzzy forecast interval
$P^*_{X_{N+h}}$	conditional probability for $\mathring{\tilde{X}}_{N+h} \subseteq \tilde{x}^*_{N+h}$

1

Introduction

1.1 Application of Time Series for Forecasting in Engineering

In engineering as well as in other fields such as the natural sciences, environmental science or economics, many processes exist for which ordered sequences of observed values are available. Examples of the latter include the settlement of a bridge measured at specific points in time, traffic loads on roads, snow depths measured over many years, the height of wheat stalks, the diameter of tree trunks or the production output in industry. The observed values, i.e. settlements, traffic loads, snow depths etc., exist for a past observation period. Under certain conditions these constitute a time series.

A time series is a temporally ordered sequence of observed values. Precisely one observed value is assigned to each discrete observation time $\tau \in \mathbf{T}$, where $\mathbf{T}$ represents a set of equidistant points in time. The set of observation time points $\tau = 1, 2, ..., N$ is referred to as the observation period.

In classical time series analysis the observed values are real-valued numbers or natural numbers, i.e. variables to which a precise numerical value is assigned.

Time series comprised of precise observed values are shown in Figs. 1.1, 1.2 and 1.3.

Forecasting of the future progression of time series containing precise observed values and hence forecasting of the process described by the observed values is the subject matter of classical time series analysis [6, 8, 60]. A forecast is possible due to the fact that particular dependencies may be deduced from the significant sequence of the observed values. In order to identify dependencies and laws within a time series, two methods are applied in classical time series analysis: descriptive time series analysis and stochastic models. In descriptive time series analysis descriptive models are used to identify attributes such as trends, seasonal variations or cyclic fluctuations. An important descriptive model is the component model. Stochastic models on the other hand

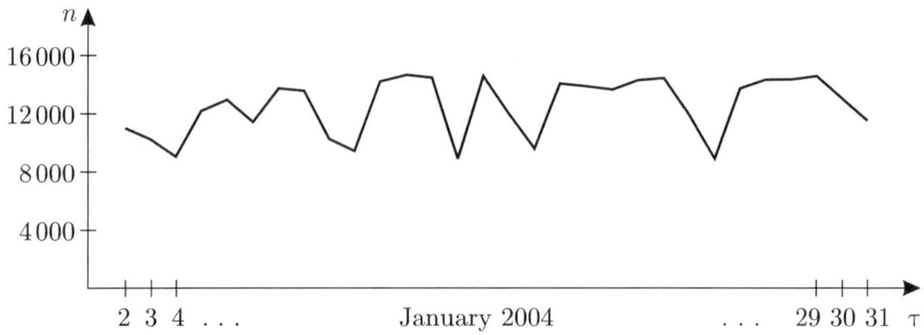

Fig. 1.1. Time series of the number of vehicles crossing the bridge 'Blaues Wunder' in Dresden [Source: Dresden Dept. of Road Construction and Public Works]

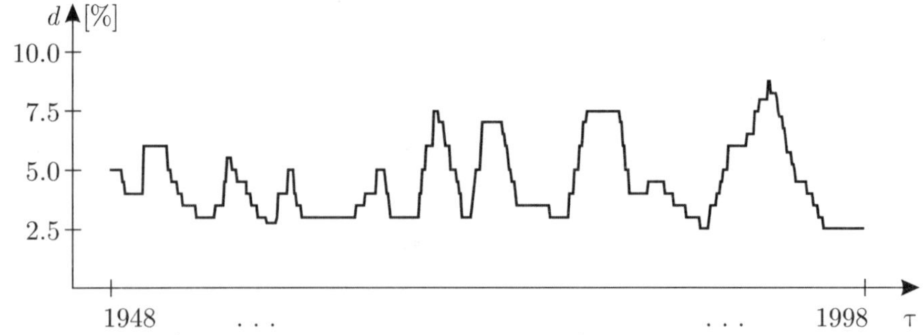

Fig. 1.2. Time series of the discount rate of the 'Deutsche Bundesbank' [Source: 'Deutsche Bundesbank']

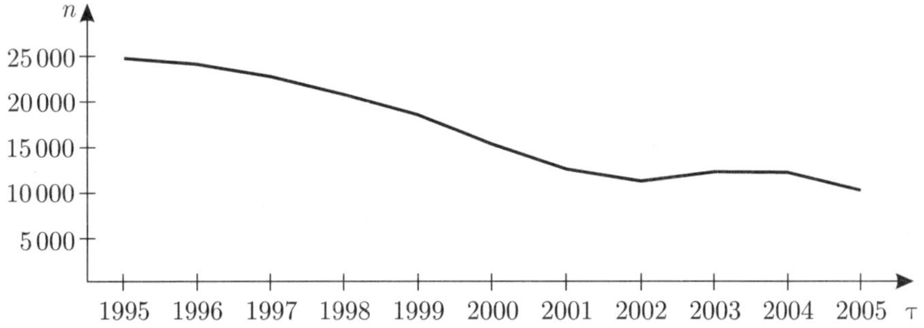

Fig. 1.3. Time series of the number of building approvals in Saxony [Source: Saxony State Office of Statistics]

assume stochastic properties, and treat the time series as the realization of a stochastic process.

If the observed values represent measured values, it is often not possible to assign precise numerical values to the observed data; they then possess data uncertainty. Data uncertainty in engineering practice is mainly due to inaccuracies in measurements, incomplete sets of observations or difficulties in performing measurements, e.g. due to local conditions. The occurrence of data uncertainty also depends on the particular observation scale adopted, i.e. whether a process is described on the microscale, mesoscale or macroscale. For example, although it is theoretically possible to precisely state the commencement of material damage on the microscale, the commencement of damage on the macroscale may be only diagnosed uncertain. Because the observation scale cannot always be chosen arbitrarily, however, the associated uncertainty must be accepted.

Measurement inaccuracy results among other things from the limited precision of a measuring device or from read errors. Geometric data in particular cannot be measured accurately in certain cases. Examples of this include water level measurements on a moving water surface, the thickness of a structural element with a very rough surface or the transport of bed material in a river. The stipulation of some sort of average value, however, means that important information may be lost.

Incomplete sets of observations signalize an information deficit due for example to gaps in a series of measurements resulting from the malfunctioning of measuring devices, irregularities in the reading of measurements or inadequate planning of the measurement regime. The measurement of parameters within a medium or construction is often extremely difficult. The corrosion behavior of steel reinforcement or the position of steel reinforcement in an RC structural element, for example, cannot be measured with absolute certainty. The same applies to crack formation in concrete elements or the quantity of water transported through a flow cross-section.

Sequences of observations may also consist of *linguistic estimates.* Examples of this include a description of concrete flaking on bridges (see Fig. 1.4), a description of the degree of discoloration of a surface or the extent of cloud cover. Linguistic estimates are a priori imprecise, as they express the subjective opinion of an expert. On the other hand, time series of linguistic observations do in fact open up new fields of application in forecasting.

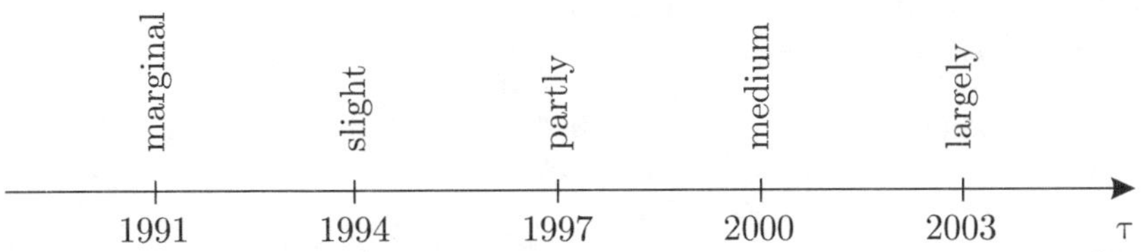

Fig. 1.4. Time series comprised of linguistic estimates of concrete flaking

Fig. 1.5 shows a comparison between a time series comprised of precise observed values and a time series consisting of uncertain observed values. The uncertainty in this case is described by an interval. This is a very simple uncertainty model. In this book the more informative uncertainty models fuzziness and fuzzy randomness are used to describe imprecise data, i.e. uncertain data. An overview of these uncertainty models is given in the following section (Sect. 1.2). New forms of representation of these uncertainty models suitable for time series analysis are derived in Sect. 2.

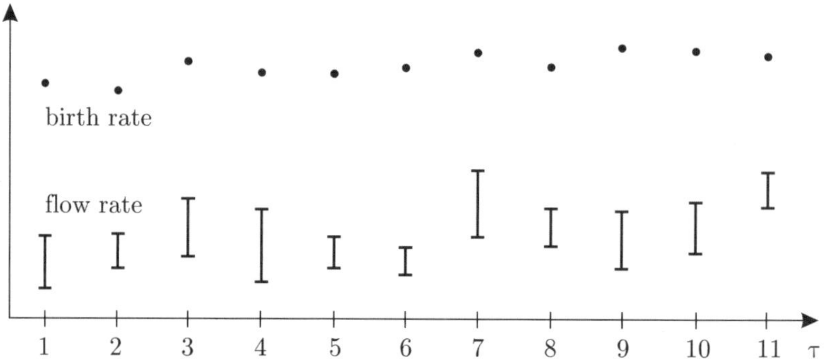

Fig. 1.5. Time series containing precise data versus time series containing uncertain data

By means of the introduced uncertainty models it is possible extend the methods of classical time series analysis in such a way as to permit the forecasting of future uncertain results under due consideration of data uncertainty. By this means it is possible to dispense with the artificial idealization of real data, the forecasting of which may lead to unrealistic results. The decision as to whether the methods of classical time series analysis or the extended methods presented in this book should be applied depends on the particular problem in question and the existing data base.

The subject matter of this book concerns time series comprised of imprecise, i.e. uncertain, observed values. This implies that an individual observed value may be uncertain. By this means it is possible to realistically model the observed values in important practical cases. Because the forecasted values are also uncertain, forecasts are obtained with higher predictive capability.

Three methods are described in the book for forecasting time series comprised of uncertain observed values:

- the fuzzy component model (Sect. 3.2) as an extension of descriptive methods,
- the fuzzy random process as an extension of stochastic models (Sect. 4.2) and

- artificial neural networks for uncertain data as an extension of artificial neural networks for real-valued data (Sect. 4.3)

1.2 Data Uncertainty and Fuzzy Time Series

If it is not possible to assign a precise numerical value to an observed value, the observed value in question possesses uncertainty. How can this uncertainty be described mathematically?

A variety of methods exist for classifying and distinguishing uncertainty. Decisive in this respect are the causes of uncertainty. If the cause is purely random, the uncertainty is referred to as aleatoric uncertainty. This is described with the aid of conventional and highly-developed stochastic models. If the uncertainty is a result of objective and subjective factors, it is then referred as epistemic uncertainty. Models for describing epistemic uncertainty include, among others, fuzziness and intervals. If it is necessary to take account of both aleatoric and epistemic effects, uncertainty is accounted for by the model fuzzy randomness.

In the case of time series the uncertainty of the individual observed values as well as the interpretation of a sequence of uncertain observed values are of interest.

The uncertainty of a single observed value is always epistemic. The uncertain observed value is thus modeled as a fuzzy variable, as illustrated in Fig. 1.6. The major causes of this type of uncertainty have already been outlined in Sect. 1.1.

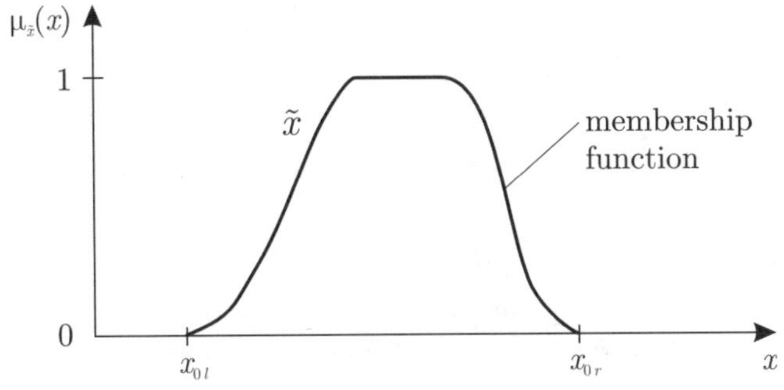

Fig. 1.6. Fuzzy variable $\tilde{x}$

The fuzzy variable $\tilde{x}$ may take on values in the interval $[x_{0\,l};\, x_{0\,r}]$. The values are assessed between zero and unity by means of a membership function $\mu_{\tilde{x}}(x)$. This assessment reflects the subjective and objective causes of the existing uncertainty, and may be used to describe the uncertain observed value. Fuzzy variables contain intervals and real numbers as special cases, as illustrated in Fig.1.7.

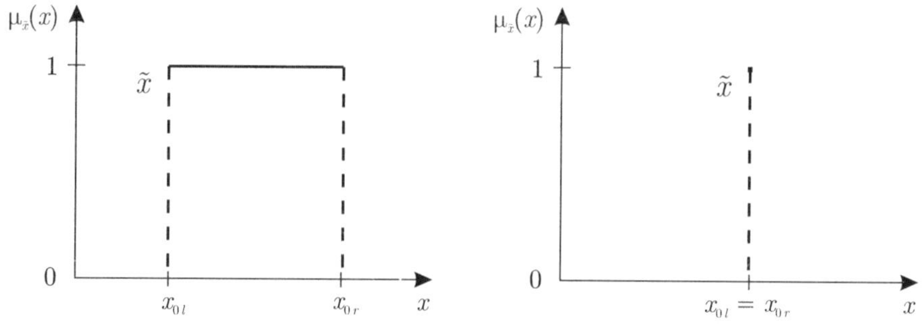

Fig. 1.7. Interval and real-valued number as special cases of a fuzzy variable $\tilde{x}$

Modeling of the individual observed values as fuzzy variables results in so-called fuzzy time series, as shown by way of example in Fig. 1.8. Starting from the uncertain observed values, the aim is to forecast future uncertain values. For this purpose the dependencies existing in the sequence of uncertain observed values are analyzed and modeled.

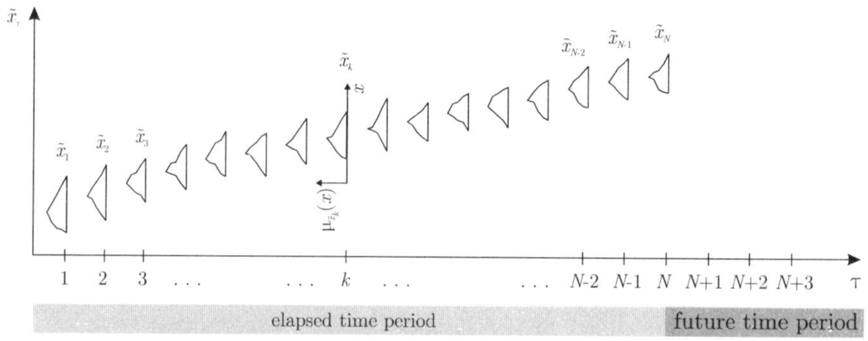

Fig. 1.8. Time series containing fuzzy variables

Modeling as a fuzzy random process. Forecasts are possible if it may be assumed that the fuzzy time series may be modeled with the aid of a fuzzy random process. A fuzzy random process is defined as a family of fuzzy random variables $\tilde{X}_\tau$.

Fuzzy random variables, as introduced in Sect. 2.2, belong to the uncertainty model fuzzy randomness. A time series of fuzzy data may be viewed as a random realization of a fuzzy random process. The realizations of this process are uncertain and thus referred to as fuzzy variables.

Only one sequence of uncertain observed values is available for determining the underlying fuzzy random process. Methods for specifying the fuzzy random process in any given case are developed in Sect. 3.5. A knowledge of this process is a precondition for the forecast. The required forecasting methods are formulated in Sect. 4.2. By means of a new incremental discretization of the fuzzy variables and fuzzy random variables the uncertainty is fully retained in the forecast. The uncertainty is also not artificially increased. This incremental representation is absolutely necessary for a *direct* description as well as for modeling and forecasting purposes. 'direct' implies that the sequence of the fuzzy variables is retained during the description, modeling and forecasting phases. No form of defuzzification or refuzzification is undertaken.

Modeling using Artificial Neural Networks. As an alternative to fuzzy random processes, methods for modeling and forecasting fuzzy time series using Artificial Neural Networks are developed in Sect. 3.6 and Sect. 4.3, respectively. The conventional multilayer perceptrons associated with the latter are extended in such a way that they may be applied to time series for fuzzy variables. A precondition for this extension is again the new incremental discretization mentioned in the foregoing. An Artificial Neural Network is first trained in an optimization process. Training is carried out on the basis of the particular fuzzy time series concerned. Different forecasting strategies are developed for forecasting purposes.

The use of Artificial Neural Networks does not require an explicit determination of the underlying fuzzy random process.

1.3 Examples of Fuzzy Time Series

The practical relevance of fuzzy time series is demonstrated by the following two examples. Further examples are given in Sect. 5.

The total global ozone change between 1965 and 2000 is shown in Fig. 1.9. The time series from 1980 onwards reflects the uncertainty of the measured data. The reason for this uncertainty is due to inaccuracies in measurements. The uncertainty is hence epistemic in nature and may be modeled by fuzziness.

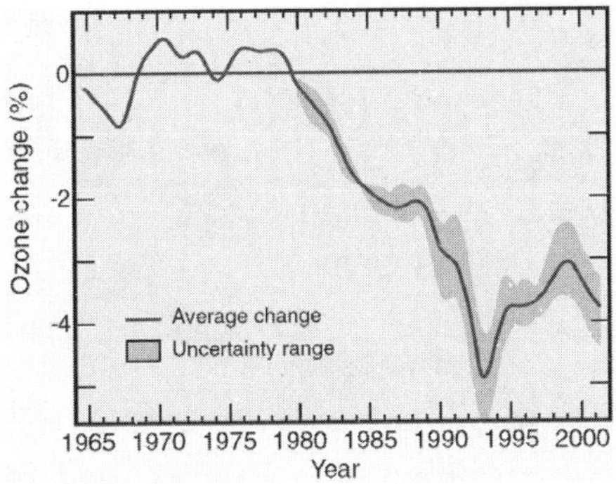

Fig. 1.9. Uncertain time series of total global ozone change [69]

The second example (Fig. 1.10) concerns measurements of the earth pressure acting on a wall. Several closely arranged pressure transducers are installed on the wall. The measured values differ from one pressure transducer to the next. The different observed values signalize uncertainty. Instead of computing an average value, this uncertainty is taken into consideration. Fuzzy variables are constructed for the measured values at each point in time. Fig. 1.10 shows a cut-out segment of the obtained fuzzy time series. The complete time series begins with measurements made in 1999.

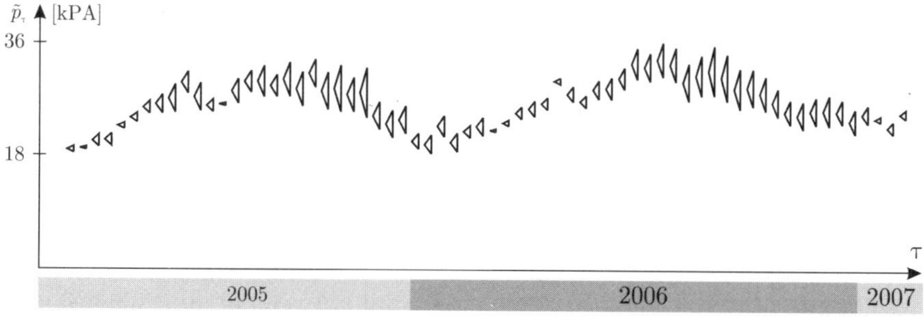

Fig. 1.10. Uncertain time series of earth pressure measurements (cut-out segment) [14]

Fuzzy time series analogous to those presented in the above examples are frequently encountered in engineering and environmental science. These share the common feature of measurable observed values. Forecasting of the latter is possible using the forecasting strategies developed in Sect. 4. If the forecasts of measurable observed values are combined with a computational model, it is also possible to forecast non-measurable observed values such as the damage state of a structure. Model-based forecasting strategies for this purpose are developed and demonstrated by way of examples in Sect. 5.

2

Mathematical Description of Uncertain Data

In this chapter fuzzy variables and fuzzy random variables for the mathematical description of uncertain data are introduced and new forms of their numerical representation especially suitable for uncertainty forecasting are developed. The mathematical description of uncertain data is limited to these basic concepts, which are essential for forecasting by means of fuzzy random processes and artificial neural networks Fuzzy set theory forms the mathematical basis for fuzzy variables. Several introductory books exist which deal with fuzzy set theory, see e.g. [5, 11, 36, 71]. Fuzzy random variables couple the uncertainty models of fuzziness and randomness. The underlying principles of this theory are presented in [5, 27, 28, 36, 44, 53, 66]. The numerical representation is based on the new $l_\alpha r_\alpha$-discretization and is a prerequisite for the exact numerical reproduction of the uncertain values of a time series.

2.1 Fuzzy Variables

Definition 2.1. *A fuzzy variable $\tilde{x}$ is defined as an uncertain subset of the fundamental set* $\mathbf{X}$.

$$\tilde{x} = \{x,\, \mu_{\tilde{x}}(x) \,|\, x \in \mathbf{X}\} \tag{2.1}$$

◆

The uncertainty is assessed by the membership function $\mu_{\tilde{x}}(x)$. A fuzzy variable $\tilde{x}$ and its membership function $\mu_{\tilde{x}}(x)$ are shown in Fig. 2.1.

Definition 2.2. *A normalized membership function $\mu_{\tilde{x}}(x)$ is defined as follows:*

$$0 \leqslant \mu_{\tilde{x}}(x) \leqslant 1 \;\; \forall\, x \in \mathbb{R} \tag{2.2}$$

$$\exists\; x_l,\; x_r \;\; \text{with} \;\; \mu_{\tilde{x}}(x) = 1 \;\; \forall\; x \in [x_l; x_r]\,. \tag{2.3}$$

◆

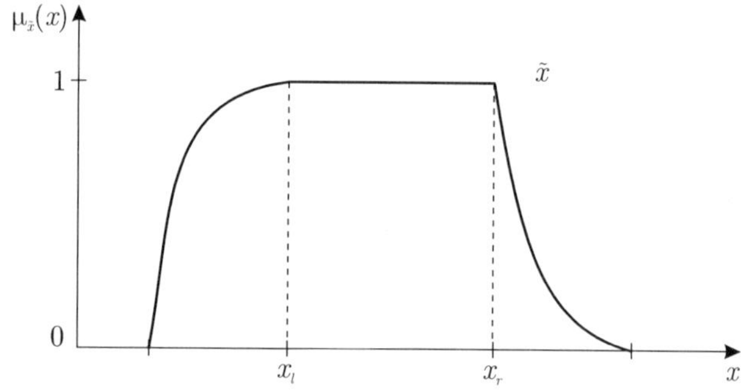

Fig. 2.1. Fuzzy variable $\tilde{x}$ and its membership function $\mu_{\tilde{x}}(x)$

A fuzzy variable $\tilde{x}$ is referred to as convex if its membership function $\mu_{\tilde{x}}(x)$ monotonically decreases on each side of the maximum value, i.e. if

$$\mu_{\tilde{x}}(x_2) \geqslant \min\left[\mu_{\tilde{x}}(x_1);\ \mu_{\tilde{x}}(x_3)\right] \quad \forall\, x_1, x_2, x_3 \in \mathbb{R} \ \text{ with } \ x_1 \leqslant x_2 \leqslant x_3 \quad (2.4)$$

applies.

The membership function may be continuous or discrete. Piecewise continuous membership functions are dealt with in the following.

Definition 2.3. *A convex fuzzy variable $\tilde{x}$ is referred to as a fuzzy number if its membership function $\mu_{\tilde{x}}(x)$ is piecewise continuous and if it has the functional value $\mu_{\tilde{x}}(x) = 1$ at precisely one of the x values with $x = x_l = x_r$ according to Eq. (2.5).*

$$\begin{aligned} x_l &= \min\left[x \in \mathbb{R} \mid \mu_{\tilde{x}}(x) = 1\right] \\ \text{and} \quad x_r &= \max\left[x \in \mathbb{R} \mid \mu_{\tilde{x}}(x) = 1\right] \end{aligned} \quad (2.5)$$

In the case $x_l < x_r$ the fuzzy variable $\tilde{x}$ constitutes a fuzzy interval. The point x_l is referred to as the peak point of the fuzzy variable. ◆

In order to describe a fuzzy number $\tilde{x}_Z$ or a fuzzy interval $\tilde{x}_I$ the so-called LR fuzzy number or the LR fuzzy interval may be used (see amongst others [5, 11, 36, 71]). The membership function of an LR fuzzy number or LR fuzzy interval is described by an $L(\cdot)$-function and an $R(\cdot)$-function. If the functions $L(\cdot)$ and $R(\cdot)$ fulfill the following four conditions:

$$L(\cdot) \text{ and } R(\cdot) \text{ are piecewise continuous from the left} \quad (2.6)$$

$$L(\cdot) \text{ and } R(\cdot) \text{ are monotonically non-increasing} \quad (2.7)$$

$$L(0) = R(0) = 1 \quad (2.8)$$

$$L(y) = 0 \text{ and } R(z) = 0, \quad \forall\, y > 1 \text{ and } \forall\, z > 1\,. \quad (2.9)$$

the membership function $\mu_{\tilde{x}}(x)$ of a fuzzy variable $\tilde{x}$ may be formulated according to Eq. (2.10).

$$\mu_{\tilde{x}}(x) = \begin{cases} L\left(\frac{x_l - x}{a}\right) & \text{if } x < x_l \\ 1 & \text{if } x_l \leqslant x \leqslant x_l + s \\ R\left(\frac{x - x_l - s}{b}\right) & \text{if } x > x_l + s \end{cases} \tag{2.10}$$

The variable x_l constitutes the peak point. The variables a, b and $s = x_r - x_l$ are parameters of the membership function as shown in Fig. 2.2.

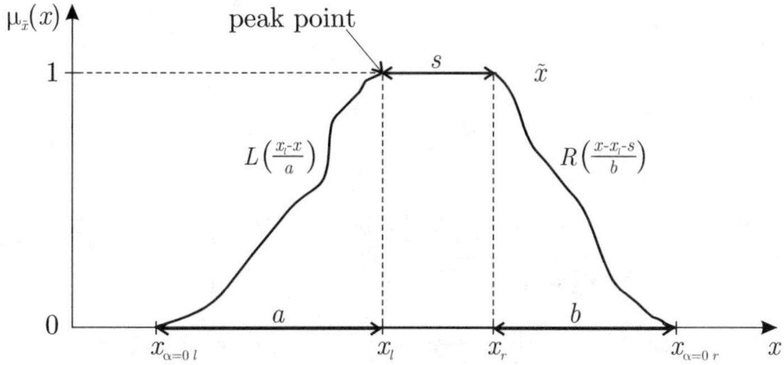

Fig. 2.2. LR-representation of a fuzzy variable $\tilde{x}$

In the special case $a = 0$ or $b = 0$ the left or right boundary functions $L(\cdot)$ and $R(\cdot)$ are defined by Eq. (2.11) and Eq. (2.12), respectively.

$$L\left(\frac{x_l - x}{a}\right) = 0 \;\text{ if } x < x_l \tag{2.11}$$

$$R\left(\frac{x - x_l - s}{b}\right) = 0 \;\text{ if } x > x_l + s \tag{2.12}$$

Under the implicit convention that the boundary functions $L(\cdot)$ and $R(\cdot)$ are known, the common abridged notation $\tilde{x}_I = (x_{\alpha=0\,l}; x_l; x_r; x_{\alpha=0\,r})_{LR}$ is used for describing a fuzzy interval $\tilde{x}_I$. In order to describe a fuzzy number $\tilde{x}_Z$ with $x_r = x_l$ the abridged notation $\tilde{x}_Z = (x_{\alpha=0\,l}; x_l; x_{\alpha=0\,r})_{LR}$ is used.

2.1.1 Classical and Incremental Discretization of Fuzzy Variables

α-Discretization of Fuzzy Variables

For all $\alpha \in (0, 1]$ closed finite intervals $[x_{\alpha l}; x_{\alpha r}]$ may be extracted from a convex fuzzy variable $\tilde{x}$. The boundaries $x_{\alpha l}$ and $x_{\alpha r}$ of the intervals are given by Eqs. (2.13) and (2.14), respectively.

$$x_{\alpha l} = \min[x \in \mathbb{R} \mid \mu_{\tilde{x}}(x) \geqslant \alpha] \tag{2.13}$$

$$x_{\alpha r} = \max[x \in \mathbb{R} \,|\, \mu_{\tilde{x}}(x) \geqslant \alpha] \qquad (2.14)$$

These intervals are referred to as α-level sets X_α. The set

$$S_{\tilde{x}} = \{x \in \mathbb{R} \,|\, \mu_{\tilde{x}}(x) > 0\} \qquad (2.15)$$

of a fuzzy variable $\tilde{x}$ is referred to as the support. The support $S_{\tilde{x}}$ of a fuzzy variable $\tilde{x}$ is referred to as an α-level set X_α with $\alpha = 0$ notwithstanding Eqs. (2.13) and (2.14). The interval boundaries $x_{\alpha l}$ and $x_{\alpha r}$ of the α-level set $X_{\alpha=0}$ are then given by Eqs. (2.16) and (2.17).

$$x_{\alpha l} = \lim_{\alpha' \to +0} \left[\min\left[x \in \mathbb{R} \,|\, \mu_{\tilde{x}}(x) > \alpha'\right]\right] \quad \text{for} \quad \alpha = 0 \qquad (2.16)$$

$$x_{\alpha r} = \lim_{\alpha' \to +0} \left[\max\left[x \in \mathbb{R} \,|\, \mu_{\tilde{x}}(x) > \alpha'\right]\right] \quad \text{for} \quad \alpha = 0 \qquad (2.17)$$

For different α-level sets of the same fuzzy variable $\tilde{x}$ the following holds:

$$X_{\alpha_k} \subseteq X_{\alpha_i} \quad \forall\, \alpha_i, \alpha_k \in [0;1] \quad \text{with} \quad \alpha_i \leqslant \alpha_k \,. \qquad (2.18)$$

Thus a convex fuzzy variable $\tilde{x}$ may be characterized by a family of α-level sets X_α according to Eq. (2.19).

$$\tilde{x} = (X_\alpha = [x_{\alpha l}, x_{\alpha r}] \,|\, \alpha \in [0,1]) \qquad (2.19)$$

If the number of α-level sets is denoted by n, then for $i = 1, 2, ..., n-1$ the following holds provided that $n \geqslant 2$:

$$0 \leqslant \alpha_i \leqslant \alpha_{i+1} \leqslant 1 \qquad (2.20)$$

$$\alpha_1 = 0 \quad \text{and} \quad \alpha_n = 1 \qquad (2.21)$$

$$X_{\alpha_{i+1}} \subseteq X_{\alpha_i} \,. \qquad (2.22)$$

Example 2.4. A convex fuzzy variable $\tilde{x}$ characterized by $n = 4$ α-level sets X_α is shown in Fig. 2.3. ◆

$l_\alpha r_\alpha$-Discretization of Fuzzy Variables

The α-level sets X_{α_i} are now considered separately for $i = 1, 2, ..., n-1$ and $i = n$.

For $i = 1, 2, ..., n-1$ each α-level set X_{α_i} of a convex fuzzy variable $\tilde{x}$ may be specified as the union of the α-level set $X_{\alpha_{i+1}}$ and the set $X^*_{\alpha_i}$ according to Eq. (2.23).

$$X_{\alpha_i} = X_{\alpha_{i+1}} \cup X^*_{\alpha_i} \quad \forall\, \alpha_i, \alpha_{i+1} \in [0;1] \quad \text{with} \quad \alpha_i \leqslant \alpha_{i+1} \qquad (2.23)$$

The set $X^*_{\alpha_i}$ is defined by two closed finite intervals $[x_{\alpha_i ll}; x_{\alpha_i lr}]$ and $[x_{\alpha_i rl}; x_{\alpha_i rr}]$. The interval boundaries of the set $X^*_{\alpha_i}$ are given by Eqs. (2.24) and (2.25).

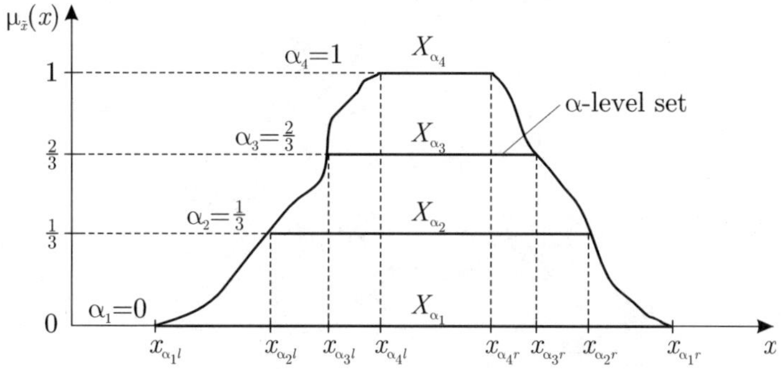

Fig. 2.3. α-discretization of a fuzzy variable $\tilde{x}$

$$x_{\alpha_i ll} = x_{\alpha_i l} \qquad \text{and} \qquad x_{\alpha_i lr} = x_{\alpha_{i+1} l} \tag{2.24}$$

$$x_{\alpha_i rl} = x_{\alpha_{i+1} r} \qquad \text{and} \qquad x_{\alpha_i rr} = x_{\alpha_i r} \tag{2.25}$$

With the aid of these definitions the interval boundaries of the α-level set X_{α_i} may be expressed by the following equations.

$$x_{\alpha_i l} = x_{\alpha_{i+1} l} - \Delta x_{\alpha_i l} \quad \text{with} \quad \Delta x_{\alpha_i l} = x_{\alpha_i lr} - x_{\alpha_i ll} \tag{2.26}$$

$$x_{\alpha_i r} = x_{\alpha_{i+1} r} + \Delta x_{\alpha_i r} \quad \text{with} \quad \Delta x_{\alpha_i r} = x_{\alpha_i rr} - x_{\alpha_i rl} \tag{2.27}$$

The terms $\Delta x_{\alpha_i l}$ and $\Delta x_{\alpha_i r}$ are referred to as $\boldsymbol{l_\alpha r_\alpha}$**-increments** and permit the $l_\alpha r_\alpha$-discretization of a fuzzy variable.

For $i = n$ the following equations hold, whereby the term $\Delta x_{\alpha_n l}$ is assigned to the peak point x_l.

$$x_{\alpha_n l} = \Delta x_{\alpha_n l} \quad \text{with} \quad \Delta x_{\alpha_n l} = x_l \tag{2.28}$$

$$x_{\alpha_n r} = x_{\alpha_n l} + \Delta x_{\alpha_n r} \quad \text{with} \quad \Delta x_{\alpha_n r} = x_r - x_l \tag{2.29}$$

The α-level sets must fulfill Eq. (2.18). For this reason the $l_\alpha r_\alpha$-increments of a convex fuzzy variable $\tilde{x}$ must be non-negative according to Eqs. (2.30) and (2.31). The requirement of non-negativity must not be fulfilled at the peak point.

$$\Delta x_{\alpha_i l} \geqslant 0 \text{ for } i = 1, 2, ..., n-1 \tag{2.30}$$

$$\Delta x_{\alpha_i r} \geqslant 0 \text{ for } i = 1, 2, ..., n \tag{2.31}$$

By way of Eqs. (2.26) to (2.31) the $l_\alpha r_\alpha$-discretization is introduced. Fig. 2.4 illustrates the $l_\alpha r_\alpha$-discretization of a fuzzy variable $\tilde{x}$ for $n = 4$.

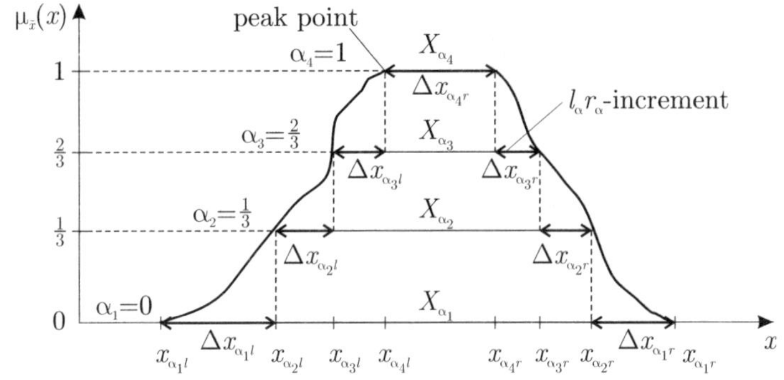

Fig. 2.4. $l_\alpha r_\alpha$-discretization of a fuzzy variable $\tilde{x}$

The $l_\alpha r_\alpha$-discretization permits an alternative, discrete representation of a fuzzy variable $\tilde{x}$ in the form of a column matrix according to Eq. (2.32), whereby the terms Δx_1, Δx_2, ..., Δx_{2n} are abridged notations of the $l_\alpha r_\alpha$-increments $\Delta x_{\alpha_1 l}$, $\Delta x_{\alpha_2 l}$, ..., $\Delta x_{\alpha_1 r}$.

$$\tilde{x} = \begin{bmatrix} \Delta x_{\alpha_1 l} \\ \Delta x_{\alpha_2 l} \\ \vdots \\ \Delta x_{\alpha_n l} \\ \Delta x_{\alpha_n r} \\ \vdots \\ \Delta x_{\alpha_2 r} \\ \Delta x_{\alpha_1 r} \end{bmatrix} = \begin{bmatrix} \Delta x_1 \\ \Delta x_2 \\ \vdots \\ \Delta x_n \\ \Delta x_{n+1} \\ \vdots \\ \Delta x_{2n-1} \\ \Delta x_{2n} \end{bmatrix} \tag{2.32}$$

Remark 2.5. If it is necessary to add a subscript to a fuzzy variable $\tilde{x}$, e.g. $\tilde{x}_j$, the following notation is adopted.

$$\tilde{x}_j = \begin{bmatrix} \Delta x_{\alpha_1 l}(j) \\ \vdots \\ \Delta x_{\alpha_1 r}(j) \end{bmatrix} = \begin{bmatrix} \Delta x_1(j) \\ \vdots \\ \Delta x_{2n}(j) \end{bmatrix} \tag{2.33}$$

◆

With the introduction of the $l_\alpha r_\alpha$-increments $\Delta x_{\alpha_i l}$ and $\Delta x_{\alpha_i r}$ the enhancement of the classical LR-representation of a fuzzy variable $\tilde{x}$ follows. Provided that Eqs. (2.20) to (2.22) hold, the membership function $\mu_{\tilde{x}}(x)$ of a convex fuzzy variable $\tilde{x}$ may then be expressed by Eq. (2.34).

$$\mu_{\tilde{x}}(x) = \begin{cases} L_{\alpha_{i,i+1}}\left(\frac{x_{\alpha_{i+1}l}-x}{\Delta x_{\alpha_i l}}\right) \cdot (\alpha_{i+1} - \alpha_i) + \alpha_i & \text{if } x_{\alpha_i l} \leqslant x < x_{\alpha_{i+1} l} \\ 1 & \text{if } x_l \leqslant x \leqslant x_l + s \\ R_{\alpha_{i,i+1}}\left(\frac{x-x_{\alpha_{i+1}r}}{\Delta x_{\alpha_i r}}\right) \cdot (\alpha_{i+1} - \alpha_i) + \alpha_i & \text{if } x_{\alpha_{i+1} r} < x \leqslant x_{\alpha_i r} \\ 0 & \text{else} \end{cases} \tag{2.34}$$

The gradient of the membership function between the α-levels α_i and α_{i+1} $(i = 1, 2, ..., n-1)$ is determined by the left and right boundary functions $L_{\alpha_{i,i+1}}(\cdot)$ and $R_{\alpha_{i,i+1}}(\cdot)$. The boundary functions $L_{\alpha_{i,i+1}}(\cdot)$ and $R_{\alpha_{i,i+1}}(\cdot)$ must likewise comply with the conditions given by Eqs. (2.6) to (2.9).

Example 2.6. The $l_\alpha r_\alpha$-discretization of a fuzzy variable $\tilde{x}$ and the enhanced LR-representation according to Eq. (2.34) are demonstrated by way of the fuzzy number shown in Fig. 2.5.

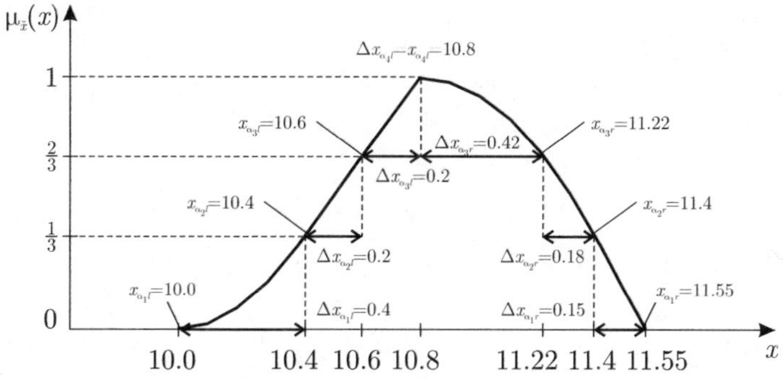

Fig. 2.5. Exemplary fuzzy number $\tilde{x}$

The corresponding left and right boundary functions $L_{\alpha_{i,i+1}}(\cdot)$ and $R_{\alpha_{i,i+1}}(\cdot)$ are given in Table 2.1.

Table 2.1. Left and right boundary functions $L_{\alpha_{i,i+1}}(\cdot)$ and $R_{\alpha_{i,i+1}}(\cdot)$

$L_{\alpha_{3,4}}(y) = 1 - y$	$R_{\alpha_{3,4}}(z) = 1 - z^2$
$L_{\alpha_{2,3}}(y) = 1 - y$	$R_{\alpha_{2,3}}(z) = \frac{4}{3} - \frac{1}{3}(z+1)^2$
$L_{\alpha_{1,2}}(y) = (y-1)^2$	$R_{\alpha_{1,2}}(z) = 1 - z$
with $y = \frac{x_{\alpha_{i+1}l}-x}{\Delta x_{\alpha_i l}}$	with $z = \frac{x-x_{\alpha_{i+1}r}}{\Delta x_{\alpha_i r}}$

The membership function $\mu_{\tilde{x}}(x)$ according to Eq. (2.34) is obtained by insertion of the $l_\alpha r_\alpha$-increments $\Delta x_{\alpha_i l}$ and $\Delta x_{\alpha_i r}$ and the interval boundaries $x_{\alpha_i l}$ and $x_{\alpha_i r}$ given in Fig. 2.5. ◆

The $l_\alpha r_\alpha$-discretization of a fuzzy variable $\tilde{x}$ presented in this section forms the basis for the numerical methods developed for analyzing and forecasting of fuzzy time series.

2.1.2 Incremental Fuzzy Arithmetic

In the context of the analysis and forecasting of fuzzy time series a new fuzzy arithmetic based on the $l_\alpha r_\alpha$-increments is required. For this purpose it is presupposed that all fuzzy variables are given in $l_\alpha r_\alpha$-increment representation according Eq. (2.32). The following operators are introduced.

Definition 2.7. *The $l_\alpha r_\alpha$-multiplication of a real-valued $[2n, 2n]$ matrix $\underline{A}$ by a fuzzy variable $\tilde{x}$ represented by n α-level sets and $2n$ $l_\alpha r_\alpha$-increments is defined by the operator $\odot$ according to Eq. (2.35).*

$$\underline{A} \odot \tilde{x} = \tilde{z} \tag{2.35}$$

The arithmetic operation constitutes the matrix product according to Eq. (2.36) and results in the $l_\alpha r_\alpha$-increments Δz_j $(j = 1, 2, ..., 2n)$ of the fuzzy result variable $\tilde{z}$.

$$\begin{bmatrix} a_{1,1} & a_{1,2} & \cdots & a_{1,2n-1} & a_{1,2n} \\ a_{2,1} & a_{2,2} & \cdots & a_{2,2n-1} & a_{2,2n} \\ \vdots & \vdots & \ddots & \vdots & \vdots \\ a_{2n-1,1} & a_{2n-1,2} & \cdots & a_{2n-1,2n-1} & a_{2n-1,2n} \\ a_{2n,1} & a_{2n,2} & \cdots & a_{2n,2n-1} & a_{2n,2n} \end{bmatrix} \begin{bmatrix} \Delta x_1 \\ \Delta x_2 \\ \vdots \\ \Delta x_{2n-1} \\ \Delta x_{2n} \end{bmatrix} = \begin{bmatrix} \Delta z_1 \\ \Delta z_2 \\ \vdots \\ \Delta z_{2n-1} \\ \Delta z_{2n} \end{bmatrix} \tag{2.36}$$

◆

Remark 2.8. Real-valued $[2n, 2n]$ matrices are processed e.g. by the parameter specification of fuzzy ARMA processes, see Sect. 3.5.5. ◆

The fact that the fuzzy result variable $\tilde{z}$ must comply with Eq. (2.18) means that Eq. (2.37) must be satisfied for $j = 1, 2, ..., n-1, n+1, ..., 2n$.

$$\Delta z_j = a_{j,1} \Delta x_1 + ... + a_{j,2n} \Delta x_{2n} \geqslant 0 \tag{2.37}$$

The requirement of non-negativity must not be fulfilled for $j = n$, i.e. at the peak point $\Delta z_n = \Delta z_{\alpha_n l} = z_{\alpha_n l}$.

Remark 2.9. If Eq. (2.37) is not fulfilled, the fuzzy result variable $\tilde{z}$ is understood to be a fuzzy variable in the improper sense. Fuzzy variables in the improper sense are only permitted as intermediate results of arithmetic operations. ◆

In the special case that the matrix $\underline{A}$ is a diagonal matrix with identical elements the $l_\alpha r_\alpha$-multiplication by a fuzzy variable $\tilde{x}$ is defined in a simplified form by Eq. (2.38).

$$\underline{A} \odot \tilde{x} = \begin{bmatrix} a & \cdots & 0 \\ \vdots & \ddots & \vdots \\ 0 & \cdots & a \end{bmatrix} \odot \tilde{x} = a\tilde{x} \tag{2.38}$$

Furthermore, special fuzzy addition and subtraction operators are required.

Definition 2.10. *The symbols $\oplus$ and $\ominus$ represent the $l_\alpha r_\alpha$-addition and $l_\alpha r_\alpha$-subtraction of the fuzzy variables $\tilde{x}$ and $\tilde{y}$ according to Eq. (2.39).*

$$\tilde{z} = \tilde{x} \oplus \tilde{y} \quad \text{and} \quad \tilde{z} = \tilde{x} \ominus \tilde{y} \tag{2.39}$$

The fuzzy result variable $\tilde{z}$ must also comply with Eq. (2.18). The corresponding conditions are given by Eq. (2.40), whereby the upper operators are applied for the $l_\alpha r_\alpha$-addition and the lower for the $l_\alpha r_\alpha$-subtraction. The requirement of non-negativity must likewise be fulfilled except at the peak point.

$$\Delta z_i = \Delta x_i \pm \Delta y_i \geqslant 0 \quad \text{for} \quad i = 1,\, 2,\, ...,\, n-1,\, n+1,\, ...,\, 2n \tag{2.40}$$

$$\Delta z_i = \Delta x_i \pm \Delta y_i \qquad \text{for} \quad i = n \tag{2.41}$$

◆

Thus the interval boundaries $[z_{\alpha_i l}; z_{\alpha_i r}]$ of the fuzzy variable $\tilde{z}$ are obtained for each α-level α_i successively according to Eqs. (2.42) and (2.43),

$$z_{\alpha_i l} = z_{\alpha_{i+1} l} - \Delta x_{\alpha_i l} \mp \Delta y_{\alpha_i l} \tag{2.42}$$

with $\quad z_{\alpha_n l} = x_{\alpha_n l} \pm y_{\alpha_n l} \quad$ and

$$z_{\alpha_i r} = z_{\alpha_{i+1} r} + \Delta x_{\alpha_i r} \pm \Delta y_{\alpha_i r} \tag{2.43}$$

with $\quad z_{\alpha_n r} = z_{\alpha_n l} + \Delta x_{\alpha_n r} \pm \Delta y_{\alpha_n r}$.

Remark 2.11. If Eqs. (2.40) and (2.41) are not fulfilled, the fuzzy result variable $\tilde{z}$ is understood to be a fuzzy variable in the improper sense (compare remark 2.9), which is only permitted as the intermediate results of arithmetic operations. ◆

Applying the $l_\alpha r_\alpha$-addition and $l_\alpha r_\alpha$-subtraction given by Eq. (2.39), the assigned LR-representation of the fuzzy result variable $\tilde{z}$ depends on the functions $L_{\alpha_{i,i+1}}(\cdot)$ and $R_{\alpha_{i,i+1}}(\cdot)$ of the summed fuzzy variables $\tilde{x}$ and $\tilde{y}$. If the types of the functions $L_{\alpha_{i,i+1}}(\cdot)$ and $R_{\alpha_{i,i+1}}(\cdot)$ of the fuzzy variables $\tilde{x}$ and $\tilde{y}$

are identical for each α-level α_i, the same functions are obtained for the enhanced LR-representation of the fuzzy result variable. In the case of different types the functions $L_{z,\alpha_{i,i+1}}(\cdot)$ and $R_{z,\alpha_{i,i+1}}(\cdot)$ of the fuzzy result variable $\tilde{z}$ are obtained according to Eqs. (2.44) and (2.45), respectively.

$$L_{z,\alpha_{i,i+1}}(\cdot) = \left(\Delta x_{\alpha_i l} L^{-1}_{x,\alpha_{i,i+1}}(\cdot) \pm \Delta y_{\alpha_i l} L^{-1}_{y,\alpha_{i,i+1}}(\cdot)\right)^{-1} \tag{2.44}$$

$$R_{z,\alpha_{i,i+1}}(\cdot) = \left(\Delta x_{\alpha_i r} R^{-1}_{x,\alpha_{i,i+1}}(\cdot) \pm \Delta y_{\alpha_i r} R^{-1}_{y,\alpha_{i,i+1}}(\cdot)\right)^{-1} \tag{2.45}$$

Taking into consideration the priority rule ($\odot$ precedes $\oplus$), it is possible to combine the introduced operators according to Eq. (2.46).

$$\tilde{z} = \underline{A} \odot \tilde{x} \oplus \ldots \ominus \ldots \oplus \ldots \oplus \underline{B} \odot \tilde{y} \tag{2.46}$$

The arithmetic operations are carried out separately with the $l_\alpha r_\alpha$-increments of each α-level. The usual calculation rules for real-valued numbers thereby hold, in particular the compliance with calculation hierarchy. Only the final $l_\alpha r_\alpha$-increments Δz_i must be non-negative, whereas negative intermediate results arising from the application of the associative law are permitted. Such intermediate fuzzy variables are regarded as fuzzy variables in the improper sense according to remark 2.9 and remark 2.11.

$$\Delta z_i \geqslant 0 \quad \text{for } i = 1,\, 2,\, \ldots,\, n-1,\, n+1,\, \ldots,\, 2n \tag{2.47}$$

The requirement according to Eq. (2.47) also represents a boundary condition for the description and modeling of fuzzy time series.

Remark 2.12. As an abbreviated notation of a sum of fuzzy variables $\tilde{x}_1 \oplus \tilde{x}_2 \oplus \ldots \oplus \tilde{x}_k$, the $l_\alpha r_\alpha$-summation operator according to Eq. (2.48) is introduced.

$$\tilde{x}_1 \oplus \tilde{x}_2 \oplus \ldots \oplus \tilde{x}_k = \bigoplus_{j=1}^{k} \tilde{x}_j \tag{2.48}$$

◆

2.1.3 Subtraction of Fuzzy Variables

The introduced $l_\alpha r_\alpha$-subtraction according to Eq. (2.39) is now compared with the HUKUHARA difference presented in [20] as well as the subtraction according the extension principle (see [5, 36, 71]), and the different properties are discussed.

Definition 2.13. *According to [20] the* HUKUHARA *difference $\tilde{x} \ominus_H \tilde{y}$ between two fuzzy variables $\tilde{x}$ and $\tilde{y}$ is defined as the solution $\tilde{z}$ of the equation $\tilde{y} \oplus \tilde{z} = \tilde{x}$:*

$$\tilde{x} \ominus_H \tilde{y} = \tilde{z} \Leftrightarrow \tilde{y} \oplus \tilde{z} = \tilde{x}\,.$$

The HUKUHARA *difference is not defined if the subtrahend $\tilde{y}$ is characterized by a higher degree of uncertainty (e.g. a wider support) than the minuend $\tilde{x}$.* ◆

If the HUKUHARA difference exists, the $l_\alpha r_\alpha$-subtraction yields the same results as the HUKUHARA difference. Both of the arithmetic operations result in fuzzy variables $\tilde{z}$ characterized by a lower uncertainty than the minuend. This type of subtraction differs from the substraction according the extension principle (see [5, 36, 71]).

The extension principle represents an alternative mathematical basis for a mapping $\tilde{z} = f(\tilde{x}, \tilde{y}, ...)$ of the fuzzy variables $\tilde{x}$, $\tilde{y}$, ... onto $\tilde{z}$.

Definition 2.14. *The fuzzy result variable $\tilde{z} = f(\tilde{x}, \tilde{y}, ...)$ according the extension principle is determined by*

$$\tilde{z} = \{z,\, \mu_{\tilde{z}}(z) \,|\, z = f(x,\, y,\, ...);\; z \in \mathbf{Z};\; (x,\, y,\, ...) \in \mathbf{X} \times \mathbf{Y} \times ...\} \tag{2.49}$$

with the fundamental sets $\mathbf{X}$, $\mathbf{Y}$, ..., $\mathbf{Z}$ *and the membership function*

$$\mu_{\tilde{z}}(z) = \begin{cases} \sup \min_{z=f(x,\, y,\, ...)} [\mu_{\tilde{x}}(x),\, \mu_{\tilde{y}}(y),\, ...] & \text{if} \quad \exists\; z = f(x,\, y,\, ...) \\ 0 & \text{otherwise} \end{cases} \tag{2.50}$$

The mapping function $f(\cdot)$ in the foregoing may be arbitrary. ◆

Applying the extension principle to the substraction, the interval boundaries $[z_{\alpha l}; z_{\alpha r}]$ of the α-level sets Z_α of the fuzzy result variable $\tilde{z} = \tilde{x}\,\text{minus}\,\tilde{y}$ are given by Eqs. (2.51) and (2.52), respectively.

$$z_{\alpha l} = x_{\alpha l} - y_{\alpha l} \qquad \forall \quad \alpha \in [0,1] \tag{2.51}$$

$$z_{\alpha r} = x_{\alpha r} + y_{\alpha r} - 2y_l \quad \forall \quad \alpha \in [0,1] \tag{2.52}$$

The subtraction according the extension principle thus leads to a higher degree of uncertainty of the fuzzy result variable $\tilde{z}$. Unlike the $l_\alpha r_\alpha$-subtraction, the $l_\alpha r_\alpha$-addition according to Eq. (2.39) and the addition according to the extension principle yield the same fuzzy result variable.

The basic distinction between the $l_\alpha r_\alpha$-subtraction and the HUKUHARA difference lies in the treatment of differences, i.e. which subtrahend possesses a higher degree of uncertainty (e.g. a wider support) than the minuend. In this case the HUKUHARA difference is not defined, whereas the $l_\alpha r_\alpha$-subtraction yields fuzzy result variables with negative $l_\alpha r_\alpha$-increments, i.e. fuzzy variables in the improper sense, which are permitted as intermediate results. In the context of fuzzy time series, however, only the final fuzzy result variables of a sequence of arithmetic operations must be fuzzy variables in the proper sense. This is a basic condition for the analysis and forecasting of fuzzy time series.

The following two examples illustrate the different kinds of subtraction of fuzzy variables.

Example 2.15. The first example demonstrates the fundamental distinction between the $l_\alpha r_\alpha$-subtraction, the HUKUHARA difference, and the subtraction according the extension principle. For a given minuend $\tilde{x} = (2.7; 3; 4)_{LR}$ and a given subtrahend $\tilde{y} = (1.4; 1.5; 2.3)_{LR}$ the different subtractions are illustrated in Fig. 2.6. The extension principle yields a fuzzy result variable $\tilde{z}$ with a higher degree of uncertainty, whereas the $l_\alpha r_\alpha$-subtraction and the HUKUHARA difference lead to a lower degree of uncertainty.

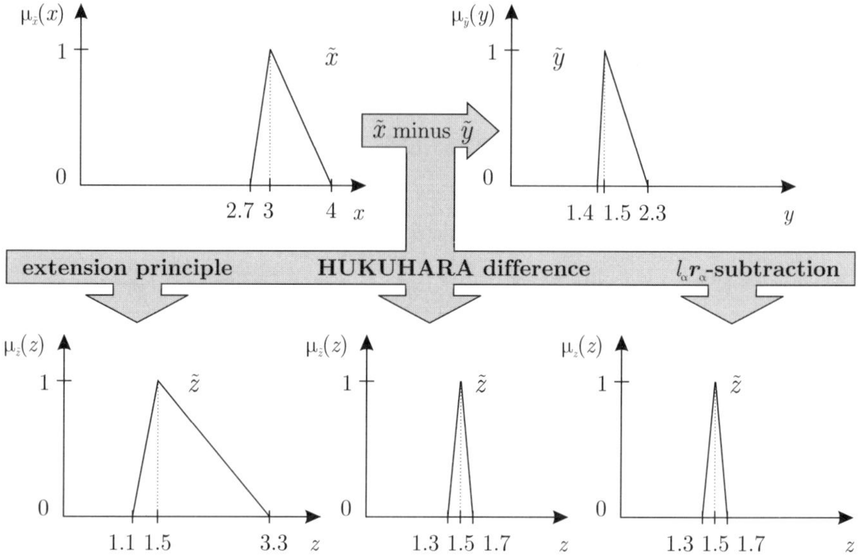

Fig. 2.6. Comparison of $l_\alpha r_\alpha$-subtraction, HUKUHARA difference, and subtraction according the extension principle – Example 2.15

◆

Example 2.16. The second example illustrates the case in which the subtrahend possesses a higher degree of uncertainty than the minuend. The minuend $\tilde{x} = (2.7; 3; 4)_{LR}$, the subtrahend $\tilde{y} = (0.5; 1.5; 2.3)_{LR}$ and the different subtractions are shown in Fig. 2.7. The subtraction according the extension principle increases the degree of uncertainty of the fuzzy result variable, whereas the HUKUHARA difference is not defined in this case. The $l_\alpha r_\alpha$-subtraction yields a fuzzy result variable in the improper sense, i.e. for the chosen $l_\alpha r_\alpha$-discretization with $n = 2$ α-levels the $l_\alpha r_\alpha$-increment $\Delta x_{\alpha_1 l}$ becomes negative. Metaphorically speaking, the $l_\alpha r_\alpha$-increment $\Delta x_{\alpha_1 l}$ 'snaps through'.

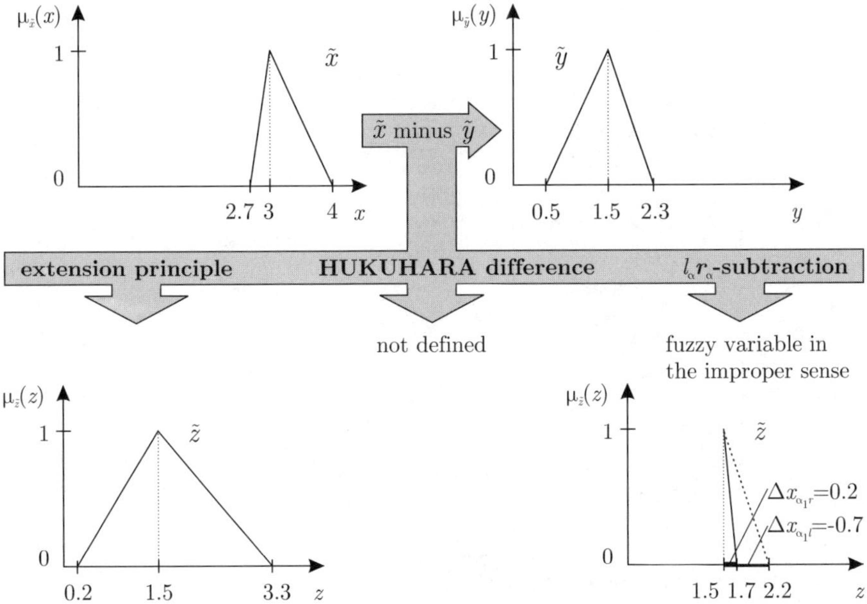

Fig. 2.7. Comparison of $l_\alpha r_\alpha$-subtraction, HUKUHARA difference, and subtraction according the extension principle – Example 2.16

◆

Moreover, the generalized HUKUHARA difference presented in [20] differs from the $l_\alpha r_\alpha$-subtraction. The generalized HUKUHARA difference yields an approximate solution, e.g. a real-valued number, if the subtrahend is more uncertain than the minuend. From this it follows that the generalized HUKUHARA difference is not applicable in equations such as (2.53), whereas the $l_\alpha r_\alpha$-subtraction may in fact be used.

$$(\tilde{x} \ominus \tilde{y}) \oplus \tilde{y} = \tilde{x} \tag{2.53}$$

Remark 2.17. The concept of a best possible difference (accordingly the generalized HUKUHARA difference) may be applied to transform a fuzzy variable in the improper sense into a fuzzy variable in the proper sense. The best possible approximation $\hat{\tilde{z}}$ is given by Eqs. (2.54) to (2.56). The following holds for $i = 1, 2, ..., n-1$ and $j = 1, 2, ..., n$:

$$\Delta\hat{z}_{\alpha_n l} = \Delta z_{\alpha_n l} \tag{2.54}$$

$$\Delta\hat{z}_{\alpha_i l} = \begin{cases} 0 & \text{if } \Delta z_{\alpha_i l} < 0 \\ 0 & \text{if } \Delta z_{\alpha_i l} \geqslant 0 \wedge \sum\limits_{t=i}^{n-1} \Delta z_{\alpha_t l} < \sum\limits_{s=i+1}^{n-1} \Delta\hat{z}_{\alpha_s l} \\ \sum\limits_{t=i}^{n-1} \Delta z_{\alpha_t l} - \sum\limits_{s=i+1}^{n-1} \Delta\hat{z}_{\alpha_s l} & \text{if } \Delta z_{\alpha_i l} \geqslant 0 \wedge \sum\limits_{t=i}^{n-1} \Delta z_{\alpha_t l} \geqslant \sum\limits_{s=i+1}^{n-1} \Delta\hat{z}_{\alpha_s l} \end{cases} \tag{2.55}$$

$$\Delta\hat{z}_{\alpha_j r} = \begin{cases} 0 & \text{if } \Delta z_{\alpha_j r} < 0 \\ 0 & \text{if } \Delta z_{\alpha_j r} \geqslant 0 \wedge \sum\limits_{t=j}^{n} \Delta z_{\alpha_t r} < \sum\limits_{s=j+1}^{n} \Delta\hat{z}_{\alpha_s r} \\ \sum\limits_{t=j}^{n} \Delta z_{\alpha_t r} - \sum\limits_{s=j+1}^{n} \Delta\hat{z}_{\alpha_s r} & \text{if } \Delta z_{\alpha_j r} \geqslant 0 \wedge \sum\limits_{t=j}^{n} \Delta z_{\alpha_t r} \geqslant \sum\limits_{s=j+1}^{n} \Delta\hat{z}_{\alpha_s r} \end{cases} \tag{2.56}$$

◆

2.1.4 Distance between Fuzzy Variables

The modeling and forecasting of fuzzy time series requires a definition of the distance $d_F(\tilde{x};\tilde{y})$ between two fuzzy variables $\tilde{x}$ and $\tilde{y}$.

Definition 2.18. *According to the metrics introduced in [24], the distance $d_F(\tilde{x};\tilde{y})$ between the fuzzy variables $\tilde{x}$ and $\tilde{y}$ is defined as the integral over the* HAUSDORFF *distance $d_H(\cdot;\cdot)$ between the α-level sets X_α and Y_α of $\tilde{x}$ and $\tilde{y}$, as given by Eq. (2.57).*

$$d_F(\tilde{x};\tilde{y}) = \int_0^1 d_H(X_\alpha;Y_\alpha)\,d\alpha \tag{2.57}$$

According to [21], the HAUSDORFF *distance $d_H(X_\alpha;Y_\alpha)$ between two non-empty closed finite α-level sets $X_\alpha;Y_\alpha \subseteq \mathbb{R}$ is defined by Eq. (2.58).*

$$d_H(X_\alpha;Y_\alpha) = \max\left\{\sup_{x\in X_\alpha}\inf_{y\in Y_\alpha} d_E(x;y)\,;\,\sup_{y\in Y_\alpha}\inf_{x\in X_\alpha} d_E(x;y)\right\} \tag{2.58}$$

The term $d_E(x;y)$ represents the EUCLID*ean distance between two real-valued variables $x, y \in \mathbb{R}$ according to Eq. (2.59).*

$$d_E(x;y) = |x-y| = \sqrt{(x-y)^2} \tag{2.59}$$

◆

Example 2.19. This is illustrated by a calculation of the distance $d_F(\tilde{x};\tilde{y})$ between the fuzzy variables $\tilde{x}$ and $\tilde{y}$, as shown in Fig. 2.8.

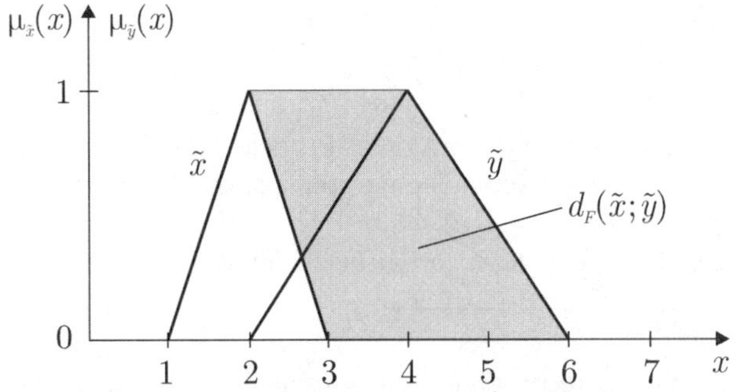

Fig. 2.8. Distance $d_F(\tilde{x};\tilde{y})$ between the fuzzy variables $\tilde{x}$ and $\tilde{y}$

The HAUSDORFF distance $d_H(X_{\alpha=0};Y_{\alpha=0})$ between the α-level sets $X_{\alpha=0}$ and $Y_{\alpha=0}$ of the support of both fuzzy variables $\tilde{x}$ and $\tilde{y}$ is given by Eq. (2.60).

$$\begin{aligned} d_H(X_{\alpha=0};Y_{\alpha=0}) &= \max\{d_E(x=1;y=2);d_E(x=3;y=6)\} \\ &= \max\{1;3\} = 3 \end{aligned} \tag{2.60}$$

The HAUSDORFF distance for the α-level $\alpha = 1$ is obviously $d_H(X_{\alpha=1};Y_{\alpha=1}) = 2$. Integration of the HAUSDORFF distance $d_H(X_\alpha;Y_\alpha)$ over all α-levels yields the distance $d_F(\tilde{x};\tilde{y})$ between the fuzzy variables $\tilde{x}$ and $\tilde{y}$. In this case the integration according to Eq. (2.57) may be reduced to a computation of the surface area of a trapezium according to Eq. (2.61). The result is also depicted in Fig. 2.8.

$$\begin{aligned} d_F(\tilde{x};\tilde{y}) &= \int_0^1 d_H(X_\alpha;Y_\alpha)\,d\alpha \\ &= \frac{d_H(X_{\alpha=0};Y_{\alpha=0}) + d_H(X_{\alpha=1};Y_{\alpha=1})}{2} = 2,5 \end{aligned} \tag{2.61}$$

◆

2.1.5 Fuzzy Functions

Definitions and basic terms relating to fuzzy functions have been introduced and enhanced by various authors, e.g. [3, 11, 36]. The definition of a fuzzy function according to [36] is presented in the following. Given are:

- the fundamental sets $\mathbf{X} \subseteq \mathbb{R}$ and $\mathbf{Y} \subseteq \mathbb{R}$,
- the set $\mathbf{F}(\mathbf{X})$ of all fuzzy variables $\tilde{x}$ on the fundamental set $\mathbf{X}$,
- the set $\mathbf{F}(\mathbf{Y})$ of all fuzzy variables $\tilde{y}$ on the fundamental set $\mathbf{Y}$.

Definition 2.20. *The mapping of* $\mathbf{F}(\mathbf{Y})$ *onto* $\mathbf{F}(\mathbf{X})$ *that assigns precisely one* $\tilde{x} \in \mathbf{F}(\mathbf{X})$ *to each* $\tilde{y} \in \mathbf{F}(\mathbf{Y})$ *is referred to as a fuzzy function denoted by*

$$\tilde{x}(\tilde{y}) : \mathbf{F}(\mathbf{Y}) \to \mathbf{F}(\mathbf{X})\,. \tag{2.62}$$

◆

For each $\tilde{y} \in \mathbf{F}(\mathbf{Y})$ the fuzzy function $\tilde{x}(\tilde{y})$ yields the fuzzy result $\tilde{x}_{\tilde{y}} = \tilde{x}(\tilde{y})$ with $\tilde{x}_{\tilde{y}} \in \mathbf{F}(\mathbf{X})$. This represents the mapping of the fuzzy variables $\tilde{y} \in \mathbf{F}(\mathbf{Y})$ onto the fuzzy variables $\tilde{x}_{\tilde{y}} \in \mathbf{F}(\mathbf{X})$.

According to $l_\alpha r_\alpha$-discretization, a fuzzy function

$$\tilde{x} = \tilde{f}(\tilde{y}) \tag{2.63}$$

may be formulated in an incremental manner. The following thus holds for the $l_\alpha r_\alpha$-increments in the general case:

$$\Delta x_j = f_j\left(\Delta y_1, \Delta y_2, ..., \Delta y_{2n}\right) \quad \text{for} \quad j = 1, 2, ..., 2n\,. \tag{2.64}$$

The deterministic functions $f_j(\cdot)$ are classical functions which are referred to as trajectories.

Taking the general incremental notation according to Eq. (2.64) as a basis, the following special cases are introduced:

- the $l_\alpha r_\alpha$-increments Δx_j only depend on the associated $l_\alpha r_\alpha$-increments Δy_j

$$\Delta x_j = f_j\left(\Delta y_j\right) \quad \text{for} \quad j = 1, 2, ..., 2n \tag{2.65}$$

- the deterministic functions $f_j(\cdot)$ are the same for each $l_\alpha r_\alpha$-increment Δx_j

$$\Delta x_j = f\left(\Delta y_1, \Delta y_2, ..., \Delta y_{2n}\right) \quad \text{for} \quad j = 1, 2, ..., 2n \tag{2.66}$$

- the $l_\alpha r_\alpha$-increments Δx_j only depend on the associated $l_\alpha r_\alpha$-increments Δy_j, and the deterministic functions $f_j(\cdot)$ are the same for each $l_\alpha r_\alpha$-increment Δx_j

$$\Delta x_j = f\left(\Delta y_j\right) \quad \text{for} \quad j = 1, 2, ..., 2n\,. \tag{2.67}$$

A further special case is when the fuzzy function $\tilde{f}(\tilde{y})$ yields the result $x = \tilde{f}(\tilde{y})$ (with $x \in \mathbb{R}$) for each $\tilde{y} \in \mathbf{F}(\mathbf{Y})$. For this case, in which the fuzzy function yields real-valued results, the notation of Eq. (2.63) reduces to the notation of Eq. (2.68) as follows.

$$x = f(\tilde{y}) \tag{2.68}$$

The deterministic function $f(\cdot)$ then maps the $l_\alpha r_\alpha$-increments Δy_j onto the result variable x according to Eq. (2.69).

$$x = f\left(\Delta y_1, \Delta y_2, ..., \Delta y_{2n}\right) \tag{2.69}$$

The special case according to Eq. (2.67) finds application for defining the fuzzy activation functions of the artificial neural networks for fuzzy variables presented in Sect. 3.6 whereas the special case according to Eq. (2.68) is used for defining the fuzzy probability distribution functions presented in Sect. 2.2.

Example 2.21. In the following example an $l_\alpha r_\alpha$-discretization is chosen with $n = 2$ α-level sets. Given are the diagonal matrices:

$$\underline{A} = \begin{bmatrix} a_{11} & 0 & 0 & 0 \\ 0 & a_{22} & 0 & 0 \\ 0 & 0 & a_{33} & 0 \\ 0 & 0 & 0 & a_{44} \end{bmatrix} \quad \text{and} \quad \underline{B} = \begin{bmatrix} b_{11} & 0 & 0 & 0 \\ 0 & b_{22} & 0 & 0 \\ 0 & 0 & b_{33} & 0 \\ 0 & 0 & 0 & b_{44} \end{bmatrix} . \tag{2.70}$$

For the fuzzy function

$$\tilde{x} = \tilde{f}(\tilde{y}) = \underline{A} \odot \tilde{y} \oplus \underline{B} \tag{2.71}$$

the $l_\alpha r_\alpha$-increments Δx_j of the fuzzy result $\tilde{x}$ are then obtained according to Eq. (2.72).

$$\Delta x_j = a_{jj}\, \Delta y_j + b_{jj} \quad \text{for} \quad j = 1, 2, ..., 2n \tag{2.72}$$

◆

2.2 Fuzzy Random Variables

The representation of fuzzy random variables $\tilde{X}$ developed in this book is based on the definition of fuzzy random variables according to [27, 28], see also [36]. The space of the random elementary events Ω is now introduced. A fuzzy realization $\tilde{X}(\omega) = \tilde{x}$ is assigned to each elementary event $\omega \in \Omega$, whereby each fuzzy realization $\tilde{X}(\omega) = \tilde{x}$ is an element of the set $\mathbf{F}(\mathbb{R})$ of all convex fuzzy variables on $\mathbb{R}$. Accordingly, a fuzzy random variable may be defined as follows.

Definition 2.22. *A fuzzy random variable* $\tilde{X}$ *is defined by the mapping*

$$\tilde{X} : \Omega \rightarrow \mathbf{F}(\mathbb{R}) \tag{2.73}$$

$$\begin{aligned} \text{with} \quad & \tilde{X}(\omega) = \tilde{x} \in \mathbf{F}(\mathbb{R}) \\ \text{and} \quad & \omega \in \Omega \, . \end{aligned}$$

◆

Each fuzzy realization $\tilde{X}(\omega) = \tilde{x}$ is defined as a convex, normalized fuzzy set, whose membership function $\mu_{\tilde{x}}(x) = \mu_{\tilde{X}(\omega)}(x)$ has to fulfill the requirements formulated in Subsection *$l_\alpha r_\alpha$-Discretization of Fuzzy Variables* on p. 12.

Example 2.23. Fig. 2.9 shows five fuzzy realizations $\tilde{X}(\omega)$ of a fuzzy random variable $\tilde{X}$.

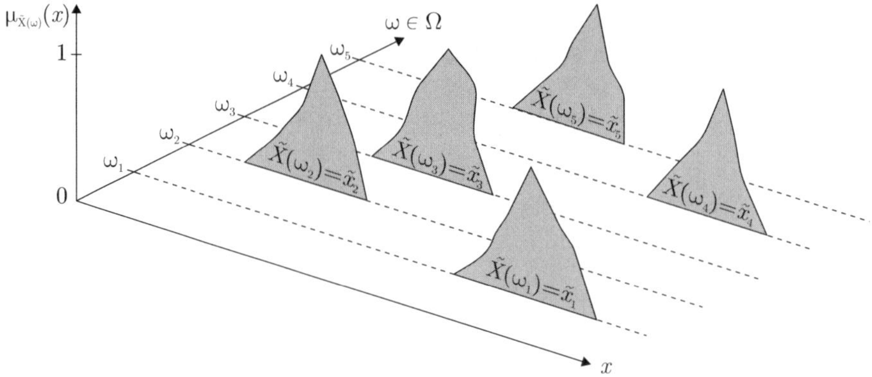

Fig. 2.9. Convex fuzzy realizations $\tilde{X}(\omega)$ of a fuzzy random variable $\tilde{X}$, e.g. as a result of uncertain measurements

◆

Definition 2.24. *A fuzzy random variable $\tilde{X}$ is referred to as a discrete fuzzy random variable if it possesses only finite or at most countable infinite different realizations $\tilde{x}_1$, $\tilde{x}_2$, ..., $\tilde{x}_m$. This means that each possible realization of a discrete fuzzy random variable $\tilde{X}$ may be assigned to a natural number bijectively (unique and reversible).* ◆

Definition 2.25. *A fuzzy random variable $\tilde{X}$ is referred to as a continuous fuzzy random variable if it possesses uncountable realizations. This means that each list $\tilde{x}_1$, $\tilde{x}_2$, $\tilde{x}_3$, ... of possible realizations of a continuous fuzzy random variable $\tilde{X}$ is incomplete.* ◆

2.2.1 Classical and Incremental Discretization of Fuzzy Random Variables

α-Discretization of Fuzzy Random Variables

Under the assumption of convex fuzzy realizations $\tilde{X}(\omega) = \tilde{x}$, a fuzzy random variable $\tilde{X}$ may be characterized by a family of **random** α-level sets X_α according to Eq. (2.74).

$$\tilde{X} = \left(X_\alpha = [X_{\alpha l}, X_{\alpha r}] \mid \alpha \in [0,1]\right) \tag{2.74}$$

In other words, closed finite random intervals $[X_{\alpha l}; X_{\alpha r}]$ are obtained, i.e. the interval boundaries $X_{\alpha l} : \Omega \to \mathbb{R}$ and $X_{\alpha r} : \Omega \to \mathbb{R}$ are real-valued random variables. The realizations $X_{\alpha l}(\omega)$ and $X_{\alpha r}(\omega)$ are assigned to each elementary event $\omega \in \Omega$. The random variables $X_{\alpha l}$ and $X_{\alpha r}$ are defined by Eqs. (2.75) and (2.76), respectively, for $\alpha \in (0,1]$. For $\alpha = 1$ the random variable $X_{\alpha l}$ is referred to as a random peak point.

$$X_{\alpha l}(\omega) = \min\left[x \in \mathbb{R} \mid \mu_{\tilde{X}(\omega)}(x) \geqslant \alpha\right] \tag{2.75}$$

$$X_{\alpha r}(\omega) = \max\left[x \in \mathbb{R} \,|\, \mu_{\tilde{X}(\omega)}(x) \geqslant \alpha\right] \tag{2.76}$$

The support of a fuzzy random variable $\tilde{X}$ is given by the random set $S_{\tilde{X}(\cdot)}$ according to Eq. (2.77).

$$S_{\tilde{X}(\omega)} = \left\{x \in \mathbb{R} \,|\, \mu_{\tilde{X}(\omega)}(x) > 0\right\} \tag{2.77}$$

The support $S_{\tilde{X}(\cdot)}$ of a fuzzy random variable $\tilde{X}$ is also referred to as a random α-level set X_α with $\alpha = 0$. The associated interval boundaries $X_{\alpha l}$ and $X_{\alpha r}$ are defined by Eqs. (2.78) and (2.79), respectively.

$$X_{\alpha l}(\omega) = \lim_{\alpha' \to +0}\left[\min\left[x \in \mathbb{R} \,|\, \mu_{\tilde{X}(\omega)}(x) > \alpha'\right]\right] \quad \text{for} \quad \alpha = 0 \tag{2.78}$$

$$X_{\alpha r}(\omega) = \lim_{\alpha' \to +0}\left[\max\left[x \in \mathbb{R} \,|\, \mu_{\tilde{X}(\omega)}(x) > \alpha'\right]\right] \quad \text{for} \quad \alpha = 0 \tag{2.79}$$

The random intervals $[X_{\alpha l}; X_{\alpha r}]$ and the random membership function $\mu_{\tilde{X}(\cdot)}(x)$ of a fuzzy random variable $\tilde{X}$ are illustrated in Fig. 2.10, which also shows assumed probability distribution functions of the random variables $X_{\alpha l}$ and $X_{\alpha r}$.

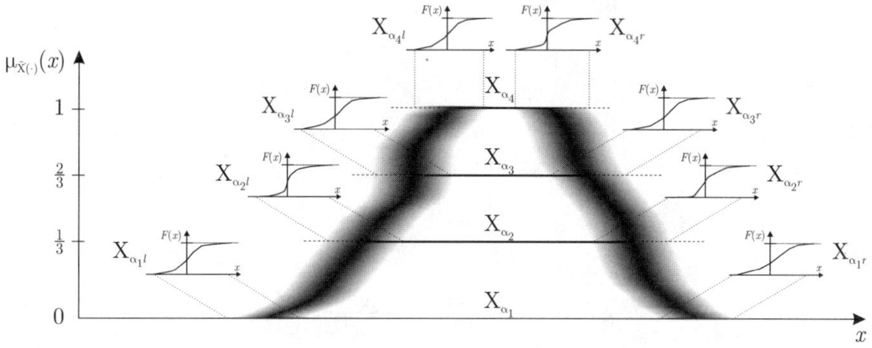

Fig. 2.10. Random α-level sets X_α and random membership function $\mu_{\tilde{X}(\cdot)}(x)$ of a fuzzy random variable $\tilde{X}$

According to Eqs. (2.75) to (2.79), a fuzzy random variable $\tilde{X}$ may be represented by its real-valued random interval boundaries $X_{\alpha l}$ and $X_{\alpha r}$ with $\alpha \in [0; 1]$. The realizations of the random variables $X_{\alpha l}$ and $X_{\alpha r}$ represent the boundaries of classical intervals. The totality of these intervals constitutes the α-level sets X_α of a convex fuzzy variable $\tilde{x}$. According to Eq. (2.73), the convex fuzzy variable $\tilde{x}$ is thus a realization of the fuzzy random variable $\tilde{X}$. Each convex realization $\tilde{x}$ of a fuzzy random variable $\tilde{X}$ must fulfill the requirement according to Eq. (2.80).

$$X_{\alpha_k} \subseteq X_{\alpha_i} \quad \forall\, \alpha_i, \alpha_k \in [0;1] \quad \text{with} \quad \alpha_i \leqslant \alpha_k \tag{2.80}$$

The requirement according to Eq. (2.80) implies that the α-level sets of a realization $\tilde{x}$ are linked interactively. The random intervals $[\mathrm{X}_{\alpha l}; \mathrm{X}_{\alpha r}]$ of a fuzzy random variable $\tilde{\mathrm{X}}$ are thus mutually dependent. The dependencies are referred to as interaction. The interaction is taken into consideration with the aid of the $l_\alpha r_\alpha$-discretization of the fuzzy random variables.

$l_\alpha r_\alpha$-Discretization of Fuzzy Random Variables

The concept of $l_\alpha r_\alpha$-discretization of fuzzy variables may be extended to fuzzy random variables.

A requirement for each random α-level set X_{α_i} of a fuzzy random variable $\tilde{\mathrm{X}}$ is that X_{α_i} may be represented as the union of the contained random α-level set $\mathrm{X}_{\alpha_{i+1}}$ and the random set $\mathrm{X}^*_{\alpha_i}$ according to Eq. (2.81).

$$\mathrm{X}_{\alpha_i} = \mathrm{X}_{\alpha_{i+1}} \cup \mathrm{X}^*_{\alpha_i} \quad \forall\, \alpha_i, \alpha_{i+1} \in (0;1] \quad \text{with} \quad \alpha_i \leqslant \alpha_{i+1} \tag{2.81}$$

The random set $\mathrm{X}^*_{\alpha_i}$ is defined by two disjoint random intervals $[\mathrm{X}_{\alpha_i ll}; \mathrm{X}_{\alpha_i lr}]$ and $[\mathrm{X}_{\alpha_i rl}; \mathrm{X}_{\alpha_i rr}]$ with real-valued random boundaries according to Eqs. (2.82) and (2.83), respectively.

$$\mathrm{X}_{\alpha_i ll} = \mathrm{X}_{\alpha_i l} \qquad \text{and} \qquad \mathrm{X}_{\alpha_i lr} = \mathrm{X}_{\alpha_{i+1} l} \tag{2.82}$$

$$\mathrm{X}_{\alpha_i rl} = \mathrm{X}_{\alpha_{i+1} r} \qquad \text{and} \qquad \mathrm{X}_{\alpha_i rr} = \mathrm{X}_{\alpha_i r} \tag{2.83}$$

The random α-level set X_{α_i} is thus characterized by the random interval boundaries $\mathrm{X}_{\alpha_i l}$ and $\mathrm{X}_{\alpha_i r}$ according Eqs. (2.84) and (2.85), respectively.

$$\mathrm{X}_{\alpha_i l} = \mathrm{X}_{\alpha_{i+1} l} - \Delta\mathrm{X}_{\alpha_i l} \quad \text{with} \quad \Delta\mathrm{X}_{\alpha_i l} = \mathrm{X}_{\alpha_i lr} - \mathrm{X}_{\alpha_i ll} \tag{2.84}$$

$$\mathrm{X}_{\alpha_i r} = \mathrm{X}_{\alpha_{i+1} r} + \Delta\mathrm{X}_{\alpha_i r} \quad \text{with} \quad \Delta\mathrm{X}_{\alpha_i r} = \mathrm{X}_{\alpha_i rr} - \mathrm{X}_{\alpha_i rl} \tag{2.85}$$

In this definition the terms $\Delta\mathrm{X}_{\alpha_i l}$ and $\Delta\mathrm{X}_{\alpha_i r}$ are correlated random variables and are referred to as **random $l_\alpha r_\alpha$-increments** of the fuzzy random variable $\tilde{\mathrm{X}}$. Eqs. (2.82) to (2.85) hold for $i = 1, 2, ..., n-1$, whereas for $i = n$, Eqs. (2.84) and (2.85) are replaced by Eqs. (2.86) and (2.87). The counter n specifies the number of α-levels.

$$\mathrm{X}_{\alpha_n l} = \Delta\mathrm{X}_{\alpha_n l} \quad \text{with} \quad \alpha_n = 1 \tag{2.86}$$

$$\mathrm{X}_{\alpha_n r} = \mathrm{X}_{\alpha_n l} + \Delta\mathrm{X}_{\alpha_n r} \quad \text{with} \quad \alpha_n = 1 \tag{2.87}$$

Recapitulating, the $l_\alpha r_\alpha$-discretization of a fuzzy random variable $\tilde{\mathrm{X}}$ is given by Eq. (2.88) with $i = 1, 2, ..., n$ for $n \geqslant 2$.

$$\tilde{X} = \begin{pmatrix} X_{\alpha_i} = [X_{\alpha_{i+1}l} - \Delta X_{\alpha_i l}; X_{\alpha_{i+1}r} + \Delta X_{\alpha_i r}] \,|\, \alpha_i \in [0,1); \\ X_{\alpha_i} = [X_{\alpha_i l}; X_{\alpha_i l} + \Delta X_{\alpha_i r}] \qquad\qquad |\, \alpha_i = 1 \end{pmatrix} \tag{2.88}$$

The $l_\alpha r_\alpha$-discretization permits an alternative, discrete representation of a fuzzy random variable $\tilde{X}$ in the form of a random column matrix given by Eq. (2.89), whereby the real-valued random variables ΔX_1, ΔX_2, ..., ΔX_{2n} are abridged notations of the random $l_\alpha r_\alpha$-increments $\Delta X_{\alpha_1 l}$, $\Delta X_{\alpha_2 l}$, ..., $\Delta X_{\alpha_1 r}$.

$$\tilde{X} = \begin{bmatrix} \Delta X_{\alpha_1 l} \\ \Delta X_{\alpha_2 l} \\ \vdots \\ \Delta X_{\alpha_n l} \\ \Delta X_{\alpha_n r} \\ \vdots \\ \Delta X_{\alpha_2 r} \\ \Delta X_{\alpha_1 r} \end{bmatrix} = \begin{bmatrix} \Delta X_1 \\ \Delta X_2 \\ \vdots \\ \Delta X_n \\ \Delta X_{n+1} \\ \vdots \\ \Delta X_{2n-1} \\ \Delta X_{2n} \end{bmatrix} \tag{2.89}$$

The random $l_\alpha r_\alpha$-increments of the fuzzy random variable $\tilde{X}$ must fulfill the requirements according to Eq. (2.90) and Eq. (2.91) in order to fulfill Eq. (2.80).

$$\Delta X_{\alpha_i l} \geqslant 0 \quad \text{for} \quad i = 1, 2, ..., n-1 \tag{2.90}$$

$$\Delta X_{\alpha_i r} \geqslant 0 \quad \text{for} \quad i = 1, 2, ..., n \tag{2.91}$$

For $i = n$ the random $l_\alpha r_\alpha$-increment $\Delta X_{\alpha_n l}$ is assigned to the random peak point $X_{\alpha_n l}$. The requirement of non-negativity must not be fulfilled at the random peak point.

The random $l_\alpha r_\alpha$-increments $\Delta X_{\alpha_i l}$ and $\Delta X_{\alpha_i r}$ of a fuzzy random variable $\tilde{X}$ are illustrated schematically in Fig. 2.11.

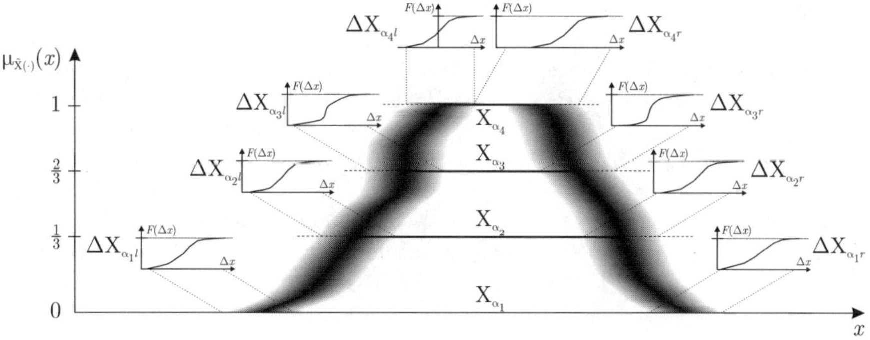

Fig. 2.11. Random $l_\alpha r_\alpha$-increments of a fuzzy random variable $\tilde{X}$

2.2.2 Fuzzy Probability Distribution Functions of Fuzzy Random Variables

A fuzzy random variable $\tilde{X}$ according to Eq. (2.73) may be described with the aid of fuzzy probability distribution functions. Different forms of fuzzy probability distribution functions have already been developed, see e.g. [36, 66]. Two forms of fuzzy probability distribution functions based on different probability measures for fuzzy random variables are presented in the following.

The first form is based on the fuzzy probability measure introduced in [36] and is referred to as the fuzzy probability distribution function form I. The fuzzy probability distribution function form I may be obtained by statistical evaluation of a concrete sample comprised of fuzzy elements. This approach, however, does not permit the precise reproduction of the underlying sample elements (e.g. by Monte Carlo simulation). The fuzzy probability distribution function form I may be advantageously applied, amongst others, in structural analysis [36, 38], in the Fuzzy Stochastic Finite Element Method (FSFEM) [35, 62] and in the safety assessment of structures [39, 63]. The fuzzy probability distribution function form I is not applicable for the analysis and forecasting of fuzzy time series.

The second form represents a new type of fuzzy probability distribution function and permits the precise reproduction of samples comprised of fuzzy elements. This is referred to as the fuzzy probability distribution function form II, which is especially suitable for the analysis of fuzzy time series [41]. Form II is based on the $l_\alpha r_\alpha$-discretization of fuzzy random variables introduced in Sect. 2.2.

Fuzzy Probability Distribution Function Form I (FPDF I)

The definition of the fuzzy probability distribution function form I $\tilde{F}_{\tilde{X}}(x)$ is based on the fuzzy probability $\tilde{P}(A)$. Only one-dimensional fuzzy random variables are considered here. Multi-variate fuzzy random variables are introduced in [36].

Definition 2.26. *The fuzzy probability $\tilde{P}(A)$ is defined according to [36] as the set of all probabilities $P(\tilde{X} \in A)$ with the membership values $\mu(P(\tilde{X} \in A))$, which take into account all states of occurrence of $\tilde{X} \in A$. The set A thereby represents a deterministic set in the EUKLIDean space $\mathbb{R}$.* ◆

Remark 2.27. The fuzzy probability space belonging to the fuzzy probability $\tilde{P}(A)$ including the concept of measurability for fuzzy random variables is described in detail in [36] and is not discussed in this application-oriented book. ◆

In order to determine the fuzzy probability $\tilde{P}(A)$ the fuzzy random variable $\tilde{X}$ is represented as family of random α-level sets X_α according to Eq. (2.92).

$$\tilde{\mathrm{X}} = (\mathrm{X}_\alpha = [\mathrm{X}_{\alpha l}, \mathrm{X}_{\alpha r}] \,|\, \alpha \in [0, 1]) \tag{2.92}$$

The random α-level sets X_α are closed finite intervals $[\mathrm{X}_{\alpha l}; \mathrm{X}_{\alpha r}]$ for each α-level, and the elements of X_α are elements of A with a certain probability. This probability is referred to as the fuzzy probability $\tilde{P}(A)$ of A, and is defined as follows by Eq. (2.93).

$$\tilde{P}(A) = (P_\alpha(A) = [P_{\alpha l}(A);\, P_{\alpha r}(A)] \,|\, \alpha \in [0;\, 1]) \tag{2.93}$$

The bounds $P_{\alpha l}(A)$ and $P_{\alpha r}(A)$ of the α-level sets $P_\alpha(A)$ are given by Eqs. (2.94) and (2.95), respectively.

$$P_{\alpha l}(A) = P(\mathrm{X}_\alpha \subseteq A) \tag{2.94}$$

$$P_{\alpha r}(A) = P(\mathrm{X}_\alpha \cap A \neq \emptyset) \tag{2.95}$$

The fuzzy probability distribution function form I $\tilde{F}_{\tilde{\mathrm{X}}}(x)$ of the fuzzy random variable $\tilde{\mathrm{X}}$ is then defined as the fuzzy probability $\tilde{P}(A)$ with $A = \{t \,|\, t < x;\, x, t \in \mathbb{R}\}$. $\tilde{F}_{\tilde{\mathrm{X}}}(x)$ thus represents a fuzzy function with the fuzzy functional values $\tilde{F}_{\tilde{\mathrm{X}}}(x)$ defined by Eq. (2.96).

$$\tilde{F}_{\tilde{\mathrm{X}}}(x) = (F_\alpha(x) = [F_{\alpha l}(x);\, F_{\alpha r}(x)] \,|\, \alpha \in [0;\, 1]) \tag{2.96}$$

$$\text{with } F_{\alpha l}(x) = P\left(\mathrm{X}_{\alpha r} < x \,|\, x \in \mathbb{R}\right)$$

$$\text{and } F_{\alpha r}(x) = P\left(\mathrm{X}_{\alpha l} < x \,|\, x \in \mathbb{R}\right)$$

According to Eq. (2.96) the empirical fuzzy probability distribution function form I may be obtained for a given sample comprised of fuzzy data.

Definition 2.28. *For a given sample comprised of s fuzzy variables $\tilde{x}_1$, $\tilde{x}_2$, ..., $\tilde{x}_s$ the empirical fuzzy probability distribution function form I $\hat{\tilde{F}}_s(x)$ is defined by Eq. (2.97).*

$$\hat{\tilde{F}}_s(x) = \left(\hat{F}_\alpha(x) = [\hat{F}_{\alpha l}(x);\, \hat{F}_{\alpha r}(x)] \,|\, \alpha \in [0;\, 1]\right) \tag{2.97}$$

$$\text{with } \hat{F}_{\alpha l}(x) = \frac{1}{s} \sum_{k=1}^{s} I_{(-\infty,x]}\left(x_{\alpha l}(k)\right)$$

$$\text{and } \hat{F}_{\alpha r}(x) = \frac{1}{s} \sum_{k=1}^{s} I_{(-\infty,x]}\left(x_{\alpha r}(k)\right)$$

The function $I_{(-\infty,x]}(\cdot)$ is the indicator function according to Eq. (2.98), and $x_{\alpha l}(k)$ and $x_{\alpha r}(k)$ are the interval boundaries of the α-level set $X_\alpha(k)$ of the kth sample element $\tilde{x}_k$.

$$I_{(-\infty,x]}(z) = \begin{cases} 1 \text{ if } & z \in (-\infty, x] \\ 0 \text{ if } & z \notin (-\infty, x] \end{cases} \tag{2.98}$$

◆

Example 2.29. For the exemplary sample of $s = 4$ fuzzy variables shown in Fig. 2.12 the empirical fuzzy probability distribution function form I $\hat{\tilde{F}}_4(x)$ is computed according to Eq. (2.97).

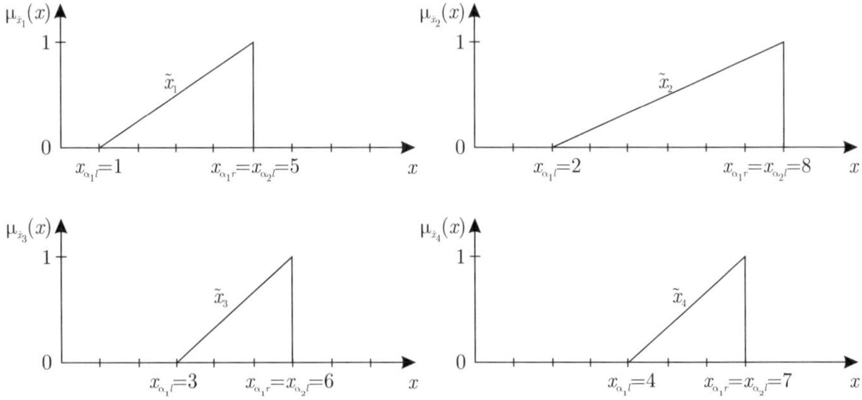

Fig. 2.12. Exemplary sample of fuzzy variables

The result is illustrated in Fig. 2.13. The empirical fuzzy probability distribution function form I $\hat{\tilde{F}}_4(x)$ is computed by separate analysis of the interval boundaries of the α-level sets of the fuzzy variables. As no account is taken of interaction between the different α-level sets, reproduction of the underlying fuzzy sample elements is not possible. ◆

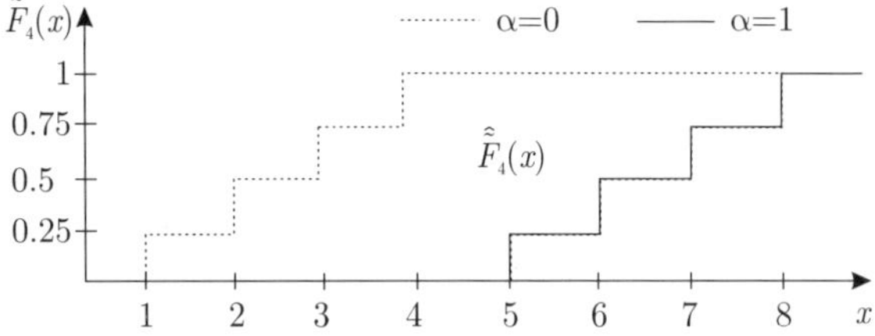

Fig. 2.13. Empirical fuzzy probability distribution function form I $\hat{\tilde{F}}_4(x)$ of the exemplary sample

Hence the empirical fuzzy probability distribution function form I $\hat{\tilde{F}}_s(x)$ does not yield a one to one description of the realizations of a fuzzy random variable $\tilde{X}$. The reproduction of underlying empirical sample elements is only possible for specific samples. The modeling of fuzzy time series, however, requires a one to one description of the realizations of fuzzy random variables $\tilde{X}$. An suitable solution is presented in the next section.

For the numerical treatment of fuzzy random variables $\tilde{X}$ the fuzzy probability distribution function form I $\tilde{F}_{\tilde{X}}(x)$ may be advantageously represented by means of fuzzy bunch parameters. For this purpose the so-called originals of fuzzy random variables are introduced.

Definition 2.30. *The realization x of a real-valued random variable X as well as the fuzzy realization $\tilde{x}$ of a fuzzy random variable $\tilde{X}$ may be assigned to an elementary event $\omega \in \Omega$. If the x are contained in $\tilde{x}$, i.e. $x \in \tilde{x}$ for all elementary events $\omega \in \Omega$, then the x constitute an original X_j of $\tilde{X}$. The original X_j is referred to as completely contained in $\tilde{X}$. Each real-valued random variable X that is completely contained in $\tilde{X}$ thus possesses the property of an original X_j. From this it follows that the fuzzy random variable $\tilde{X}$ may also be defined as a fuzzy set of all possible originals X_j contained in $\tilde{X}$. Thus $\tilde{X}$ may be represented by the assessed bunch of their originals, which may be specified by means of fuzzy bunch parameters $\underline{\tilde{s}}$.*

$$\tilde{X} = X(\underline{\tilde{s}}) \tag{2.99}$$

◆

The bunch of originals may be assessed with the aid of fuzzy bunch parameters $\underline{\tilde{s}}$. Each vector of fuzzy bunch parameters $\underline{s}_j \in \underline{\tilde{s}}$ with the membership value $\mu(\underline{s}_j)$ definitely determines one original with $X_j = X(\underline{s}_j)$. A fuzzy random variable $\tilde{X}$ may thus be defined as a family of the originals $X_j \in \tilde{X}$ with $\mu(X_j) = \mu(X(\underline{s}_j)) = \mu(\underline{s}_j)$.

$$\tilde{X} = X(\underline{\tilde{s}}) = \big(X_j = X(\underline{s}_j) \,|\, \mu(X_j) = \mu(\underline{s}_j) \quad \forall \quad \underline{s}_j \in \underline{\tilde{s}}\big) \tag{2.100}$$

The originals X_j represent real-valued random variables for which the real-valued probability distribution functions $F_X(\underline{s}_j, x)$ exist. Using these real-valued probability distribution functions $F_X(\underline{s}_j, x)$ the fuzzy probability distribution function form I $\tilde{F}_{\tilde{X}}(x)$ of the fuzzy random variable $X(\underline{\tilde{s}})$ may then be expressed by $\tilde{F}_{\tilde{X}}(x) = F_{\tilde{X}}(\underline{\tilde{s}}, x)$. This is the bunch parameter representation of the fuzzy probability distribution function. The real-valued probability distribution functions $F_X(\underline{s}_j, x)$ are also referred to as trajectories of $F_{\tilde{X}}(\underline{\tilde{s}}, x)$.

$$F_{\tilde{X}}(\underline{\tilde{s}}, x) = \big(F_\alpha(\underline{\tilde{s}}, x) = [\inf(F_X(\underline{s}_j, x)); \sup(F_X(\underline{s}_j, x))] \\ \,|\, \underline{s}_j \in \underline{\tilde{s}}, \alpha \in [0;1]\big) \tag{2.101}$$

Fuzzy Probability Distribution Function Form II (FPDF II)

The fuzzy probability distribution function form II is based on the deterministic probability measure $P(F(A))$, whereby $F(A)$ is a finite, countable infinite or uncountable infinite subset of fuzzy variables contained in the deterministic set A. The set $F(A)$ is referred to as a discrete set if it contains only a finite or countable infinite number of fuzzy variables. This means that each fuzzy element of a discrete set $F(A)$ of fuzzy variables may be assigned to a natural number bijectively (uniquely and reversibly). If the set $F(A)$ contains an uncountable number of fuzzy variables, it is referred to as a continuous set. This means that each list $\tilde{x}_1$, $\tilde{x}_2$, $\tilde{x}_3$, ... of fuzzy elements of a continuous set $F(A)$ of fuzzy variables. is incomplete.

Definition 2.31. *The probability measure $P(F(A))$ expresses the probability with which a fuzzy random variable $\tilde{X}$ takes a value $\tilde{x}$ belonging to the (discrete or continuous) set $F(A)$ of fuzzy variables, i.e. $\tilde{X} = \tilde{x} \in F(A) \subseteq \mathbf{F}(\mathbb{R})$. $\mathbf{F}(\mathbb{R})$ denotes the set of all fuzzy variables in the EUKLIDean space $\mathbb{R}$.* ◆

Remark 2.32. The probability space belonging to the probability $P(F(A))$ including the concept of measurability for fuzzy random variables is equivalent to the probability space of random matrices (see amongst others [10]), because each fuzzy random variable may be represented by a random column matrix according to Eq. (29). The probability space is thus not discussed in this application-oriented book. ◆

Example 2.33. Let $F(A)$ be a discrete subset of three fuzzy variables $\tilde{a}_1$, $\tilde{a}_2$ and $\tilde{a}_3$ contained in the deterministic set A. Fig. 2.14 shows four realizations $\tilde{x}_1$, $\tilde{x}_2$, $\tilde{x}_3$ and $\tilde{x}_4$ of a fuzzy random variable $\tilde{X}$ with $\tilde{x}_2$, $\tilde{x}_4 \in F(A)$ and $\tilde{x}_1$, $\tilde{x}_3 \notin F(A)$. ◆

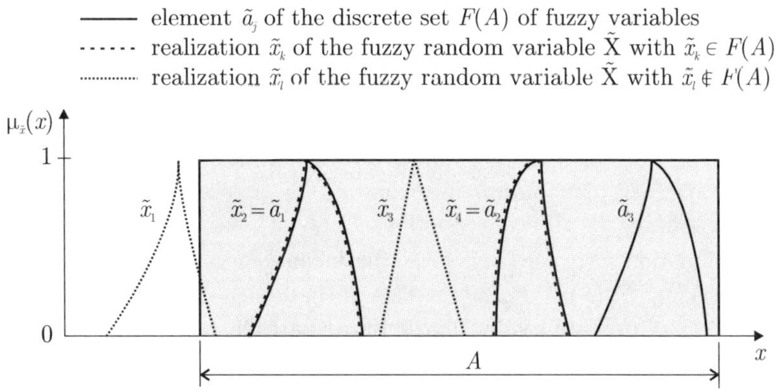

Fig. 2.14. Deterministic set A and discrete set $F(A)$ of three fuzzy variables $\tilde{a}_j$ $(j = 1, 2, 3)$ showing realizations $\tilde{x}_k$ $(k = 2, 4)$ and $\tilde{x}_l$ $(l = 1, 3)$ of a fuzzy random variable $\tilde{X}$

Compared with the fuzzy probability $\tilde{P}(A)$ according to Eq. (2.93), it is not the uncertain probability of the event $\tilde{X} \in A$ that is described in this case but rather the deterministic probability of the (complete) membership of realizations $\tilde{x}$ of a fuzzy random variable $\tilde{X}$ to a (finite, countable infinite or uncountable infinite) set $F(A)$ of fuzzy variables. The probability with which a realization $\tilde{x}$ of a fuzzy random variable $\tilde{X}$ is an element of the set $F(A)$ of fuzzy variables is denoted by $P(F(A))$ of $F(A)$ and is defined by Eq. (2.102).

$$P(F(A)) = \left\{ P(\tilde{X}) \,|\, \tilde{X} = \tilde{x} \in F(A) \right\} \tag{2.102}$$

The fuzzy probability distribution function form II may thus be defined in two different coordinate systems, as presented in the following.

FPDF II in the coordinate system of the interval bounds. For the numerical computation of the probability $P(F(A))$ the fuzzy random variable $\tilde{X}$ is represented as a family of random α-level sets X_α according to Eq. (2.103).

$$\tilde{X} = (X_\alpha = [X_{\alpha l}, X_{\alpha r}] \,|\, \alpha \in [0,1]) \tag{2.103}$$

The bounds of the closed, finite random intervals $[X_{\alpha l}; X_{\alpha r}]$ which determine the α-level sets X_α are real-valued, interactively linked random variables. Each of the random variables $X_{\alpha l}$ and $X_{\alpha r}$ may be described by a real-valued probability distribution function $F_{X_{\alpha l}}(x_{\alpha l})$ and $F_{X_{\alpha r}}(x_{\alpha r})$, respectively. The realizations of the random variables $X_{\alpha l}$ and $X_{\alpha r}$ are denoted by $x_{\alpha l}$ and $x_{\alpha r}$, respectively. According to Sect. 2.1.1 the realizations $x_{\alpha l}$ and $x_{\alpha r}$ represent the interval boundaries of the α-level sets of the realizations $\tilde{x}$ of the fuzzy random variable $\tilde{X}$. Realizations $\tilde{x}$ of the fuzzy random variable $\tilde{X}$ are thus represented by realizations $x_{\alpha l}$ and $x_{\alpha r}$ of the random variables $X_{\alpha l}$ and $X_{\alpha r}$, respectively. The probability $P(F(A))$ according to Eq. (2.102) may hence be expressed by the probability with which realizations $x_{\alpha l}$ and $x_{\alpha r}$ correspond to the interval boundaries $a_{\alpha l}$ and $a_{\alpha r}$ of an element $\tilde{a}$ of $F(A)$.

For the numerical computation the fuzzy random variable $\tilde{X}$ is represented by n α-level sets. The realizations $x_{\alpha_1 l}$, ..., $x_{\alpha_n l}$ and $x_{\alpha_1 r}$, ..., $x_{\alpha_n r}$ of the random interval boundaries $X_{\alpha_1 l}$, ..., $X_{\alpha_n l}$ and $X_{\alpha_1 r}$, ..., $X_{\alpha_n r}$, respectively, are thus regarded as the coordinates of the $2n$-dimensional coordinate system of the interval bounds.

Definition 2.34. *The fuzzy probability distribution function form II $F_{\tilde{X}}(\tilde{x})$ of the fuzzy random variable $\tilde{X}$, which is discretized by $n \geqslant 2$ random α-level sets $[X_{\alpha_i l}; X_{\alpha_i r}]$, is defined as the $2n$-dimensional probability distribution function of the random variables $X_{\alpha_i l}$ and $X_{\alpha_i r}$ according to Eq. (2.104), whereby $x_{\alpha_1 l}$, ..., $x_{\alpha_n l}$ and $x_{\alpha_1 r}$, ..., $x_{\alpha_n r}$ are the interval boundaries of the α-level sets X_α of the fuzzy variable $\tilde{x}$. These form the coordinates in the coordinate system of the interval bounds.*

$$F_{\tilde{X}}(\tilde{x}) = P\left(\{\omega \,|\, X_{\alpha_1 l}(\omega) \leqslant x_{\alpha_1 l}, \ldots, X_{\alpha_n l}(\omega) \leqslant x_{\alpha_n l}, \right. \\ \left. X_{\alpha_1 r}(\omega) \leqslant x_{\alpha_1 r}, \ldots, X_{\alpha_n r}(\omega) \leqslant x_{\alpha_n r}\}\right) \qquad (2.104)$$

$$= P\left(\{X_{\alpha_1 l} \leqslant x_{\alpha_1 l}, \ldots, X_{\alpha_n l} \leqslant x_{\alpha_n l}, \right. \\ \left. X_{\alpha_1 r} \leqslant x_{\alpha_1 r}, \ldots, X_{\alpha_n r} \leqslant x_{\alpha_n r}\}\right)$$

◆

For each random variable $X_{\alpha_i l}$ and $X_{\alpha_i r}$ ($i = 1, 2, \ldots, n$) of a fuzzy random variable $\tilde{X}$ the real-valued probability distribution functions $F_{X_{\alpha_i l}}(x_{\alpha_i l})$ and $F_{X_{\alpha_i r}}(x_{\alpha_i r})$ are given by Eqs. (2.105) and (2.106), respectively. Both of these correspond with the marginal distribution function of the $2n$-dimensional probability distribution function $F_{\tilde{X}}(\tilde{x})$ according to Eq. (2.104).

$$F_{X_{\alpha_i l}}(x_{\alpha_i l}) = \lim_{\substack{x_{\alpha_j l}, x_{\alpha_k r} \to \infty \\ j,k=1,2,\ldots,n \\ j \neq i}} F_{\tilde{X}}(\tilde{x}) \qquad (2.105)$$

$$F_{X_{\alpha_i r}}(x_{\alpha_i r}) = \lim_{\substack{x_{\alpha_j l}, x_{\alpha_k r} \to \infty \\ j,k=1,2,\ldots,n \\ k \neq i}} F_{\tilde{X}}(\tilde{x}) \qquad (2.106)$$

According to the definition 2.34 the empirical fuzzy probability distribution function form II may be derived in the coordinate system of the interval bounds. This presupposes that a sample of s fuzzy variables $\tilde{x}_1, \tilde{x}_2, \ldots, \tilde{x}_s$ is available.

Definition 2.35. *The empirical fuzzy probability distribution function form II $\hat{F}_s(\tilde{x})$ for a sample of s fuzzy variables $\tilde{x}_1, \tilde{x}_2, \ldots, \tilde{x}_s$ is defined in the coordinate system of the interval bounds by Eq. (2.107).*

$$\hat{F}_s(\tilde{x}) = \frac{\#\left\{\tilde{x}_j \,\middle|\, \begin{array}{l} x_{\alpha_1 1}(j) \leqslant x_{\alpha_1 1}, \ldots, x_{\alpha_n 1}(j) \leqslant x_{\alpha_n 1} \\ x_{\alpha_1 r}(j) \leqslant x_{\alpha_1 r}, \ldots, x_{\alpha_n r}(j) \leqslant x_{\alpha_n r} \end{array}, j = 1, 2, \ldots, s\right\}}{s} \qquad (2.107)$$

◆

The coordinates $x_{\alpha_1 l}, \ldots, x_{\alpha_n l}$ and $x_{\alpha_1 r}, \ldots, x_{\alpha_n r}$ are the interval bounds of the α-level sets X_α of the fuzzy variable $\tilde{x}$, and $x_{\alpha_1 l}(j), \ldots, x_{\alpha_n l}(j)$ and $x_{\alpha_1 r}(j), \ldots, x_{\alpha_n r}(j)$ are the interval bounds of the fuzzy variables $\tilde{x}_1, \tilde{x}_2, \ldots, \tilde{x}_s$. The symbol $\#\{\cdot\}$ denotes the number of fuzzy variables $\tilde{x}_j$ ($j = 1, 2, \ldots, s$) for which the requirements $x_{\alpha_1 1}(j) \leqslant x_{\alpha_1 1}, \ldots, x_{\alpha_n 1}(j) \leqslant x_{\alpha_n 1}$ and $x_{\alpha_1 r}(j) \leqslant x_{\alpha_1 r}, \ldots, x_{\alpha_n r}(j) \leqslant x_{\alpha_n r}$ are fulfilled. According to Eq. (2.107) the empirical fuzzy probability distribution function form II $\hat{F}_s(\tilde{x})$ is a monotonically non-decreasing $2n$-dimensional step function.

Example 2.36. The computation of $\hat{F}_s(\tilde{x})$ is demonstrated by means of the exemplary sample of 4 fuzzy variables, as shown in Fig. 2.12. In this case the peak point of each fuzzy sample element is identical with the right boundary of the associated support interval. The multidimensional empirical probability distribution function $\hat{F}_s(\tilde{x})$ according to Eq. (2.107) may hence be depicted two-dimensionally. Discretization of the fuzzy variables is thus applied for two α-levels ($\alpha_1 = 0$ and $\alpha_2 = 1$).

Applying Eq. (2.107), it is necessary to determine the number $\#\{\cdot\}$ of fuzzy variables $\tilde{x}_j$ for which the requirements $x_{\alpha_1 1}(j) \leqslant x_{\alpha_1 1}$, ..., $x_{\alpha_n 1}(j) \leqslant x_{\alpha_n 1}$ and $x_{\alpha_1 r}(j) \leqslant x_{\alpha_1 r}$, ..., $x_{\alpha_n r}(j) \leqslant x_{\alpha_n r}$ are fulfilled. The functional values of $\hat{F}_4(\tilde{x})$ for a selected number of coordinates $x_{\alpha_1 l}$, $x_{\alpha_2 l}$, $x_{\alpha_1 r}$ and $x_{\alpha_2 r}$ are shown in Table 2.2.

Table 2.2. Computation of the empirical fuzzy probability distribution function form II $\hat{F}_4(\tilde{x})$ in the coordinate system of the interval bounds

Coordinates				Fuzzy variables					$\#\{\cdot\}$	$\hat{F}_4(\tilde{x})$
$x_{\alpha_1 1}$	$x_{\alpha_2 1}$	$x_{\alpha_1 r}$	$x_{\alpha_2 r}$	$\tilde{x}_j$	$x_{\alpha_1 1}(j)$	$x_{\alpha_2 1}(j)$	$x_{\alpha_1 r}(j)$	$x_{\alpha_2 r}(j)$		
1	4	4	4	–					0	0
1	5	5	5	$\tilde{x}_1$	1	5	5	5	1	0.25
3	6	6	6	$\tilde{x}_1$	1	5	5	5	2	0.5
				$\tilde{x}_3$	3	6	6	6		
3	8	8	8	$\tilde{x}_1$	1	5	5	5	3	0.75
				$\tilde{x}_2$	2	8	8	8		
				$\tilde{x}_3$	3	6	6	6		
4	8	8	8	$\tilde{x}_1$	1	5	5	5	4	1
				$\tilde{x}_2$	2	8	8	8		
				$\tilde{x}_3$	3	6	6	6		
				$\tilde{x}_4$	4	7	7	7		
⋮										⋮

The empirical fuzzy probability distribution function form II $\hat{F}_4(\tilde{x})$ obtained for this sample of fuzzy variables is shown in Fig. 2.15 together with the empirical fuzzy probability distribution function form I according to Eq. (2.96). This depiction illustrates the interrelation between the fuzzy probability distribution function form II according to Eq. (2.104) and the fuzzy probability distribution function form I according to Eq. (2.96). The marginal distributions of the multidimensional probability distribution function according to Eq. (2.104) correspond to the left and right boundary functions of the fuzzy probability distribution function form I. The uncoupled treatment of the marginal distributions does not take account of the dependencies between the different α-level sets. The interaction between the α-level sets of the fuzzy realizations of a fuzzy random variable is only taken into account using the fuzzy probability distribution function form II. For this reason it is possible to reproduce the underlying fuzzy sample elements. ◆

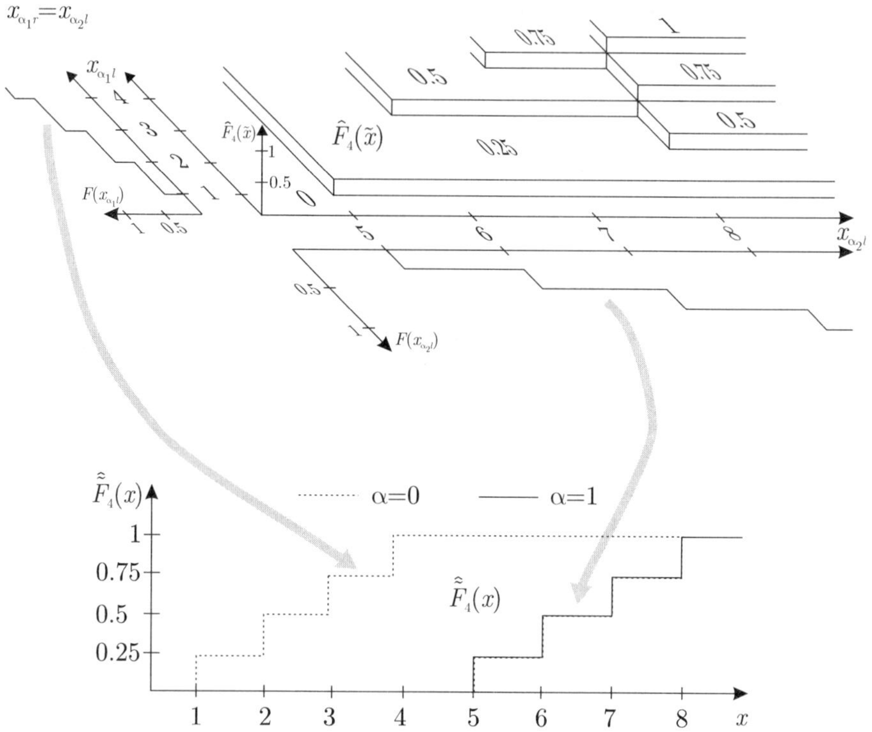

Fig. 2.15. Empirical fuzzy probability distribution function form II $\hat{F}_4(\tilde{x})$ in the coordinate sytem of the interval bounds together with the empirical fuzzy probability distribution function form I $\hat{\hat{F}}_4(x)$ for the exemplary sample

Remark 2.37. The fuzzy probability distribution function form I may be regarded as a simplified representation of form II, with the restriction that the dependencies between the different α-level sets are not taken into consideration. ◆

The simulation of a fuzzy random variable $\tilde{\mathrm{X}}$ requires that the realizations $x_{\alpha_1 l}$, ..., $x_{\alpha_n l}$ and $x_{\alpha_1 r}$, ..., $x_{\alpha_n r}$ of the interval boundaries must fulfill Eq. (2.80). A simulation in the coordinate system of the interval bounds and thus the fulfillment of Eq. (2.80) is exceedingly difficult, however. For this reason the fuzzy random variable $\tilde{\mathrm{X}}$ is expressed with the aid of $l_\alpha r_\alpha$-discretization. The realizations $\Delta x_{\alpha_1 l}$, ..., $\Delta x_{\alpha_n l}$ and $\Delta x_{\alpha_1 r}$, ..., $\Delta x_{\alpha_n r}$ of the random $l_\alpha r_\alpha$-increments $\Delta \mathrm{X}_{\alpha_1 l}$, ..., $\Delta \mathrm{X}_{\alpha_n l}$ and $\Delta \mathrm{X}_{\alpha_1 r}$, ..., $\Delta \mathrm{X}_{\alpha_n r}$, respectively, are thus regarded as the coordinates of the $2n$-dimensional coordinate system of the increments. The random $l_\alpha r_\alpha$-increments must fulfill the requirements according to Eqs. (2.90) and (2.91). The requirements are fulfilled easily if the coordinates are restricted to the set of positive real numbers in $\mathbb{R}^n$.

FPDF II in the coordinate system of the increments. For each α-level the random $l_\alpha r_\alpha$-increments $\Delta X_{\alpha_i l}$ and $\Delta X_{\alpha_i r}$ are real-valued random variables which are linked interactively. The realizations $\Delta x_{\alpha_i l}$ and $\Delta x_{\alpha_i r}$ of the random $l_\alpha r_\alpha$-increments $\Delta X_{\alpha_i l}$ and $\Delta X_{\alpha_i r}$, respectively, correspond to the $l_\alpha r_\alpha$-increments $a_{\alpha_i l}$ and $a_{\alpha_i r}$ of a fuzzy element $\tilde{a}$ of $F(A)$ with a specific probability. This probability is expressed by the probability $P(F(A))$.

Definition 2.38. *The fuzzy probability distribution function form II* $F_{\tilde{X}}(\tilde{x})$ *of a fuzzy random variable* $\tilde{X}$ *discretized by* $2n$ *random* $l_\alpha r_\alpha$*-increments* ΔX_1, ΔX_2, ..., ΔX_{2n} *is defined as the* $2n$*-dimensional probability distribution function* ${}_{lr}F_{\tilde{X}}(\tilde{x})$ *in the coordinate system of the increments according to Eq. (2.108), whereby* Δx_1, Δx_2, ..., Δx_{2n} *are the* $l_\alpha r_\alpha$*-increments of the fuzzy variable* $\tilde{x}$*. These form the coordinates in the coordinate system of the increments.*

$$
\begin{aligned}
{}_{lr}F_{\tilde{X}}(\tilde{x}) &= P\left(\{\omega \,|\, \Delta X_1(\omega) \leqslant \Delta x_1, \,...,\, \Delta X_{2n}(\omega) \leqslant \Delta x_{2n}\}\right) \qquad (2.108) \\
&= P\left(\{\Delta X_1 \leqslant \Delta x_1, \,...,\, \Delta X_{2n} \leqslant \Delta x_{2n}\}\right)
\end{aligned}
$$

◆

Furthermore, a real-valued probability distribution function $F_{\Delta X_i}(\Delta x_i)$ according to Eq. (2.109) exists for each random $l_\alpha r_\alpha$-increment ΔX_i $(i = 1, 2, ..., 2n)$ of a fuzzy random variable $\tilde{X}$. The probability distribution functions $F_{\Delta X_i}(\Delta x_i)$ given by Eq. (2.109) correspond to the marginal distribution functions of the $2n$-dimensional probability distribution function ${}_{lr}F_{\tilde{X}}(\tilde{x})$ according to Eq. (2.108).

$$
F_{\Delta X_i}(\Delta x_i) = \lim_{\substack{\Delta x_j \to \infty \\ j=1,2,...,2n \\ j \neq i}} {}_{lr}F_{\tilde{X}}(\tilde{x}) \qquad (2.109)
$$

Owing to the requirement of non-negativity for the random $l_\alpha r_\alpha$-increments ΔX_i according to Eqs. (2.90) and (2.91) the following equation holds for the marginal distribution functions $F_{\Delta X_i}(\Delta x_i)$.

$$
F_{\Delta X_i}(\Delta x_i) = 0 \mid \Delta x_i < 0 \quad \text{for} \quad i = 1, 2, ..., n-1, n+1, ..., 2n \qquad (2.110)
$$

The requirement according to Eq. (2.110) does not hold for the probability distribution function $F_{\Delta X_n}(\Delta x_n)$ at the random peak point $\Delta X_n = X_{\alpha_n l}$ due to the fact that the requirement of non-negativity according to Eq. (2.90) does not apply to $X_{\alpha_n l}$.

Remark 2.39. Normally, the multi-dimensionality of the both fuzzy probability distribution functions form II $F_{\tilde{X}}(\tilde{x})$ and ${}_{lr}F_{\tilde{X}}(\tilde{x})$ does not permit the graphical representation of $F_{\tilde{X}}(\tilde{x})$ and ${}_{lr}F_{\tilde{X}}(\tilde{x})$. It is possible to graphically illustrate the marginal distributions, however, without consideration of the dimensionality, i.e. this is independent of the number of chosen α-levels. Although this permits the use of the fuzzy probability distribution function form I as a simplified graphical representation of form II, it is not able to completely characterize a fuzzy random variable. For practical applications a tabular representation or functional description of $F_{\tilde{X}}(\tilde{x})$ and ${}_{lr}F_{\tilde{X}}(\tilde{x})$ is recommended. ◆

For the special case of a discrete fuzzy random variable $\tilde{X}$ (see definition 2.24) a discrete fuzzy probability distribution function form II is obtained. A discrete fuzzy random variable $\tilde{X}$ possesses a finite or countable infinite number of realizations $\tilde{x}_1, \tilde{x}_2, ..., \tilde{x}_m$. Each of these realizations appears with a certain probability $P(\tilde{X} = \tilde{x}_j) = P_j$. The definition of the fuzzy probability distribution function form II ${}_{lr}F_{\tilde{X}}(\tilde{x})$ of a discrete fuzzy random variable $\tilde{X}$ is thus given by Eq. (2.111), whereby $j_1, ..., j_{2n} = 1, 2, ..., m$ holds.

$$
{}_{lr}F_{\tilde{X}}(\tilde{x}) = \sum_{\substack{\Delta x_1(j_1) \leqslant \Delta x_1 \\ \vdots \\ \Delta x_{2n}(j_{2n}) \leqslant \Delta x_{2n}}} P\left(\{\Delta X_1 = \Delta x_1(j_1), \, ..., \, \Delta X_{2n} = \Delta x_{2n}(j_{2n})\}\right) \quad (2.111)
$$

Example 2.40. A discrete fuzzy random variable $\tilde{X}$ may adopt ten different realizations $\tilde{x}_1, \tilde{x}_2, ..., \tilde{x}_{10}$. For the $l_\alpha r_\alpha$-increments $\Delta x_i(j)$ of the possible realizations $\tilde{x}_j$ $(j = 1, 2, ..., 10)$ Eq. (2.112) holds.

$$
\Delta x_i(j) < \Delta x_i(k) \quad \text{for} \quad j < k \quad \forall\, j, k = 1, 2, ..., 10 \qquad (2.112)
$$
$$
i = 1, 2, ..., 2n
$$

Each of the ten realizations appears with a probability of $P(\tilde{X} = \tilde{x}_j) = \frac{1}{10}$. Selected functional values of the associated fuzzy probability distribution function form II ${}_{lr}F_{\tilde{X}}(\tilde{x})$ (evaluated according to Eq. (2.111)) are given in Table 2.3.

Table 2.3. Selected functional values of a discrete fuzzy probability distribution function form II ${}_{lr}F_{\tilde{X}}(\tilde{x})$ in the coordinate system of the increments

${}_{lr}F_{\tilde{X}}(\tilde{x}_1) = \frac{1}{10}$	${}_{lr}F_{\tilde{X}}(\tilde{x}_6) = \frac{6}{10}$
${}_{lr}F_{\tilde{X}}(\tilde{x}_2) = \frac{2}{10}$	${}_{lr}F_{\tilde{X}}(\tilde{x}_7) = \frac{7}{10}$
${}_{lr}F_{\tilde{X}}(\tilde{x}_3) = \frac{3}{10}$	${}_{lr}F_{\tilde{X}}(\tilde{x}_8) = \frac{8}{10}$
${}_{lr}F_{\tilde{X}}(\tilde{x}_4) = \frac{4}{10}$	${}_{lr}F_{\tilde{X}}(\tilde{x}_9) = \frac{9}{10}$
${}_{lr}F_{\tilde{X}}(\tilde{x}_5) = \frac{5}{10}$	${}_{lr}F_{\tilde{X}}(\tilde{x}_{10}) = 1$

◆

The fuzzy probability density functions form II ${}_{lr}f_{\tilde{\mathrm{X}}}(\tilde{x})$ of continuous fuzzy random variables $\tilde{\mathrm{X}}$ according to Definition 2.25 are obtained (analogous to classical probability theory) by partial differentiation of the fuzzy probability distribution function form II ${}_{lr}F_{\tilde{\mathrm{X}}}(\tilde{x})$ according to Eq. (2.113).

$$ {}_{lr}f_{\tilde{\mathrm{X}}}(\tilde{x}) = \frac{\partial^{2n}\,{}_{lr}F_{\tilde{\mathrm{X}}}(\tilde{x})}{\partial \Delta x_1 \cdots \partial \Delta x_{2n}} \tag{2.113} $$

The fuzzy probability distribution function form II ${}_{lr}F_{\tilde{\mathrm{X}}}(\tilde{x})$ of a continuous fuzzy random variable $\tilde{\mathrm{X}}$ may thus be obtained by inversion of the differentiation given by Eq. (2.113), i.e. ${}_{lr}F_{\tilde{\mathrm{X}}}(\tilde{x})$ may be obtained by integration of the fuzzy probability density function form II ${}_{lr}f_{\tilde{\mathrm{X}}}(\tilde{x})$ according to Eq. (2.114).

$$ {}_{lr}F_{\tilde{\mathrm{X}}}(\tilde{x}) = \int\limits_{0}^{\Delta x_1} \cdots \int\limits_{-\infty}^{\Delta x_n} \cdots \int\limits_{0}^{\Delta x_{2n}} {}_{lr}f_{\tilde{\mathrm{X}}}(\tilde{t})\; d\Delta t_1 \cdots d\Delta t_n \cdots d\Delta t_{2n} \tag{2.114} $$

According to the definition 2.38 the empirical fuzzy probability distribution function form II in the coordinate system of the increments may also be derived. This presupposes that a sample of s fuzzy variables is available.

Definition 2.41. *The empirical fuzzy probability distribution function form II ${}_{lr}\hat{F}_s(\tilde{x})$ for a sample of s fuzzy variables $\tilde{x}_1$, $\tilde{x}_2$, ..., $\tilde{x}_s$ is defined in the coordinate system of the increments by Eq. (2.115).*

$$ {}_{lr}\hat{F}_s(\tilde{x}) = \frac{\#\left\{\tilde{x}_j \,|\, \Delta x_1(j) \leqslant \Delta x_1,\, ...,\, \Delta x_{2n}(j) \leqslant \Delta x_{2n},\, j = 1, 2, ..., s\right\}}{s} \tag{2.115} $$

◆

The coordinates Δx_1, Δx_2, ..., Δx_{2n} are the $l_\alpha r_\alpha$-increments of the fuzzy variable $\tilde{x}$, and $\Delta x_1(j)$, $\Delta x_2(j)$, ..., $\Delta x_{2n}(j)$ are the $l_\alpha r_\alpha$-increments of the fuzzy variables $\tilde{x}_1$, $\tilde{x}_2$, ..., $\tilde{x}_s$. The symbol $\#\{\cdot\}$ denotes the number of fuzzy variables $\tilde{x}_j$ ($j = 1, 2, ..., s$) for which the requirements $\Delta x_1(j) \leqslant \Delta x_1$, ..., $\Delta x_{2n}(j) \leqslant \Delta x_{2n}$ are fulfilled. According to Eq. (2.115) the empirical fuzzy probability distribution function form II ${}_{lr}\hat{F}_s(\tilde{x})$ is a monotonically non-decreasing $2n$-dimensional step function.

Example 2.42. The computation of ${}_{lr}\hat{F}_s(\tilde{x})$ according to Eq. (2.115) is demonstrated by way of the exemplary sample of 4 fuzzy variables, as shown in Fig. 2.12. The fuzzy variables are discretized for two α-levels ($\alpha_1 = 0$ and $\alpha_2 = 1$). The interval boundaries of the α-level sets and the $l_\alpha r_\alpha$-increments of the fuzzy sample elements are shown in Fig. 2.16.

Eq. (2.115) is now applied to determine the number $\#\{\cdot\}$ of fuzzy variables $\tilde{x}_j$ ($j = 1, 2, 3, 4$) for which the requirements $\Delta x_1(j) \leqslant \Delta x_1$, ..., $\Delta x_{2n}(j) \leqslant \Delta x_{2n}$ are fulfilled. The functional values of ${}_{lr}\hat{F}_s(\tilde{x})$ are also shown in Table 2.4 for a selected number of coordinates.

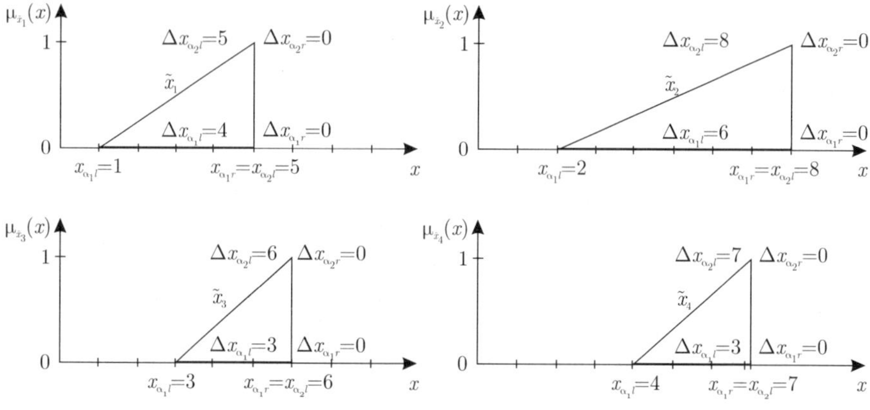

Fig. 2.16. Exemplary sample of fuzzy variables

Table 2.4. Computation of the empirical fuzzy probability distribution function form II ${}_{lr}\hat{F}_4(\tilde{x})$ in the coordinate system of the increments

Coordinates				Fuzzy variables					$\#\{\cdot\}$	${}_{lr}\hat{F}_4(\tilde{x})$
Δx_1	Δx_2	Δx_3	Δx_4	$\tilde{x}_j$	$\Delta x_1(j)$	$\Delta x_2(j)$	$\Delta x_3(j)$	$\Delta x_4(j)$		
3	5	0	0	–					0	0
3	6	0	0	$\tilde{x}_3$	3	6	0	0	1	0.25
5	6	0	0	$\tilde{x}_1$	4	5	0	0	2	0.5
				$\tilde{x}_3$	3	6	0	0		
5	8	0	0	$\tilde{x}_1$	4	5	0	0	3	0.75
				$\tilde{x}_3$	3	6	0	0		
				$\tilde{x}_4$	3	7	0	0		
6	8	0	0	$\tilde{x}_1$	4	5	0	0	4	1
				$\tilde{x}_2$	6	8	0	0		
				$\tilde{x}_3$	3	6	0	0		
				$\tilde{x}_4$	3	7	0	0		
⋮										⋮

Due to the fact that the peak point of each fuzzy sample element coincides with the right-hand boundary of the associated support interval, i.e. $\Delta x_{\alpha_1 r} = 0$ and $\Delta x_{\alpha_2 r} = 0$, the four-dimensional empirical probability distribution function ${}_{lr}\hat{F}_4(\tilde{x})$ according to Eq. (2.115) may be represented two-dimensionally. The resulting empirical fuzzy probability distribution function form II ${}_{lr}\hat{F}_4(\tilde{x})$ is shown in Fig. 2.17. ◆

Empirical frequency distribution. For a concrete sample of fuzzy variables an empirical frequency distribution may also be evaluated. In so doing, it is necessary to draw a distinction between samples comprised of discrete or

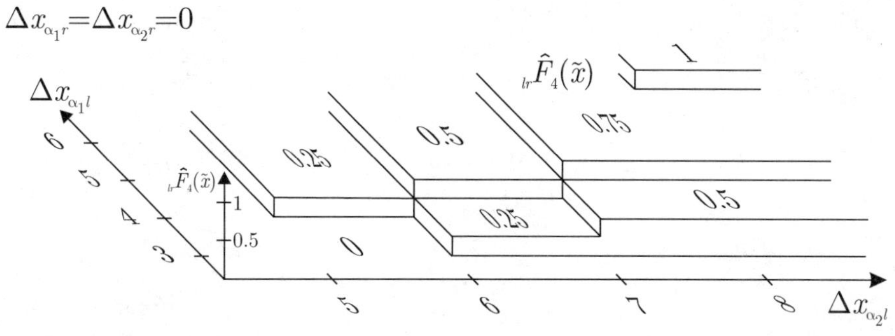

Fig. 2.17. Empirical fuzzy probability distribution function form II $_{lr}\hat{F}_s(\tilde{x})$ in the coordinate system of the increments for the exemplary sample

continuous fuzzy random variables. This procedure, which is explained in the following, requires an evaluation of the relative frequencies of the realizations of the fuzzy random variable.

Under the condition that the realizations of the discrete fuzzy random variable $\tilde{X}$ are restricted to the set $\tilde{a}_1$, $\tilde{a}_2$, ..., $\tilde{a}_m$ of m different fuzzy variables, the following definition holds.

Definition 2.43. *Considering s realizations $\tilde{x}_1$, $\tilde{x}_2$, ..., $\tilde{x}_s$ of the discrete fuzzy random variable $\tilde{X}$, the relative frequency $r_s(\tilde{a}_k)$ of the fuzzy variable $\tilde{a}_k$ is defined by Eq. (2.116) as follows:*

$$r_s(\tilde{a}_k) = \frac{h_s(\tilde{a}_k)}{s} \quad \text{with} \quad k = 1, 2, ..., m\,. \tag{2.116}$$

◆

In this equation, $h_s(\tilde{a}_k)$ denotes the absolute frequency of $\tilde{a}_k$, i.e. the number of realizations (sample elements) $\tilde{x}$ corresponding to $\tilde{a}_k$.

Example 2.44. Let us consider 60 realizations $\tilde{x}_1$, $\tilde{x}_2$, ..., $\tilde{x}_{60}$ of a discrete fuzzy random number $\tilde{X}$, which may be assigned to ten different fuzzy variables $\tilde{a}_1$, $\tilde{a}_2$, ..., $\tilde{a}_{10}$. Assuming that 18 fuzzy sample elements correspond to the fuzzy variable $\tilde{a}_1$, the absolute frequency of $\tilde{a}_1$ is thus $h_{60}(\tilde{a}_1) = 18$. In accordance with Eq. (2.116) the relative frequency $r_{60}(\tilde{a}_1)$ of the fuzzy variable $\tilde{a}_1$ is hence $r_{60}(\tilde{a}_1) = \frac{18}{60} = 0.3$. ◆

Let $\tilde{x}_1$, $\tilde{x}_2$, ..., $\tilde{x}_s$ be a concrete sample of s realizations of a continuous fuzzy random variable. In order to evaluate the frequencies of the sample it is necessary to classify the $l_\alpha r_\alpha$-increments of the fuzzy sample elements. For this purpose $2n$ intervals are constructed for the $2n$ random $l_\alpha r_\alpha$-increments, which include all corresponding $l_\alpha r_\alpha$-increments of the sample elements in each case. These intervals are subdivided into several classes, whereby the individual classes are denoted by $K_{a_1}(1)$, $K_{a_2}(2)$, ..., $K_{a_{2n}}(2n)$. Taking m_i to be the number of classes for the ith random $l_\alpha r_\alpha$-increment, $a_i = 1, 2, ..., m_i$ holds.

Definition 2.45. *Let* $\tilde{x}_1$, $\tilde{x}_2$, ..., $\tilde{x}_s$ *be s realizations of a continuous fuzzy random variable* $\tilde{X}$. *The relative class frequency* $r_s(K_{a_1}(1), K_{a_2}(2), ..., K_{a_{2n}}(2n))$ *of the sample is thus defined by Eq. (2.117).*

$$r_s(K_{a_1}(1), K_{a_2}(2), ..., K_{a_{2n}}(2n)) = \frac{h_s(K_{a_1}(1), K_{a_2}(2), ..., K_{a_{2n}}(2n))}{s} \quad (2.117)$$

$$\text{with} \quad a_i = 1,\, 2,\, ...,\, m_i$$
$$\text{and} \quad i = 1,\, 2,\, ...,\, 2n$$

◆

In the above equation, $h_s(K_{a_1}(1), K_{a_2}(2), ..., K_{a_{2n}}(2n))$ is the absolute class frequency of $K_{a_1}(1)$, $K_{a_2}(2)$, ..., $K_{a_{2n}}(2n)$, i.e. the number of fuzzy sample elements whose $l_\alpha r_\alpha$-increments are all contained in the classes $K_{a_1}(1)$, $K_{a_2}(2)$, ..., $K_{a_{2n}}(2n)$.

Example 2.46. Let $\tilde{x}_1$, $\tilde{x}_2$, ..., $\tilde{x}_{80}$ be a sample of 80 realizations of the continuous fuzzy random variable $\tilde{X}$. The fuzzy variables are discretized exemplarily by $n = 2$ α-level sets. For $\alpha = 0$ all $l_\alpha r_\alpha$-increments $\Delta x_{\alpha_1 l}$ and $\Delta x_{\alpha_1 r}$ are contained in the intervals [3.2; 4.1] and [1.8; 2.3]. For $\alpha = 1$ all $l_\alpha r_\alpha$-increments $\Delta x_{\alpha_2 l}$ and $\Delta x_{\alpha_2 r}$ are contained in the intervals [−50.3; −43.1] and [1.3; 1.5]. These intervals are subdivided in each case into ten classes $K_1(i)$, $K_2(i)$, ..., $K_{10}(i)$. The counter $i = 1, 2, 3, 4$ denotes the number of the subdivided interval. For the purpose of demonstration the classes $K_1(1)$, $K_1(2)$, $K_1(3)$ and $K_1(4)$ are analyzed. The $l_\alpha r_\alpha$-increments of four fuzzy sample elements are completely contained in the classes $K_1(1)$, $K_1(2)$, $K_1(3)$ and $K_1(4)$, i.e. the absolute class frequency of $K_1(1)$, $K_1(2)$, $K_1(3)$ and $K_1(4)$ is given by $h_{80}(K_1(1),K_1(2),K_1(3),K_1(4)) = 4$. In accordance with Eq. (2.117) the relative class frequency is thus computed to be $r_{80}(K_1(1), K_1(2), K_1(3), K_1(4)) = \frac{4}{80} = 0.05$. ◆

Remark 2.47. The introduced relative class frequency of a sample of continuous fuzzy random variables may be approximately interpreted as an empirical fuzzy probability density function form II $_{lr}\hat{f}_s(\tilde{x})$. In this formulation the relative class frequencies are replaced by the probabilities of occurence of the fuzzy random variable. The interpretation of the relative frequency of a discrete fuzzy random variable as a fuzzy probability density function form II is not fully correct, however, as it is not a density function by definition. ◆

2.2.3 Characteristic Moments

As in the case of random variables, fuzzy random variables are characterized using the first and second order moments. The definitions of the characteristic moments on the basis of $l_\alpha r_\alpha$-discretization are presented in the following.

Definition 2.48. *The first order moment of a fuzzy random variable* $\tilde{X}$ *is the fuzzy expected value* $E[\tilde{X}] = \tilde{m}_{\tilde{X}}$ *given by Eq. (2.118). The fuzzy expected value* $E[\tilde{X}] = \tilde{m}_{\tilde{X}}$ *is a fuzzy variable which may be represented by* $l_\alpha r_\alpha$*-discretization according to Eq. (2.32).*

$$E[\tilde{X}] = \tilde{m}_{\tilde{X}} = \int_0^\infty \cdots \int_{-\infty}^\infty \cdots \int_0^\infty {}_{lr}f_{\tilde{X}}(\tilde{x})\, \tilde{x}\; d\Delta x_1 \cdots d\Delta x_n \cdots d\Delta x_{2n} \quad (2.118)$$

The $l_\alpha r_\alpha$*-increments* $\Delta m_{\alpha_i l}$ *and* $\Delta m_{\alpha_i r}$ *of the fuzzy expected value* $E[\tilde{X}] = \tilde{m}_{\tilde{X}}$ *of a fuzzy random variable* $\tilde{X}$ *are obtained according to Eq. (2.119).*

$$E[\tilde{X}] = \tilde{m}_{\tilde{X}} = \begin{bmatrix} \Delta m_{\alpha_1 l} \\ \vdots \\ \Delta m_{\alpha_n l} \\ \vdots \\ \Delta m_{\alpha_1 r} \end{bmatrix} = \begin{bmatrix} \int_0^\infty \Delta x_{\alpha_1 l}\, f_{\Delta X_{\alpha_1 l}}(\Delta x_{\alpha_1 l}) d\Delta x_{\alpha_1 l} \\ \vdots \\ \int_{-\infty}^\infty \Delta x_{\alpha_n l}\, f_{\Delta X_{\alpha_n l}}(\Delta x_{\alpha_n l}) d\Delta x_{\alpha_n l} \\ \vdots \\ \int_0^\infty \Delta x_{\alpha_1 r}\, f_{\Delta X_{\alpha_1 r}}(\Delta x_{\alpha_1 r}) d\Delta x_{\alpha_1 r} \end{bmatrix} \quad (2.119)$$

◆

The real-valued functions $f_{\Delta X_{\alpha_i l}}(\Delta x_{\alpha_i l})$ and $f_{\Delta X_{\alpha_i r}}(\Delta x_{\alpha_i r})$ are the probability distribution functions of the random $l_\alpha r_\alpha$-increments $\Delta X_{\alpha_i l}$ and $\Delta X_{\alpha_i r}$ ($i = 1, 2, ..., n$) of the fuzzy random variable $\tilde{X}$. It is only necessary to evaluate the integral in the above equation from $-\infty$ to $+\infty$ for the $l_\alpha r_\alpha$-increment $\Delta m_{\alpha_n l}$ of the peak point. The integration limits of the remaining integrals follow from the requirements of Eqs. (2.90) and (2.91).

Remark 2.49. Although the number n of the chosen α-levels determines the $l_\alpha r_\alpha$-increment representation of $E[\tilde{X}] = \tilde{m}_{\tilde{X}}$, the fuzzy expected value $E[\tilde{X}] = \tilde{m}_{\tilde{X}}$ is inherently independent of n. ◆

Example 2.50. The fuzzy expected value $E[\tilde{X}] = \tilde{m}_{\tilde{X}}$ of an exemplary fuzzy random variable $\tilde{X}$ is illustrated in Fig. 2.18. ◆

Definition 2.51. *Linear dependencies between the random* $l_\alpha r_\alpha$*-increments of a fuzzy random variable* $\tilde{X}$ *are quantified by the* $l_\alpha r_\alpha$*-covariance* $k^{\alpha_i l*}_{\alpha_j r*}$ *according to Eq. (2.120).*

$$k^{\alpha_i l*}_{\alpha_j r*} = E\left[\left(\Delta X_{\alpha_i l*} - \Delta m_{\alpha_i l*} \right) \left(\Delta X_{\alpha_j r*} - \Delta m_{\alpha_j r*} \right) \right] \quad (2.120)$$

◆

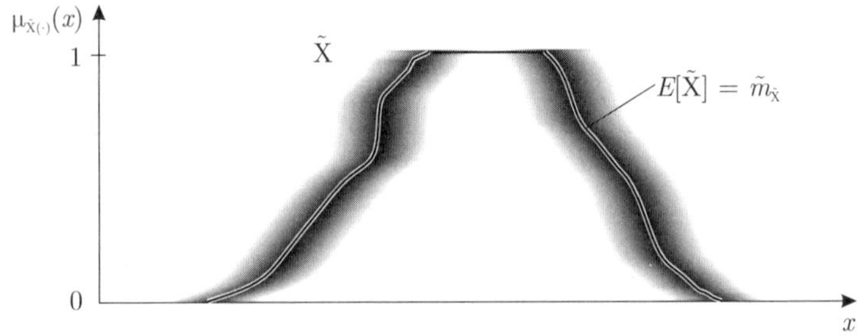

Fig. 2.18. Fuzzy expected value $E[\tilde{X}] = \tilde{m}_{\tilde{X}}$ of a fuzzy random variable $\tilde{X}$

The subscripts α_i and α_j refer to the different α-levels, whereas $l^* = l, r$ and $r^* = l, r$ denote the left-hand and right-hand branches of the memberhip functions, respectively. The $l_\alpha r_\alpha$-covariance $k_{\alpha_j r*}^{\alpha_i l*}$ is computed for $i, j = 1, 2, ..., n$ by solving the integral given by Eq. (2.121).

$$k_{\alpha_j r*}^{\alpha_i l*} = \int_{-\infty}^{\infty} \int_{-\infty}^{\infty} \left(\Delta x_{\alpha_i l*} - \Delta m_{\alpha_i l*}\right) \left(\Delta x_{\alpha_j r*} - \Delta m_{\alpha_j r*}\right) ... \tag{2.121}$$

$$...f\left(\Delta x_{\alpha_i l*}, \Delta x_{\alpha_j r*}\right) d\Delta x_{\alpha_i l*} d\Delta x_{\alpha_j r*}$$

The function $f\left(\Delta x_{\alpha_i l*}, \Delta x_{\alpha_j r*}\right)$ is the joint probability density function of the random $l_\alpha r_\alpha$-increments $\Delta \mathrm{X}_{\alpha_i l*}$ and $\Delta \mathrm{X}_{\alpha_i r*}$ according to definition 2.38. The values of the $l_\alpha r_\alpha$-covariances $k_{\alpha_j r*}^{\alpha_i l*}$ are arranged in the $l_\alpha r_\alpha$-covariance matrix ${}_{lr}\underline{K}_{\tilde{X}}$ according to Eq. (2.122).

$${}_{lr}\underline{K}\left[\tilde{X}\right] = {}_{lr}\underline{K}_{\tilde{X}} = \begin{bmatrix} k_{\alpha_1 l}^{\alpha_1 l} & k_{\alpha_1 l}^{\alpha_2 l} & \cdots & k_{\alpha_1 l}^{\alpha_2 r} & k_{\alpha_1 l}^{\alpha_1 r} \\ k_{\alpha_2 l}^{\alpha_1 l} & k_{\alpha_2 l}^{\alpha_2 l} & \cdots & k_{\alpha_2 l}^{\alpha_2 r} & k_{\alpha_2 l}^{\alpha_1 r} \\ \vdots & \vdots & \ddots & \vdots & \vdots \\ k_{\alpha_2 r}^{\alpha_1 l} & k_{\alpha_2 r}^{\alpha_2 l} & \cdots & k_{\alpha_2 r}^{\alpha_2 r} & k_{\alpha_2 r}^{\alpha_1 r} \\ k_{\alpha_1 r}^{\alpha_1 l} & k_{\alpha_1 r}^{\alpha_2 l} & \cdots & k_{\alpha_1 r}^{\alpha_2 r} & k_{\alpha_1 r}^{\alpha_1 r} \end{bmatrix} \tag{2.122}$$

Definition 2.52. *The scale-invariant $l_\alpha r_\alpha$-correlation matrix ${}_{lr}\underline{R}_{\tilde{X}}$ given by Eq. (2.123) is obtained by element-by-element division of the $l_\alpha r_\alpha$-covariance matrix ${}_{lr}\underline{K}_{\tilde{X}}$ by the associated diagonal elements according to Eq. (2.124).*

$$
{}_{lr}\underline{R}\left[\tilde{\mathrm{X}}\right] = {}_{lr}\underline{R}_{\tilde{\mathrm{X}}} = \begin{bmatrix} r^{\alpha_1 l}_{\alpha_1 l} & r^{\alpha_2 l}_{\alpha_1 l} & \cdots & r^{\alpha_2 r}_{\alpha_1 l} & r^{\alpha_1 r}_{\alpha_1 l} \\ r^{\alpha_1 l}_{\alpha_2 l} & r^{\alpha_2 l}_{\alpha_2 l} & \cdots & r^{\alpha_2 r}_{\alpha_2 l} & r^{\alpha_1 r}_{\alpha_2 l} \\ \vdots & \vdots & \ddots & \vdots & \vdots \\ r^{\alpha_1 l}_{\alpha_2 r} & r^{\alpha_2 l}_{\alpha_2 r} & \cdots & r^{\alpha_2 r}_{\alpha_2 r} & r^{\alpha_1 r}_{\alpha_2 r} \\ r^{\alpha_1 l}_{\alpha_1 r} & r^{\alpha_2 l}_{\alpha_1 r} & \cdots & r^{\alpha_2 r}_{\alpha_1 r} & r^{\alpha_1 r}_{\alpha_1 r} \end{bmatrix} \tag{2.123}
$$

$$
r^{\alpha_i l*}_{\alpha_j l*} = \frac{k^{\alpha_i l*}_{\alpha_j l*}}{\sqrt{k^{\alpha_i l*}_{\alpha_i l*}\, k^{\alpha_j l*}_{\alpha_j l*}}} \tag{2.124}
$$

◆

Remark 2.53. The $l_\alpha r_\alpha$-covariance matrix ${}_{lr}\underline{K}_{\tilde{\mathrm{X}}}$ as well as the $l_\alpha r_\alpha$-correlation matrix ${}_{lr}\underline{R}_{\tilde{\mathrm{X}}}$ are strictly dependent on the $l_\alpha r_\alpha$-increment representation and thus dependent on the number n of chosen α-levels. The number n of chosen α-levels determines the dimension of the matrices. ◆

Definition 2.54. *The $l_\alpha r_\alpha$-variance ${}_{lr}Var[\tilde{\mathrm{X}}] = {}_{lr}\underline{\sigma}^2_{\tilde{\mathrm{X}}}$ of a fuzzy random variable $\tilde{\mathrm{X}}$ is a measure of the variance of the realizations of $\tilde{\mathrm{X}}$, and is represented in the form of a column vector with $[2n]$ elements. The elements of the vector correspond to the diagonal elements of the $l_\alpha r_\alpha$-covariance matrix ${}_{lr}\underline{K}_{\tilde{\mathrm{X}}}$ according to Eq. (2.125).*

$$
{}_{lr}Var[\tilde{\mathrm{X}}] = {}_{lr}\underline{\sigma}^2_{\tilde{\mathrm{X}}} = \begin{bmatrix} k^{\alpha_1 l}_{\alpha_1 l} \\ k^{\alpha_2 l}_{\alpha_2 l} \\ \vdots \\ k^{\alpha_n l}_{\alpha_n l} \\ k^{\alpha_n r}_{\alpha_n r} \\ \vdots \\ k^{\alpha_2 r}_{\alpha_2 r} \\ k^{\alpha_1 r}_{\alpha_1 r} \end{bmatrix} = \begin{bmatrix} {}_{lr}\sigma^2_1 \\ {}_{lr}\sigma^2_2 \\ \vdots \\ {}_{lr}\sigma^2_n \\ {}_{lr}\sigma^2_{n+1} \\ \vdots \\ {}_{lr}\sigma^2_{2n-1} \\ {}_{lr}\sigma^2_{2n} \end{bmatrix} \tag{2.125}
$$

◆

Remark 2.55. The $l_\alpha r_\alpha$-variance ${}_{lr}\underline{\sigma}^2_{\tilde{\mathrm{X}}}$ (i.e. the dimension of the vector and its elements) is thus dependent on the number n of chosen α-levels. Generally speaking, however, the $l_\alpha r_\alpha$-variance ${}_{lr}\underline{\sigma}^2_{\tilde{\mathrm{X}}}(n_1)$ based on the $l_\alpha r_\alpha$-discretization with n_1 α-levels is not directly transformable into the $l_\alpha r_\alpha$-variance ${}_{lr}\underline{\sigma}^2_{\tilde{\mathrm{X}}}(n_2)$ based on n_2 discrete α-levels. The transformation is performed with the aid of the $l_\alpha r_\alpha$-standard deviation. The $l_\alpha r_\alpha$-standard deviation ${}_{lr}\underline{\sigma}_{\tilde{\mathrm{X}}}$ is obtained by extracting the positive square root for each element of the $l_\alpha r_\alpha$-variance

${}_{lr}\underline{\sigma}^2_{\tilde{\mathrm{X}}}$ of a fuzzy random vector $\tilde{\mathrm{X}}$ according to Eq. (2.126). The $l_\alpha r_\alpha$-standard deviation ${}_{lr}\underline{\sigma}_{\tilde{\mathrm{X}}}$ is thus a $[2n]$ column vector with the vector elements

$$ {}_{lr}\sigma_i = +\sqrt{{}_{lr}\sigma_i^2}\,. \tag{2.126}$$

In specific cases the $l_\alpha r_\alpha$-standard deviation ${}_{lr}\underline{\sigma}_{\tilde{\mathrm{X}}}(n_1)$ based on the $l_\alpha r_\alpha$-discretization with n_1 α-levels may be transformed directly into the $l_\alpha r_\alpha$-standard deviation ${}_{lr}\underline{\sigma}_{\tilde{\mathrm{X}}}(n_2)$ with n_2 underlying α-levels. For the specific case of a fuzzy random variable $\tilde{\mathrm{X}}$ whose realizations are in the form of fuzzy triangular numbers or fuzzy intervals the transformation is given exemplarily by Eq. (2.127).

$$ {}_{lr}\sigma_i(n_2) = \begin{cases} \frac{1}{n_2-1} \sum\limits_{j=1}^{n_1-1} {}_{lr}\sigma_j(n_1) & \text{for} \quad i = 1, 2, ..., n_2 - 1 \\ {}_{lr}\sigma_i(n_1) & \text{for} \quad i = n_2, n_2 + 1 \\ \frac{1}{n_2-1} \sum\limits_{j=n_1+2}^{2n_1} {}_{lr}\sigma_j(n_1) & \text{for} \quad i = n_2 + 2, n_2 + 3, ..., 2n_2 \end{cases} \tag{2.127}$$

◆

Linear dependencies between the random $l_\alpha r_\alpha$-increments of a fuzzy random variable $\tilde{\mathrm{X}}$ and the random $l_\alpha r_\alpha$-increments of a fuzzy random variable $\tilde{\mathrm{Y}}$ are quantified by the $l_\alpha r_\alpha$-covariance matrix ${}_{lr}\underline{K}[\tilde{\mathrm{X}}, \tilde{\mathrm{Y}}] = {}_{lr}\underline{K}_{\tilde{\mathrm{X}}\tilde{\mathrm{Y}}}$ and the $l_\alpha r_\alpha$-correlation matrix ${}_{lr}\underline{R}[\tilde{\mathrm{X}}, \tilde{\mathrm{Y}}] = {}_{lr}\underline{R}_{\tilde{\mathrm{X}}\tilde{\mathrm{Y}}}$, respectively. Both ${}_{lr}\underline{K}_{\tilde{\mathrm{X}}\tilde{\mathrm{Y}}}$ and ${}_{lr}\underline{R}_{\tilde{\mathrm{X}}\tilde{\mathrm{Y}}}$ correspond the $[2n, 2n]$ matrices according Eqs. (2.122) and (2.123). The elements of the $l_\alpha r_\alpha$-covariance matrix ${}_{lr}\underline{K}_{\tilde{\mathrm{X}}\tilde{\mathrm{Y}}}$ are defined by Eq. (2.128), whereby $i, j = 1, 2, ..., n$ holds. The subscripts α_i and α_j likewise refer to the different α-levels, whereas $l^* = l, r$ and $r^* = l, r$ specify the left-hand and right-hand branches of the memberhip functions, respectively.

$$ k^{\alpha_i l*}_{\alpha_j r*}[\tilde{\mathrm{X}}\tilde{\mathrm{Y}}] = \int\limits_{-\infty}^{\infty} \int\limits_{-\infty}^{\infty} \left(\Delta x_{\alpha_i l*} - \Delta m_{\alpha_i l*}(\tilde{\mathrm{X}})\right)\left(\Delta y_{\alpha_j r*} - \Delta m_{\alpha_j r*}(\tilde{\mathrm{Y}})\right) ... \tag{2.128}$$

$$ ... f\left(\Delta x_{\alpha_i l*}, \Delta y_{\alpha_j r*}\right) d\Delta x_{\alpha_i l*} d\Delta y_{\alpha_j r*}$$

Analogous to the application of Eq. (2.124), the elements of the scale-invariant $l_\alpha r_\alpha$-correlation matrix ${}_{lr}\underline{R}_{\tilde{\mathrm{X}}\tilde{\mathrm{Y}}}$ are obtained by dividing the $l_\alpha r_\alpha$-covariance matrix ${}_{lr}\underline{K}_{\tilde{\mathrm{X}}\tilde{\mathrm{Y}}}$ by the associated diagonal elements of the $l_\alpha r_\alpha$-covariance matrices ${}_{lr}\underline{K}_{\tilde{\mathrm{X}}}$ and ${}_{lr}\underline{K}_{\tilde{\mathrm{Y}}}$ element-by-element according to Eq. (2.129).

$$ r^{\alpha_i l*}_{\alpha_j l*}[\tilde{\mathrm{X}}\tilde{\mathrm{Y}}] = \frac{k^{\alpha_i l*}_{\alpha_j l*}[\tilde{\mathrm{X}}\tilde{\mathrm{Y}}]}{\sqrt{k^{\alpha_i l*}_{\alpha_i l*}[\tilde{\mathrm{X}}]\; k^{\alpha_j l*}_{\alpha_j l*}[\tilde{\mathrm{Y}}]}} \tag{2.129}$$

The correlation between the random $l_\alpha r_\alpha$-increments of a fuzzy random variable $\tilde{X}$ may be impaired by the influence of fuzzy random variables $\tilde{Y}_1$, $\tilde{Y}_2$, ..., $\tilde{Y}_m$. In order to uncouple the influence of the fuzzy random variables $\tilde{Y}_1$, $\tilde{Y}_2$, ..., $\tilde{Y}_m$ on a fuzzy random variable $\tilde{X}$ the partial $l_\alpha r_\alpha$-correlation matrix ${}_{lr}\underline{P}_{\tilde{X}/\tilde{Y}}$ is defined.

Definition 2.56. *After eliminating the influence of the fuzzy random variables $\tilde{Y}_1$, $\tilde{Y}_2$, ..., $\tilde{Y}_m$, the partial $l_\alpha r_\alpha$-correlation matrix ${}_{lr}\underline{P}_{\tilde{X}/\tilde{Y}}$ of a fuzzy random variable is given by Eq. (2.130).*

$$ {}_{lr}\underline{P}_{\tilde{X}/\tilde{Y}} = {}_{lr}\underline{R}\left[\tilde{X} \ominus \check{\tilde{X}}\right] \tag{2.130}$$

◆

The term $\check{\tilde{X}}$ hereby represents the best linear approximation of the fuzzy random variable $\tilde{X}$ by the fuzzy random variables $\tilde{Y}_1$, $\tilde{Y}_2$, ..., $\tilde{Y}_m$, i.e. it holds that

$$\check{\tilde{X}} = \underline{A}_1 \odot \tilde{Y}_1 \oplus \underline{A}_2 \odot \tilde{Y}_2 \oplus \ldots \oplus \underline{A}_m \odot \tilde{Y}_m \tag{2.131}$$

with the requirement according to Eq. (2.132).

$$\sum_{i=1}^{2n} E\left[\Delta X_i - \sum_{j=1}^{2n}\sum_{k=1}^{m} a_k[i,j]\,\Delta Y_j(k)\right] \stackrel{!}{=} \min \tag{2.132}$$

The terms $a_k[i,j]$ are the elements of the $[2n, 2n]$ coefficient matrices $\underline{A}_k$. Compliance with Eq. (2.90) and Eq. (2.91) is not required, however, as the $l_\alpha r_\alpha$-increments which arise during the computation of the partial $l_\alpha r_\alpha$-correlation matrix ${}_{lr}\underline{P}_{\tilde{X}/\tilde{Y}}$ only serve as intermediate results.

Furthermore, the correlations between the random $l_\alpha r_\alpha$-increments of two fuzzy random variables $\tilde{X}$ and $\tilde{Z}$ may be influenced by fuzzy random variables $\tilde{Y}_1$, $\tilde{Y}_2$, ..., $\tilde{Y}_m$. In order to eliminate the influence of the fuzzy random variables $\tilde{Y}_1$, $\tilde{Y}_2$, ..., $\tilde{Y}_m$ on the correlations between $\tilde{X}$ and $\tilde{Z}$ the partial $l_\alpha r_\alpha$-correlation matrix ${}_{lr}\underline{P}_{\tilde{X}\tilde{Z}/\tilde{Y}}$ is defined.

Definition 2.57. *After eliminating the influence of the fuzzy random variables $\tilde{Y}_1$, $\tilde{Y}_2$, ..., $\tilde{Y}_m$, the partial $l_\alpha r_\alpha$-correlation matrix ${}_{lr}\underline{P}_{\tilde{X}\tilde{Z}/\tilde{Y}}$ of two fuzzy random variables $\tilde{X}$ and $\tilde{Z}$ is given by Eq. (2.133).*

$$ {}_{lr}\underline{P}_{\tilde{X}\tilde{Z}/\tilde{Y}} = {}_{lr}\underline{R}\left[\tilde{X} \ominus \check{\tilde{X}}, \tilde{Z} \ominus \check{\tilde{Z}}\right] \tag{2.133}$$

◆

The best linear approximations $\check{\tilde{X}}$ and $\check{\tilde{Z}}$ of the fuzzy random variables $\tilde{X}$ and $\tilde{Z}$ by the fuzzy random variables $\tilde{Y}_1$, $\tilde{Y}_2$, ..., $\tilde{Y}_m$ are defined analogously by Eqs. (2.131) and (2.132).

Remark 2.58. An estimation of the characteristic moments of a fuzzy random variable by computing the empirical moments based on a concrete sample of fuzzy variables is presented in Sect. 3.5.4 in the context of the description and modeling of fuzzy time series. ◆

2.2.4 Monte Carlo Simulation of Fuzzy Random Variables

The numerical simulation of a fuzzy random variable $\tilde{\mathrm{X}}$, i.e. the generation of realizations $\tilde{\mathrm{x}}$, is based on a Monte Carlo simulation. The Monte Carlo simulation depends on the characteristic of the fuzzy random variable $\tilde{\mathrm{X}}$, i.e. whether $\tilde{\mathrm{X}}$ is a continuous or a discrete fuzzy random variable.

Continuous fuzzy random variable. If the fuzzy random variable $\tilde{\mathrm{X}}$ is continuous, it is computed as a one-to-one mapping of the uniformly distributed fuzzy random variable $\tilde{\mathrm{Y}}$ according to Eq. (2.134).

$$\tilde{\mathrm{X}} = f_c(\tilde{\mathrm{Y}}) \tag{2.134}$$

The random $l_\alpha r_\alpha$-increments $\Delta \mathrm{Y}_j$ ($j = 1, 2, ..., 2n$) of the fuzzy random variable $\tilde{\mathrm{Y}}$ are in the interval $[0, 1]$ uniformly distributed uncorrelated random variables. The random $l_\alpha r_\alpha$-increments $\Delta \mathrm{X}_j$ of the fuzzy random variable $\tilde{\mathrm{X}}$ are correlated according the $l_\alpha r_\alpha$-covariance matrix ${}_{lr}\underline{K}_{\tilde{\mathrm{X}}}$. The mapping according to Eq. (2.134) is thus nontrivial and requires the transformation of the fuzzy random variables $\tilde{\mathrm{X}}$ and $\tilde{\mathrm{Y}}$ into the correlated GAUSSian space. The numerical procedure is described in the following.

The uniformly distributed uncorrelated random $l_\alpha r_\alpha$-increments $\Delta \mathrm{Y}_j$ of the fuzzy random variable $\tilde{\mathrm{Y}}$ in the interval $[0, 1]$ are simulated with the aid of pseudo random numbers (see e.g. [50]) or low-discrepancy numbers (see e.g. [46]). The simulation yields realizations Δy_j of the random $l_\alpha r_\alpha$-increments $\Delta \mathrm{Y}_j$. By applying the inverse probability distribution function method, the realizations Δy_j ($j = 1, 2, ..., 2n$) are transformed into the uncorrelated GAUSSian space according to Eq. (2.135).

$$\Delta u_j = \Phi^{-1}(\Delta y_j) \tag{2.135}$$

The Δu_j are $l_\alpha r_\alpha$-increments of a realization $\tilde{u}$ of the fuzzy random variable $\tilde{\mathrm{U}}$. The random $l_\alpha r_\alpha$-increments $\Delta \mathrm{U}_j$ of $\tilde{\mathrm{U}}$ are uncorrelated GAUSSian distributed random variables with the standard normal probability distribution function $\Phi(\cdot)$. The fuzzy expected value and the $l_\alpha r_\alpha$-variance of the fuzzy random variable $\tilde{U}$ are given by:

$$E[\tilde{\mathrm{U}}] = 0 \tag{2.136}$$

$${}_{lr}Var[\tilde{\mathrm{U}}] = (1, 1, ..., 1)^T \, . \tag{2.137}$$

The realizations $\tilde{u}$ of the fuzzy random variable $\tilde{\mathrm{U}}$ are thus fuzzy variables in the improper sense (see remark 2.11).

The realizations $\tilde{u}$ are transformed into the correlated GAUSSian space according to Eq. (2.138) (see also [57]).

$$\tilde{s} = {}_{lr}C_{\tilde{\mathrm{S}}} \odot \tilde{u} \tag{2.138}$$

The fuzzy variables $\tilde{s}$ in the improper sense are realizations of the fuzzy random variable $\tilde{\mathrm{S}}$. The random $l_\alpha r_\alpha$-increments $\Delta \mathrm{S}_j$ of $\tilde{\mathrm{S}}$ are now correlated GAUSSian distributed random variables with the standard normal probability distribution function $\Phi(\cdot)$. The matrix $_{lr}\underline{C}_{\tilde{\mathrm{S}}}$ is obtained by CHOLESKY decomposition of the $l_\alpha r_\alpha$-covariance matrix $_{lr}\underline{K}_{\tilde{\mathrm{S}}}$ of the fuzzy random variable $\tilde{\mathrm{S}}$ according to Eq. (2.139).

$$_{lr}\underline{K}_{\tilde{\mathrm{S}}} = {_{lr}\underline{C}_{\tilde{\mathrm{S}}}}\ {_{lr}\underline{C}_{\tilde{\mathrm{S}}}}^T \tag{2.139}$$

The real-valued matrix $_{lr}\underline{C}_{\tilde{\mathrm{S}}}$ represents the lower triangular matrix according to Eq. (2.140).

$$_{lr}\underline{C}_{\tilde{\mathrm{S}}} = \begin{bmatrix} c_{1,1} & 0 & \cdots & 0 \\ c_{2,1} & c_{2,2} & \cdots & 0 \\ \vdots & \vdots & \ddots & \vdots \\ c_{2n,1} & c_{2n,2} & \cdots & c_{2n,2n} \end{bmatrix} \tag{2.140}$$

The elements $k_{u,v}(\tilde{\mathrm{S}})$ of the $l_\alpha r_\alpha$-covariance matrix $_{lr}\underline{K}_{\tilde{\mathrm{S}}}$ are given implicitly by Eq. (2.141) (see also [15]). For each (given) element $k_{u,v}(\tilde{\mathrm{X}})$ of the $l_\alpha r_\alpha$-covariance matrix $_{lr}\underline{K}_{\tilde{\mathrm{X}}}$ the associated element $k_{u,v}(\tilde{\mathrm{S}})$ is obtained iteratively by numerical solution of the integral given by Eq. (2.141).

$$k_{u,v}(\tilde{\mathrm{X}}) = \int_{-\infty}^{\infty} \int_{-\infty}^{\infty} (F_{\Delta \mathrm{X}_u}{}^{-1}(\Phi(\Delta s_u)) - \Delta m_u)(F_{\Delta \mathrm{X}_v}{}^{-1}(\Phi(\Delta s_v)) - \Delta m_v) \ldots \tag{2.141}$$

$$\ldots \phi_{\Delta \mathrm{S}_u \Delta \mathrm{S}_v}(\Delta s_u, \Delta s_v, k_{u,v}(\tilde{\mathrm{S}}))\ d\Delta s_u\, d\Delta s_u$$

$$\text{with} \quad u, v = 1,\, 2\,, \ldots,\, 2n$$

In the above equation, $F_{\Delta \mathrm{X}_u}(\cdot)$ and $F_{\Delta \mathrm{X}_v}(\cdot)$ are the given probability distribution functions of the random $l_\alpha r_\alpha$-increments $\Delta \mathrm{X}_u$ and $\Delta \mathrm{X}_v$, $\Phi(\cdot)$ is the standard normal probability distribution, and $\phi_{\Delta \mathrm{S}_u \Delta \mathrm{S}_v}(\cdot)$ is the two-dimensional standard normal probability density function of the GAUSSian distributed random $l_\alpha r_\alpha$-increments $\Delta \mathrm{S}_u$ and $\Delta \mathrm{S}_v$. The terms Δm_u and Δm_v are the $l_\alpha r_\alpha$-increments of the fuzzy expected value $E[\tilde{\mathrm{X}}] = \tilde{m}_{\tilde{\mathrm{X}}}$.

The sought realizations $\tilde{x}$ of the fuzzy random variable $\tilde{\mathrm{X}}$ are obtained by increment-by-increment transformation of the fuzzy variables $\tilde{s}$ according to Eq. (2.142)

$$\Delta x_j = F_{\Delta \mathrm{X}_j}{}^{-1}\left(\Phi(\Delta s_j)\right), \quad j = 1,\, 2,\, \ldots,\, 2n\,, \tag{2.142}$$

whereby the Δs_j are the increments of the realizations $\tilde{s}$ (see Eq. (2.138)).

Discrete fuzzy random variable. If the fuzzy random variable $\tilde{\mathrm{X}}$ is discrete, a simplified procedure is adopted for the Monte Carlo simulation. Let us consider the possible realizations $\tilde{x}_1$, $\tilde{x}_2$, ..., $\tilde{x}_m$ of the discrete fuzzy random variable $\tilde{X}$ with the given probabilities of occurrence $P(\tilde{x}_1)$, $P(\tilde{x}_2)$, ..., $P(\tilde{x}_m)$. In order to generate realizations from the set $\tilde{x}_1$, $\tilde{x}_2$, ..., $\tilde{x}_m$ a uniformly distributed random variable Y in the interval $[0, 1]$ is simulated and transformed according to Eq. (2.143).

$$\tilde{X} = f_d(Y) \tag{2.143}$$

The mapping according to Eq. (2.143) is performed numerically by a Monte Carlo simulation of the random variable Y using pseudo random numbers or low-discrepancy numbers. The simulation yields realizations y of Y. The sought realizations $\tilde{x}_1$, $\tilde{x}_2$, ..., $\tilde{x}_m$ of the fuzzy random variable $\tilde{X}$ are obtained by applying the following equation.

$$\tilde{x} = \begin{cases} \tilde{x}_1 & \text{for} \quad 0 \leqslant y < \sum_{j=1}^{1} P(\tilde{x}_j) \\ \tilde{x}_2 & \text{for} \quad \sum_{j=1}^{1} P(\tilde{x}_j) \leqslant y < \sum_{j=1}^{2} P(\tilde{x}_j) \\ \vdots & \qquad \vdots \\ \tilde{x}_m & \text{for} \quad \sum_{j=1}^{m-1} P(\tilde{x}_j) \leqslant y \leqslant 1 \end{cases} \tag{2.144}$$

Example 2.59. A discrete fuzzy random variable $\tilde{X}$ may take $m = 4$ different realizations $\tilde{x}_1$, $\tilde{x}_2$, $\tilde{x}_3$, $\tilde{x}_4$. Each of the four realizations occurs with a probability according to Fig. 2.19, e.g. as estimated from a given sample with $s \gg m$ fuzzy elements.

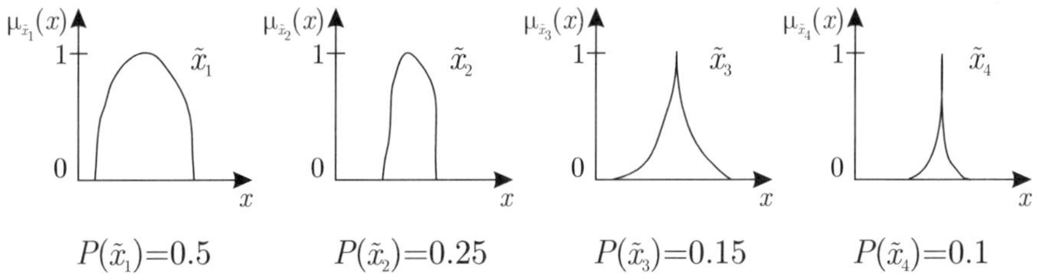

Fig. 2.19. Realizations $\tilde{x}_1$, $\tilde{x}_2$, $\tilde{x}_3$, $\tilde{x}_4$ and associated probabilities $P(\tilde{x}_1)$, $P(\tilde{x}_2)$, $P(\tilde{x}_3)$, $P(\tilde{x}_4)$ of the discrete fuzzy random variable $\tilde{X}$

Each Monte Carlo simulation yields a realization y of the uniformly distributed random variable Y. The mapping according Eqs. (2.143) and (2.144) yields a realization $\tilde{x}_1$, $\tilde{x}_2$, $\tilde{x}_3$ or $\tilde{x}_4$. The mapping is shown in Fig. 2.20 while several results are listed in Table 2.5.

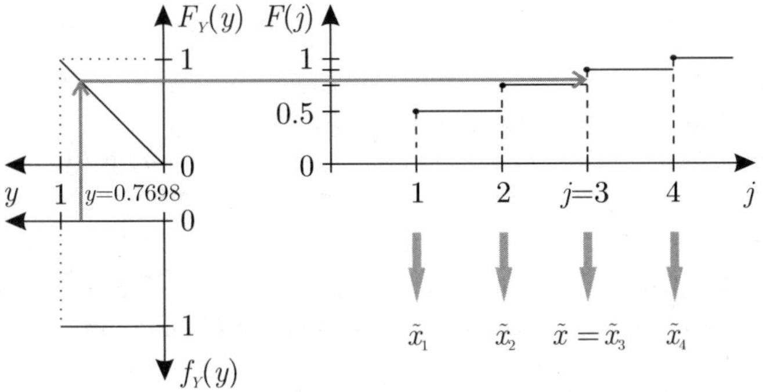

Fig. 2.20. Simplified Monte Carlo simulation of the discrete fuzzy random variable $\tilde{X}$

Table 2.5. Simplified Monte Carlo simulation of the discrete fuzzy random variable $\tilde{X}$

$y_1 = 0.7698 \rightarrow \mathrm{j}{=}3 \rightarrow \tilde{x}_3$
$y_2 = 0.4631 \rightarrow \mathrm{j}{=}1 \rightarrow \tilde{x}_1$
$y_3 = 0.0185 \rightarrow \mathrm{j}{=}1 \rightarrow \tilde{x}_1$
$y_4 = 0.6574 \rightarrow \mathrm{j}{=}2 \rightarrow \tilde{x}_2$
$\vdots \qquad \vdots$

◆

Remark 2.60. The simplified Monte Carlo simulation permits the non-parametric simulation of a fuzzy random variable by means of an existing empirical probability distribution function form II (see definitions 2.35 and 2.41). ◆

2.3 Fuzzy Random Processes

In classical time series analysis use is made of the model concept that a given sequence of deterministic observed values are random realizations of a random process. This model concept may be extended to uncertain data. In the case of a sequence of uncertain observed values this is considered to be a random realization of a fuzzy random process. This leads to the creation of a fuzzy random process model in compliance with the definition of fuzzy random variables according to Sect. 2.2.

Definition 2.61. *A fuzzy random process* $(\tilde{X}_\tau)_{\tau \in \mathbf{T}}$ *is defined as a family of fuzzy random variables* $\tilde{X}_\tau$ *over the space* $\mathbf{T}$ *of the time coordinate* τ*, and represents the fuzzy result of the mapping according to Eq. (2.145) for* $\tau \in \mathbf{T}$.

$$\tilde{X}_\tau : \Omega \to \mathbf{F}(\mathbb{R}) \tag{2.145}$$

In the foregoing Ω is the space of the random elementary events ω, and $\mathbf{F}(\mathbb{R})$ is the set of all fuzzy variables in the EUCLIDian space $\mathbb{R}$. By way of Eq. (2.145) fuzzy realizations $\tilde{X}_\tau(\omega) = \tilde{x}_\tau$ with $\tau \in \mathbf{T}$ are assigned to each random elementary event $\omega \in \Omega$. The realization of a fuzzy random process $(\tilde{X}_\tau)_{\tau\in\mathbf{T}}$ is hence a fuzzy time series $(\tilde{x}_\tau)_{\tau\in\mathbf{T}}$. ◆

A fuzzy random process $(\tilde{X}_\tau)_{\tau\in\mathbf{T}}$ is referred to as stationary in strong sense if the fuzzy random variables $\tilde{X}_\tau$ are independent of the parameter τ.

$$\tilde{X}_\tau = \tilde{X} \quad \forall \quad \tau \in \mathbf{T} \tag{2.146}$$

Example 2.62. A family of four fuzzy random variables of a fuzzy random process is shown in Fig. 2.21. A corresponding sequence of fuzzy realizations relating to the elementary event ω_i is presented in Fig. 2.22. This sequence constitutes a time series. ◆

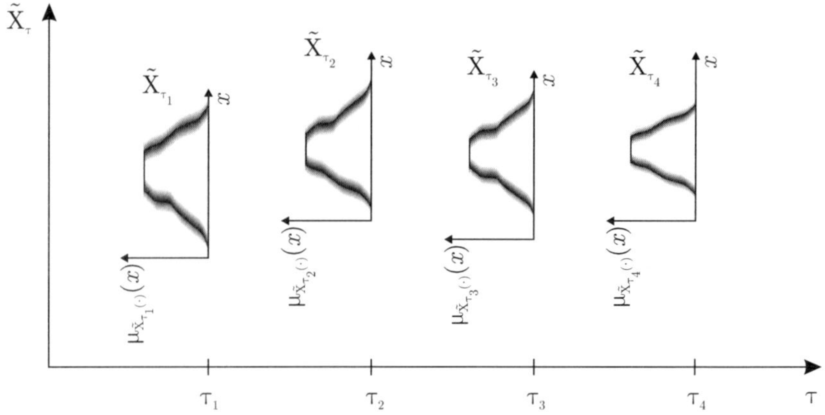

Fig. 2.21. Fuzzy random variables of a fuzzy random process

The fuzzy random variables $\tilde{X}_\tau$ of a fuzzy random process $(\tilde{X}_\tau)_{\tau\in\mathbf{T}}$ available at time τ may be represented and characterized according to Sect. 2.2. This means that a corresponding fuzzy probability distribution function form I or II may be formulated for $\tilde{X}_\tau$ at each point in time τ. Moreover, analogous to random processes, the first and second order moments may be used for characterizing fuzzy random processes. If the fuzzy random variables $\tilde{X}_\tau$ of a fuzzy random process $(\tilde{X}_\tau)_{\tau\in\mathbf{T}}$ are represented numerically by n α-levels with the aid of $l_\alpha r_\alpha$-discretization, the following definitions hold for the characteristic moments.

Definition 2.63. *For each point in time τ the fuzzy expected value $E[\tilde{X}_\tau] = \tilde{m}_{\tilde{X}_\tau}$ of a fuzzy random process $(\tilde{X}_\tau)_{\tau\in\mathbf{T}}$ is defined according to Eq. (2.147).*

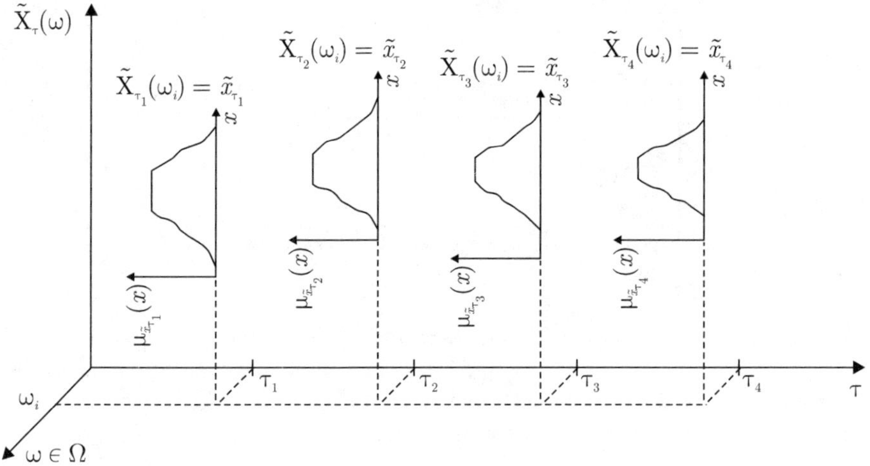

Fig. 2.22. Realization of a fuzzy random process

$$E[\tilde{X}_\tau] = \tilde{m}_{\tilde{X}_\tau} = \int_0^\infty \cdots \int_{-\infty}^\infty \cdots \int_0^\infty {}_{lr}f_{\tilde{X}_\tau}(\tilde{x})\, \tilde{x}\; d\Delta x_1 \cdots d\Delta x_n \cdots d\Delta x_{2n} \quad (2.147)$$

$E[\tilde{X}_\tau]$ is referred to as fuzzy expected value function. The $l_\alpha r_\alpha$-increments $\Delta m_{\alpha_i l}(\tau)$ and $\Delta m_{\alpha_i r}(\tau)$ of the fuzzy expected value function $E[\tilde{X}_\tau] = \tilde{m}_{\tilde{X}_\tau}$ are computed according to Eq. (2.148).

$$E[\tilde{X}_\tau] = \tilde{m}_{\tilde{X}_\tau} = \begin{bmatrix} \Delta m_{\alpha_1 l}(\tau) \\ \vdots \\ \Delta m_{\alpha_n l}(\tau) \\ \vdots \\ \Delta m_{\alpha_1 r}(\tau) \end{bmatrix} = \begin{bmatrix} \int_0^\infty \Delta x_{\alpha_1 l}\, f_{\Delta X_{\alpha_1 l}}(\Delta x_{\alpha_1 l}, \tau) d\Delta x_{\alpha_1 l} \\ \vdots \\ \int_{-\infty}^\infty \Delta x_{\alpha_n l}\, f_{\Delta X_{\alpha_n l}}(\Delta x_{\alpha_n l}, \tau) d\Delta x_{\alpha_n l} \\ \vdots \\ \int_0^\infty \Delta x_{\alpha_1 r}\, f_{\Delta X_{\alpha_1 r}}(\Delta x_{\alpha_1 r}, \tau) d\Delta x_{\alpha_1 r} \end{bmatrix} \quad (2.148)$$

The functions $f_{\Delta X_{\alpha_i l}}(\Delta x_{\alpha_i l}, \tau)$ and $f_{\Delta X_{\alpha_i r}}(\Delta x_{\alpha_i r}, \tau)$ in the foregoing are the probability density functions of the random $l_\alpha r_\alpha$-increments $\Delta X_{\alpha_i l}(\tau)$ and $\Delta X_{\alpha_i r}(\tau)$, respectively (with $i = 1, 2, ..., n$), of the fuzzy random variables $\tilde{X}_\tau$ at times τ. ◆

Definition 2.64. *Linear dependencies between the fuzzy random variables $\tilde{X}_{\tau_a}$ and $\tilde{X}_{\tau_b}$ of a fuzzy random process $(\tilde{X}_\tau)_{\tau \in \mathbf{T}}$ at times τ_a and τ_b are attributed to the dependencies between the $l_\alpha r_\alpha$-random increments $\Delta\tilde{X}_{\tau_a}$ and $\Delta\tilde{X}_{\tau_b}$. This leads to the $l_\alpha r_\alpha$-covariance matrix ${}_{lr}\underline{K}_{\tilde{X}_\tau}(\tau_a, \tau_b)$ according to Eq. (2.149). Taking into consideration arbitrary discrete points in time τ_a and τ_b, ${}_{lr}\underline{K}_{\tilde{X}_\tau}(\tau_a, \tau_b)$ is referred to as the $l_\alpha r_\alpha$-covariance function.*

$$
{}_{lr}\underline{K}_{\tilde{X}_\tau}(\tau_a,\tau_b)=\begin{bmatrix}
k_{\alpha_1 l}^{\alpha_1 l}(\tau_a,\tau_b) & k_{\alpha_1 l}^{\alpha_2 l}(\tau_a,\tau_b) & \cdots & k_{\alpha_1 l}^{\alpha_2 r}(\tau_a,\tau_b) & k_{\alpha_1 l}^{\alpha_1 r}(\tau_a,\tau_b) \\
k_{\alpha_2 l}^{\alpha_1 l}(\tau_a,\tau_b) & k_{\alpha_2 l}^{\alpha_2 l}(\tau_a,\tau_b) & \cdots & k_{\alpha_2 l}^{\alpha_2 r}(\tau_a,\tau_b) & k_{\alpha_2 l}^{\alpha_1 r}(\tau_a,\tau_b) \\
\vdots & \vdots & \ddots & \vdots & \vdots \\
k_{\alpha_2 r}^{\alpha_1 l}(\tau_a,\tau_b) & k_{\alpha_2 r}^{\alpha_2 l}(\tau_a,\tau_b) & \cdots & k_{\alpha_2 r}^{\alpha_2 r}(\tau_a,\tau_b) & k_{\alpha_2 r}^{\alpha_1 r}(\tau_a,\tau_b) \\
k_{\alpha_1 r}^{\alpha_1 l}(\tau_a,\tau_b) & k_{\alpha_1 r}^{\alpha_2 l}(\tau_a,\tau_b) & \cdots & k_{\alpha_1 r}^{\alpha_2 r}(\tau_a,\tau_b) & k_{\alpha_1 r}^{\alpha_1 r}(\tau_a,\tau_b)
\end{bmatrix} \quad (2.149)
$$

◆

The elements of the $l_\alpha r_\alpha$-covariance function ${}_{lr}\underline{K}_{\tilde{X}_\tau}(\tau_a,\tau_b)$ are determined for $i,j = 1,2,...,n$ according to Eq. (2.150). The indices α_i and α_j denote the α-levels under consideration whereas $l^* = l,r$ and $r^* = l,r$ denote the left and right branches of the membership function, respectively.

$$
k_{\alpha_j r*}^{\alpha_i l*}(\tau_a,\tau_b)=\int_{-\infty}^{\infty}\int_{-\infty}^{\infty}(\Delta x_{\alpha_i l*}-\Delta m_{\alpha_i l*}(\tau_a))(\Delta x_{\alpha_j r*}-\Delta m_{\alpha_j r*}(\tau_b))... \quad (2.150)
$$

$$
...f\left(\Delta x_{\alpha_i l*},\Delta x_{\alpha_j r*},\tau_a,\tau_b\right) d\Delta x_{\alpha_i l*}\, d\Delta x_{\alpha_j r*}
$$

The term $f\left(\Delta x_{\alpha_i l*},\Delta x_{\alpha_j r*},\tau_a,\tau_b\right)$ in the foregoing represents the joint probability density function of the $l_\alpha r_\alpha$-random increments $\Delta \mathrm{X}_{\alpha_i l*}(\tau_a)$ and $\Delta \mathrm{X}_{\alpha_j r*}(\tau_b)$.

Definition 2.65. *The element by element division of the $l_\alpha r_\alpha$-covariance function ${}_{lr}\underline{K}_{\tilde{X}_\tau}(\tau_a,\tau_b)$ by the corresponding leading diagonal elements according to Eq. (2.152) yields the scale-invariant $l_\alpha r_\alpha$-correlation function ${}_{lr}\underline{R}_{\tilde{X}_\tau}(\tau_a,\tau_b)$ given by Eq. (2.151).*

$$
{}_{lr}\underline{R}_{\tilde{X}_\tau}(\tau_a,\tau_b)=\begin{bmatrix}
r_{\alpha_1 l}^{\alpha_1 l}(\tau_a,\tau_b) & r_{\alpha_1 l}^{\alpha_2 l}(\tau_a,\tau_b) & \cdots & r_{\alpha_1 l}^{\alpha_2 r}(\tau_a,\tau_b) & r_{\alpha_1 l}^{\alpha_1 r}(\tau_a,\tau_b) \\
r_{\alpha_2 l}^{\alpha_1 l}(\tau_a,\tau_b) & r_{\alpha_2 l}^{\alpha_2 l}(\tau_a,\tau_b) & \cdots & r_{\alpha_2 l}^{\alpha_2 r}(\tau_a,\tau_b) & r_{\alpha_2 l}^{\alpha_1 r}(\tau_a,\tau_b) \\
\vdots & \vdots & \ddots & \vdots & \vdots \\
r_{\alpha_2 r}^{\alpha_1 l}(\tau_a,\tau_b) & r_{\alpha_2 r}^{\alpha_2 l}(\tau_a,\tau_b) & \cdots & r_{\alpha_2 r}^{\alpha_2 r}(\tau_a,\tau_b) & r_{\alpha_2 r}^{\alpha_1 r}(\tau_a,\tau_b) \\
r_{\alpha_1 r}^{\alpha_1 l}(\tau_a,\tau_b) & r_{\alpha_1 r}^{\alpha_2 l}(\tau_a,\tau_b) & \cdots & r_{\alpha_1 r}^{\alpha_2 r}(\tau_a,\tau_b) & r_{\alpha_1 r}^{\alpha_1 r}(\tau_a,\tau_b)
\end{bmatrix} \quad (2.151)
$$

$$
r_{\alpha_j l*}^{\alpha_i l*}(\tau_a,\tau_b) = \frac{k_{\alpha_j l*}^{\alpha_i l*}(\tau_a,\tau_b)}{\sqrt{k_{\alpha_i l*}^{\alpha_i l*}(\tau_a,\tau_b)\cdot k_{\alpha_j l*}^{\alpha_j l*}(\tau_a,\tau_b)}} \quad (2.152)
$$

◆

If all fuzzy expected values $E[\tilde{\mathrm{X}}_\tau]$ of a fuzzy random process $(\tilde{\mathrm{X}}_\tau)_{\tau\in\mathbf{T}}$ are constant

$$E[\tilde{X}_\tau] = \tilde{m}_{\tilde{X}_\tau} = \text{constant} \quad \forall \quad \tau \in \mathbf{T} \tag{2.153}$$

and Eq. (2.154) is satisfied, i.e. the $l_\alpha r_\alpha$-covariance function ${}_{lr}\underline{K}_{\tilde{X}_\tau}(\tau_a, \tau_b)$ or the $l_\alpha r_\alpha$-correlation function ${}_{lr}\underline{R}_{\tilde{X}_\tau}(\tau_a, \tau_b)$ is not dependent on τ_a and τ_b but only on the difference $\Delta\tau = \tau_a - \tau_b$, then $(\tilde{X}_\tau)_{\tau \in \mathbf{T}}$ is described as being weakly stationary.

$${}_{lr}\underline{K}_{\tilde{X}_\tau}(\tau_a, \tau_b) = {}_{lr}\underline{K}_{\tilde{X}_\tau}(\tau_a - \tau_b) = {}_{lr}\underline{K}_{\tilde{X}_\tau}(\Delta\tau) \quad \forall \quad \tau_a, \tau_b \in \mathbf{T} \tag{2.154}$$

The $l_\alpha r_\alpha$-variance function ${}_{lr}Var[\tilde{X}_\tau] = {}_{lr}\underline{\sigma}^2_{\tilde{X}_\tau}$ provides a measure of the fluctuation of the realizations of a fuzzy random process $(\tilde{X}_\tau)_{\tau \in \mathbf{T}}$. For each point in time τ the elements of the $l_\alpha r_\alpha$-variance correspond to the leading diagonal elements of ${}_{lr}\underline{K}_{\tilde{X}_\tau}(\tau_a, \tau_b)$ for $\tau_a = \tau_b = \tau$. For a stationary fuzzy random process $(\tilde{X}_\tau)_{\tau \in \mathbf{T}}$ the $l_\alpha r_\alpha$-variance function ${}_{lr}\underline{\sigma}^2_{\tilde{X}_\tau}$ according to Eq. (2.155) is hence constant at each point in time τ with $i = 1, 2, ..., 2n$.

$${}_{lr}\sigma_i^2(\tau) = {}_{lr}\sigma_i^2 = k_{i,i}(\tau_a - \tau_b) = k_{i,i}(\tau - \tau) = k_{i,i}(0) \tag{2.155}$$

Special fuzzy random processes include fuzzy **W**hite-**N**oise processes, fuzzy **M**oving **A**verage processes, fuzzy **A**uto**R**egressive processes and fuzzy **A**uto-**R**egressive **M**oving **A**verage processes. The modeling of fuzzy time series by means of these fuzzy random processes is presented in the following section.

3

Analysis of Time Series Comprised of Uncertain Data

The aim of time series analysis is to recognize and model structural features in a sequence of observed values. In the following chapter various commonly applied methods of classical time series analysis are extended to deal with time series comprised of fuzzy data.

Definition 3.1. *A time series comprised of fuzzy data* $(\tilde{x}_\tau)_{\tau \in \mathbf{T}}$ *is a temporally ordered sequence of fuzzy variables* $\tilde{x}_\tau$*, where* $\mathbf{T}$ *represents a set of equidistant points in time* τ*. Precisely one fuzzy variable* $\tilde{x}_\tau$ *is assigned to each discrete observation time* $\tau = 1, 2, ..., N$*. The* $l_\alpha r_\alpha$*-increments of the fuzzy variable* $\tilde{x}_\tau$ *are denoted by* $\Delta x_{\alpha_i l}(\tau)$ *and* $\Delta x_{\alpha_i r}(\tau)$*.*

$$\tilde{x}_\tau = \begin{bmatrix} \Delta x_{\alpha_1 l}(\tau) \\ \Delta x_{\alpha_2 l}(\tau) \\ \vdots \\ \Delta x_{\alpha_1 r}(\tau) \end{bmatrix} = \begin{bmatrix} \Delta x_1(\tau) \\ \Delta x_2(\tau) \\ \vdots \\ \Delta x_{2n}(\tau) \end{bmatrix} \tag{3.1}$$

◆

In the following the abridged term 'fuzzy time series' is used to express a time series comprised of fuzzy data. Time series $(x_\tau)_{\tau \in \mathbf{T}}$ of real observations x are special cases of fuzzy time series.

Definition 3.2. *A portion* $\tilde{x}_k, \tilde{x}_{k+1}, ..., \tilde{x}_l$ *with* $1 \leqslant k < l \leqslant N$ *of a fuzzy time series* $\tilde{x}_1, \tilde{x}_2, ..., \tilde{x}_N$ *is referred to as a segment.* ◆

Example 3.3. A fuzzy time series is shown by way of example in Fig. 3.1. ◆

3.1 Plot of Fuzzy Time Series

According to [59], a graphical representation or plot 'should always be the first step in the analysis of a time series'. A plot provides initial information

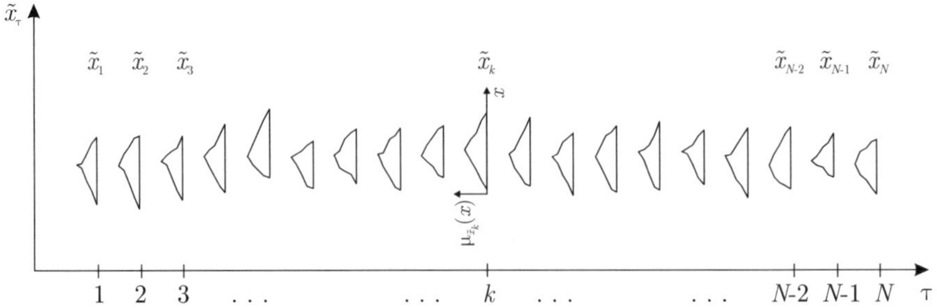

Fig. 3.1. Example of a fuzzy time series $(\tilde{x}_\tau)_{\tau \in \mathbf{T}}$

concerning the characterization of a fuzzy time series, especially regarding the existence of a fuzzy trend, cyclic fluctuations, dependencies between the discrete α-level sets X_α, and the presence of freak values. A full plot of a fuzzy time series should include the graphical representation of individual fuzzy variables $\tilde{x}_\tau$ at time points τ and selected $l_\alpha r_\alpha$-increments $\Delta x_{\alpha_i l}(\tau)$ and $\Delta x_{\alpha_i r}(\tau)$.

$\tilde{x}_\tau$-plot. In the $\tilde{x}_\tau$-plot, the fuzzy variables $\tilde{x}_\tau$ are represented as projections or in perspective at the N time points τ. This representation may be supplemented by including the polygonally connected interval bounds $[x_{\alpha_i l}(\tau); x_{\alpha_i r}(\tau)]$ of selected α-level sets $X_{\alpha_i}(\tau)$ of the fuzzy variables $\tilde{x}_\tau$. Both variants are shown in Fig. 3.2.

$l_\alpha r_\alpha$-increment plot. In the $l_\alpha r_\alpha$-increment plot, the increments $\Delta x_{\alpha_i l}(\tau)$ and $\Delta x_{\alpha_i r}(\tau)$ are polygonally connected at time points τ for selected α-levels α_i. An example of an $l_\alpha r_\alpha$-increment plot of a fuzzy time series $(\tilde{x}_\tau)_{\tau \in \mathbf{T}}$ for $n = 3$ is shown in Fig. 3.3.

The examples 3.4 and 3.5 clearly illustrate the purpose of the $\tilde{x}_\tau$-plot as well as the $l_\alpha r_\alpha$-increment plot. Both forms of representation should always be used in order to recognize the structure of a fuzzy time series.

Example 3.4. In the example shown in Fig. 3.4 the $l_\alpha r_\alpha$-discretization is carried out for $n = 5$. In the $\tilde{x}_\tau$-plot of the fuzzy time series the constancy of the interval bounds $[x_{\alpha_i l}(\tau); x_{\alpha_i r}(\tau)]$ of the α-level sets $X_{\alpha_i}(\tau)$ for $i = 1$ and $i = 3$ is clearly recognizable, i.e. no random effects are present. ◆

Example 3.5. In the example shown in Fig. 3.5 the $l_\alpha r_\alpha$-increments exhibit regularities. The $l_\alpha r_\alpha$-discretization is carried out for $n = 5$. With the aid of the $l_\alpha r_\alpha$-increment plot it is seen that the progressions of the $l_\alpha r_\alpha$-increments $\Delta x_{\alpha_1 l}(\tau)$, $\Delta x_{\alpha_2 l}(\tau)$, $\Delta x_{\alpha_3 l}(\tau)$ and $\Delta x_{\alpha_4 r}(\tau)$ are identical, as also applies to the progressions of the $l_\alpha r_\alpha$-increments $\Delta x_{\alpha_1 r}(\tau)$, $\Delta x_{\alpha_2 r}(\tau)$, $\Delta x_{\alpha_3 r}(\tau)$ and $\Delta x_{\alpha_4 l}(\tau)$. This indicates that the $l_\alpha r_\alpha$-increments $\Delta x_{\alpha_1 l}(\tau)$, $\Delta x_{\alpha_2 l}(\tau)$, $\Delta x_{\alpha_3 l}(\tau)$ and $\Delta x_{\alpha_4 r}(\tau)$ as well as $\Delta x_{\alpha_1 r}(\tau)$, $\Delta x_{\alpha_2 r}(\tau)$, $\Delta x_{\alpha_3 r}(\tau)$

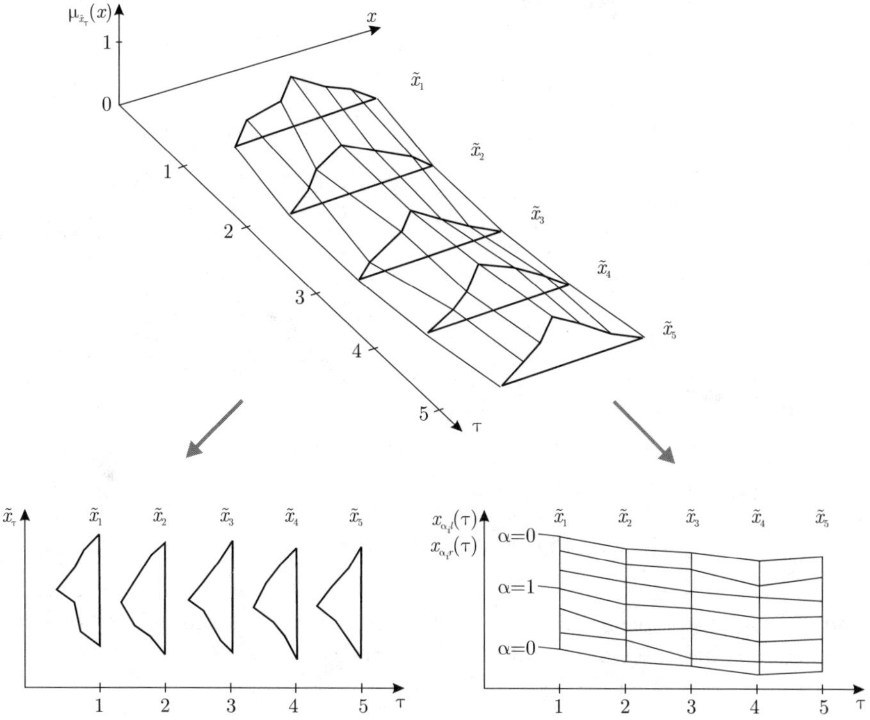

Fig. 3.2. $\tilde{x}_\tau$-plots of a fuzzy time series

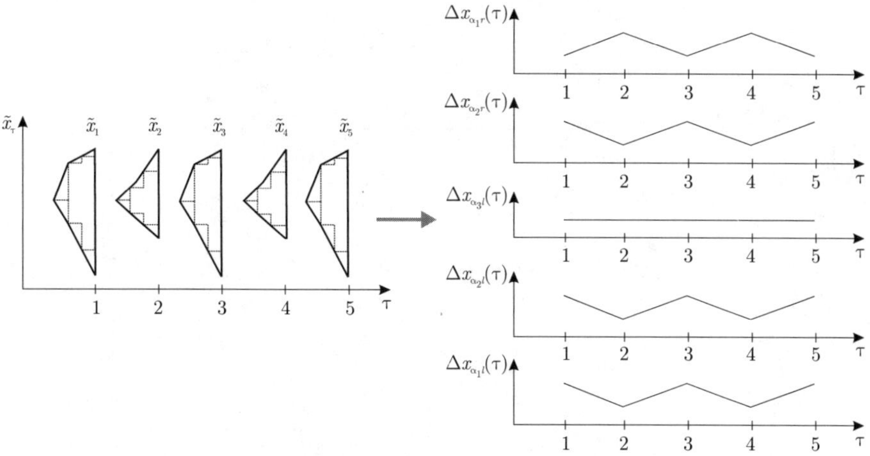

Fig. 3.3. $l_\alpha r_\alpha$-increment plot of a fuzzy time series for $n = 3$

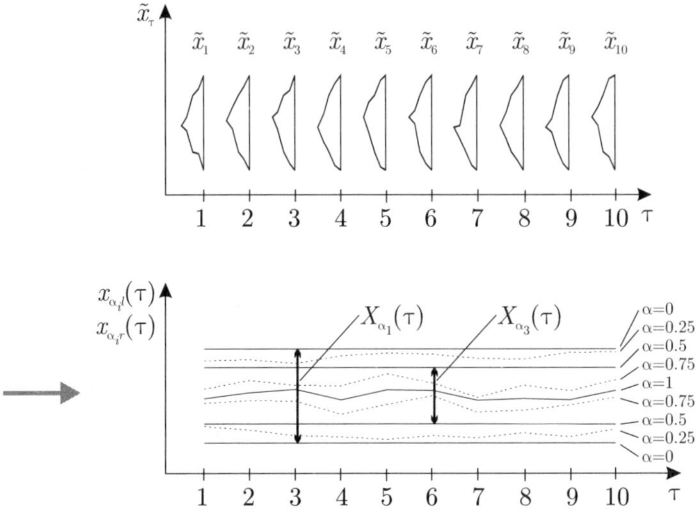

Fig. 3.4. $\tilde{x}_\tau$-plot for $n = 5$

and $\Delta x_{\alpha_4 l}(\tau)$ are positively fully correlated. Moreover, the affinity of both progressions indicates a mutual positive correlation. ◆

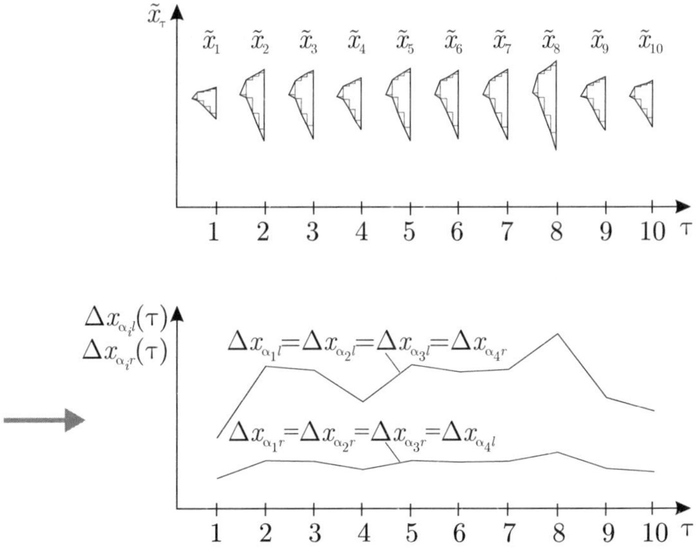

Fig. 3.5. $l_\alpha r_\alpha$-increment plot

With the aid of both plots it is possible to recognize important features of a fuzzy time series merely by visual inspection. Numerical methods for determining the characteristics of fuzzy time series are introduced in the following sections.

3.2 Fuzzy Component Model

Non-stationary fuzzy time series may be analyzed by means of the fuzzy component model. This model subdivides the given fuzzy time series additively into a trend component, a cyclic component and a fuzzy-random residual component.

Definition 3.6. *In order to describe the fuzzy time series* $(\tilde{x}_\tau)_{\tau\in\mathbf{T}}$ *the fuzzy component model according to Eq. (3.2) is introduced as an extension of the component model of classical time series analysis.*

$$\tilde{x}_\tau = \tilde{t}_\tau \oplus \tilde{z}_\tau \oplus \tilde{r}_\tau \tag{3.2}$$

◆

The fuzzy variables $\tilde{t}_\tau$ and $\tilde{z}_\tau$ are hereby functional values of a fuzzy trend function $\tilde{t}(\tau)$ or a fuzzy cycle function $\tilde{z}(\tau)$ at time τ. An introductory account of fuzzy functions is given in Sect. 2.1.5 and e.g. by [36, 45]. Fuzzy random processes are described in Sects. 2.3 and 3.5. The fuzzy residual component $\tilde{r}_\tau$ is the realization of a stationary fuzzy random interference process $(\tilde{\mathrm{R}}_\tau)_{\tau\in\mathbf{T}}$ at time point τ.

For identical $l_\alpha r_\alpha$-representation of the three fuzzy variables in Eq. (3.2) the following holds for each $l_\alpha r_\alpha$-increment $\Delta x_j(\tau)$

$$\Delta x_j(\tau) = \Delta t_j(\tau) + \Delta z_j(\tau) + \Delta r_j(\tau) \;\forall\; j = 1, 2, ..., 2n \tag{3.3}$$

under the conditions

$$\left.\begin{array}{l} \Delta x_j(\tau) \geqslant 0 \\ \Delta t_j(\tau) \geqslant 0 \\ \Delta z_j(\tau) \geqslant 0 \\ \Delta r_j(\tau) \geqslant 0 \end{array}\right\} \forall\; \tau \in \mathbf{T}, \quad j = 1, 2, ..., n-1, n+1, ..., 2n\,. \tag{3.4}$$

Because negative $l_\alpha r_\alpha$-increments of the realizations $\tilde{r}_\tau$ are excluded, it cannot be assumed (in contrast to the classical component model) that the fuzzy expected value $E[\tilde{\mathrm{R}}_\tau]$ of the fuzzy random interference process is equal to zero. The following holds for the fuzzy expected value $E[\tilde{\mathrm{R}}_\tau]$:

$$E[\tilde{\mathrm{R}}_\tau] = \tilde{m}_{\tilde{\mathrm{R}}_\tau} = \text{constant} \quad \forall\; \tau \in \mathbf{T} \tag{3.5}$$

where

$$E\left[\Delta \mathrm{R}_j(\tau)\right] \geqslant 0 \;\; \text{for } j = 1, 2, ..., n-1, n+1, ..., 2n\,. \tag{3.6}$$

According to Eq. (3.3) the determination of the fuzzy trend function $\tilde{t}(\tau)$, the fuzzy cycle function $\tilde{z}(\tau)$, and the fuzzy random interference process $(\tilde{\mathrm{R}}_\tau)_{\tau \in \mathbf{T}}$ reduces to the determination of the real-valued trend function $\Delta t_j(\tau)$, the real-valued cycle function $\Delta z_j(\tau)$, and the real-valued realizations $\Delta r_j(\tau)$ of the fuzzy random interference process$(\tilde{\mathrm{R}}_\tau)_{\tau \in \mathbf{T}}$.

The values of $\Delta t_j(\tau)$, $\Delta z_j(\tau)$ and $\Delta r_j(\tau)$ are computed according to the algorithm described in the following. This is demonstrated by way of example for the $l_\alpha r_\alpha$-increment $\Delta x_1(\tau)$ of the fuzzy time series shown in Fig. 3.6 a. The remaining $l_\alpha r_\alpha$-increments $\Delta x_j(\tau)$ are dealt with in a similar manner. The $l_\alpha r_\alpha$-discretization of the considered fuzzy time series is carried out for $n = 2$ α-levels.

Determination of the trend auxiliary functions $t_j^*(\tau)$. The $l_\alpha r_\alpha$-increment plot corresponding to each $l_\alpha r_\alpha$-increment $\Delta x_j(\tau)$ is developed from the given fuzzy time series. Fig. 3.6 b shows the plot for $\Delta x_1(\tau)$. This $l_\alpha r_\alpha$-increment function is approximated by a trend auxiliary function $t_j^*(\tau)$. The free parameters of a suitably chosen function $t_j^*(\tau)$ are determined by the method of least squares according to Eq. (3.7). The requirement according to Eq. (3.8) must hereby be fulfilled.

$$\sum_{\tau=1}^{N} \left(t_j^*(\tau) - \Delta x_j(\tau)\right)^2 \overset{!}{=} \min \tag{3.7}$$

$$t_j^*(\tau) \geqslant 0 \quad \forall \quad \tau \in \mathbf{T}, \;\; j = 1, 2, ..., n-1, n+1, ..., 2n \tag{3.8}$$

Determination of the cycle auxiliary functions $z_j^*(\tau)$. The differences $d_1(\tau) = \Delta x_1(\tau) - t_1^*(\tau)$ between the functions $\Delta x_1(\tau)$ and $t_1^*(\tau)$ are indicative of an existing cycle (see Fig. 3.6 c). The latter is approximated by means of a cycle auxiliary function $z_1^*(\tau)$. The free parameters of a suitably chosen function $z_1^*(\tau)$ are again determined by the method of least squares.

$$\sum_{\tau=1}^{N} \left[z_j^*(\tau) - \left(\Delta x_j(\tau) - t_j^*(\tau)\right)\right] \overset{!}{=} \min \tag{3.9}$$

The techniques of classical time series analysis are applied to determine the cycle and select the cycle auxiliary function $z_j^*(\tau)$. For the series of differences $d_j(\tau)$ the periodogram or sample spectrum is computed according to Eq. (3.10) and represented graphically.

$$I(\lambda_j) = N\left[\frac{1}{N}\sum_{\tau=1}^{N}\left(d_j(\tau) - \bar{d}_j\right)\cos 2\pi\lambda_j\tau\right]^2 + N\left[\frac{1}{N}\sum_{\tau=1}^{N}\left(d_j(\tau) - \bar{d}_j\right)\sin 2\pi\lambda_j\tau\right]^2 \tag{3.10}$$

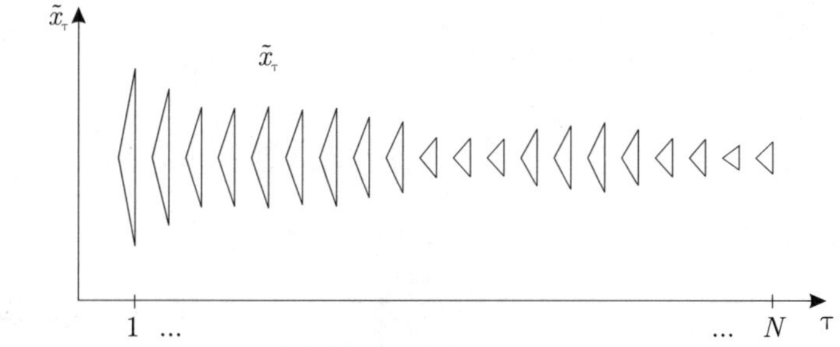

(a) Fuzzy time series

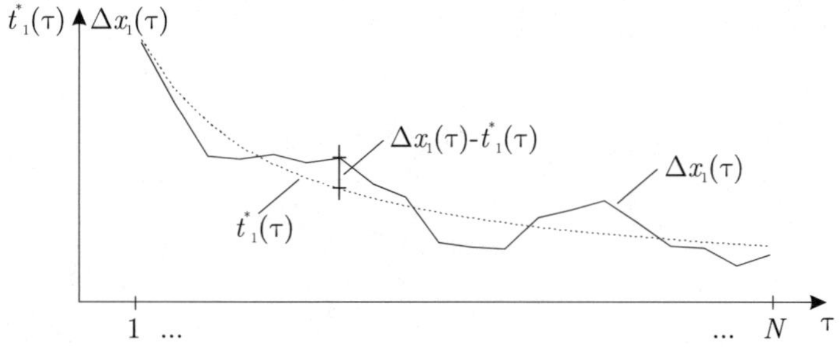

(b) Determination of the trend auxiliary function

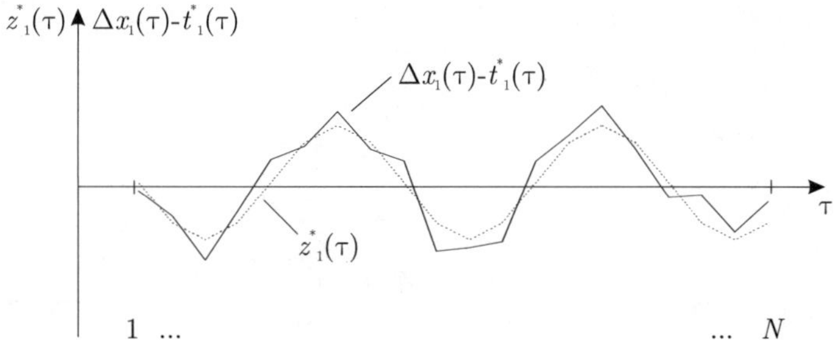

(c) Determination of the cycle auxiliary function

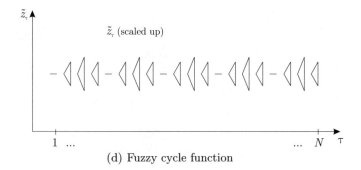

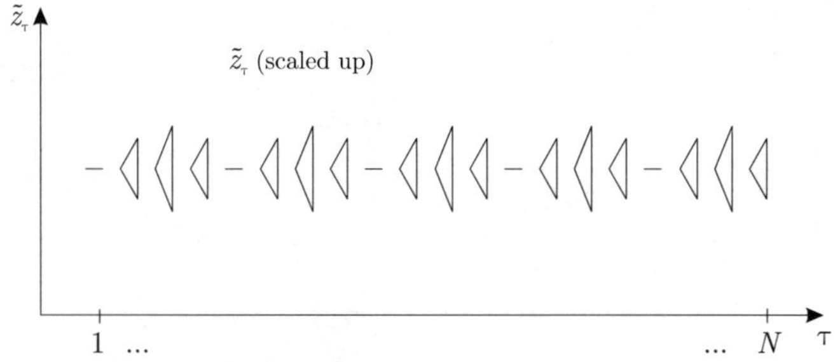

(d) Fuzzy cycle function

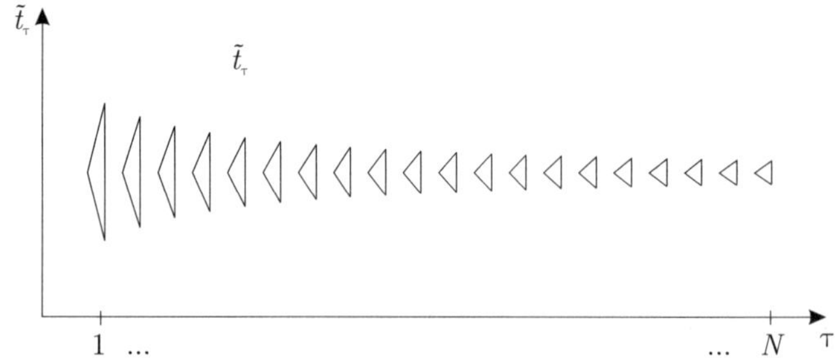

(e) Fuzzy trend function

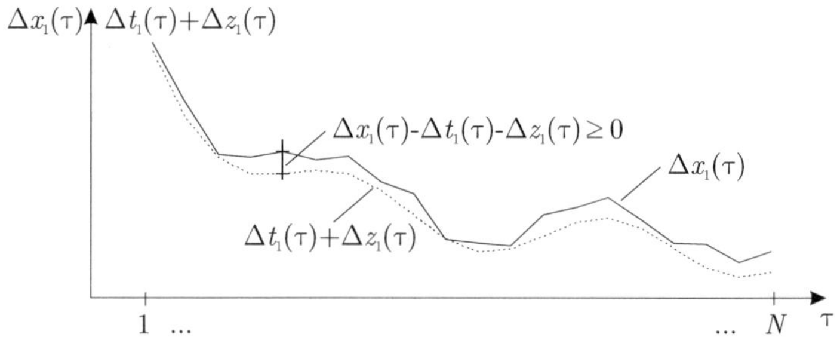

(f) Determination of the fuzzy residual component

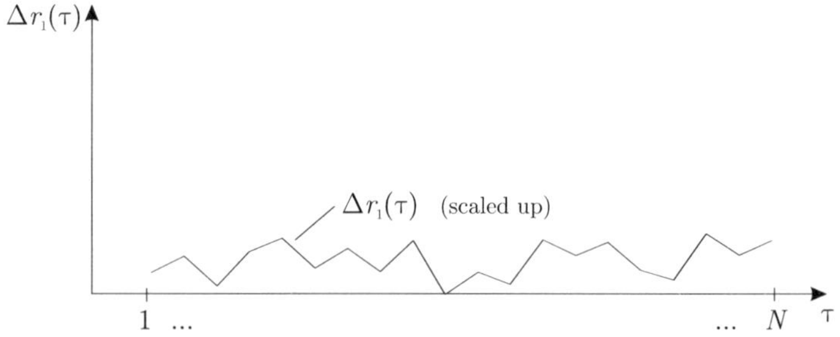

(g) Residual component

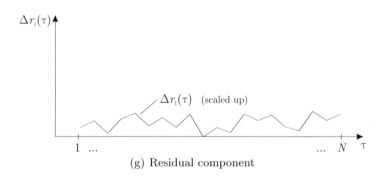

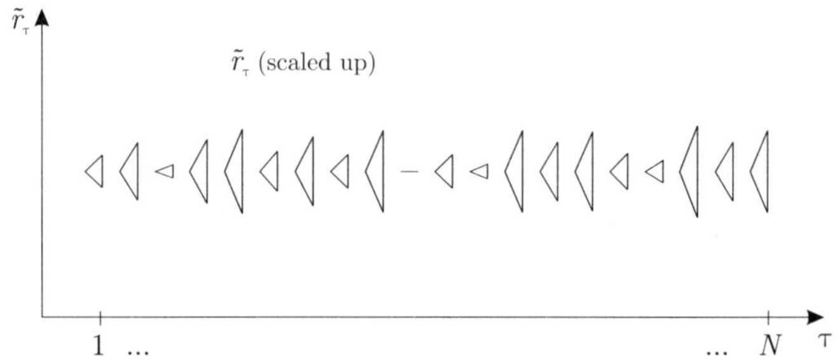

(h) Fuzzy residual component

Fig. 3.6. Fuzzy component model of a fuzzy time series

In the above equation $\overline{d}_j$ denotes the time average of the differences $d_j(\tau)$, and λ_j, the frequency. The presence of cyclic fluctuations with a period $\frac{1}{\lambda_j^m}$ may be deduced from the maximum values λ_j^m of the sample spectrum.

Example 3.7. By way of example the periodogram showing the variation of the peak points $h_l(\tau)$ of empirical humidity data $\tilde{h}_\tau$ is plotted in Fig. 3.7. The maximum value of the periodogram for $\lambda = \frac{1}{24h}$ confirms the natural period length of the humidity (Fig. 3.7 b).

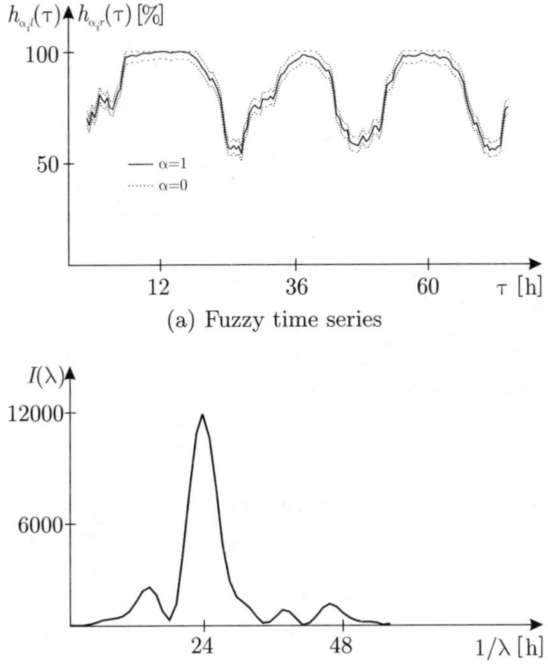

(a) Fuzzy time series

(b) Periodogram

Fig. 3.7. Humidity $\tilde{h}_\tau$ and periodogram for the mean value $h_l(\tau)$

◆

Cycle matching. The $l_\alpha r_\alpha$-increments $\Delta z_j(\tau)$ of the fuzzy cycle function $\tilde{z}(\tau)$ may now be determined for each α-level according to Eq. (3.11). The computation according to Eq. (3.11) is equivalent to a parallel displacement of the cycle auxiliary functions $z_j^*(\tau)$ into the positive range of values. The requirement according to Eq. (3.4) is thus automatically complied with.

$$\Delta z_j(\tau) = z_j^*(\tau) - \min_{a \in \mathbf{T}} \left[z_j^*(a) \right] \tag{3.11}$$

$$\text{with } j = 1, 2, ..., n-1, n+1, ..., 2n$$

The equality $\Delta z_n(\tau) = z_n^*(\tau)$ holds for $j = n$. The plot of the fuzzy cycle function $\tilde{z}(\tau)$ is shown in Fig. 3.6 d.

Trend matching. The $l_\alpha r_\alpha$-increments $\Delta t_j(\tau)$ of the fuzzy trend function $\tilde{t}_\tau(\tau)$ may now be computed by means of Eq. (3.12). The computation according to Eq. (3.12) is equivalent to a parallel displacement of the trend auxiliary functions $t_j^*(\tau)$ so that $\Delta x_j(\tau) - \Delta t_j(\tau) - \Delta z_j(\tau) \geqslant 0$ holds. In other words, the summated function $\Delta t_j(\tau) + \Delta z_j(\tau)$ always runs below $\Delta x_j(\tau)$ (see Fig. 3.6 f). In compliance with Eq. (3.4) the requirement according to Eq. (3.13) must be fulfilled.

$$\Delta t_j(\tau) = t_j^*(\tau) - \max_{a \in \mathbf{T}} \left[t_j^*(a) + \Delta z_j(a) - \Delta x_j(a) \right] \tag{3.12}$$

$$\text{with } \; j = 1, 2, ..., n-1, n+1, ..., 2n$$

$$\Delta t_j(\tau) \geqslant 0 \quad \forall \;\; \tau \in \mathbf{T}, \; j = 1, 2, ..., n-1, n+1, ..., 2n \tag{3.13}$$

The equality $\Delta t_n(\tau) = t_n^*(\tau)$ holds for $j = n$. The plot of the fuzzy trend function $\tilde{t}_\tau(\tau)$ is shown in Fig. 3.6 e.

Remark 3.8. If Eq. (3.13) is not fulfilled, then the selected fuzzy component model is unsuitable. By a critical assessment and reselection of the trend and cycle auxiliary function to be matched, it is possible to suitably modify the fuzzy component model. ◆

Determination of the fuzzy residual component $\tilde{r}_\tau$. The fuzzy residual component $\tilde{r}_\tau$ is the realization of a stationary fuzzy-random interference process $(\tilde{R}_\tau)_{\tau \in \mathbf{T}}$ at time τ. This is obtained according to Eq. (3.14) by adjusting the fuzzy observed values $\tilde{x}_\tau$ using the functional values $\tilde{t}_\tau$ and $\tilde{z}_\tau$ of the fuzzy trend function $\tilde{t}(\tau)$ and the fuzzy cycle function $\tilde{z}(\tau)$ at time τ.

$$\tilde{r}_\tau = \tilde{x}_\tau \ominus \tilde{t}_\tau \ominus \tilde{z}_\tau \tag{3.14}$$

The $l_\alpha r_\alpha$-increments $\Delta r_j(\tau)$ of the fuzzy residual component $\tilde{r}_\tau$ are thus obtained for $j = 1, 2, ..., 2n$ according to Eq. (3.15). If the requirement according to Eq. (3.13) is fulfilled, the non-negativity requirement for $\Delta r_j(\tau)$ according to Eq. (3.4) is automatically complied with. This procedure is illustrated in Fig. 3.6 f.

$$\Delta r_j(\tau) = \Delta x_j(\tau) - \Delta t_j(\tau) - \Delta z_j(\tau) \tag{3.15}$$

The plot for the $l_\alpha r_\alpha$-increment $\Delta r_1(\tau)$ is presented in Fig. 3.6 g. Fig. 3.6 h shows the plot of the fuzzy residual component $\tilde{r}_\tau$.

The fuzzy residual component $\tilde{r}_\tau$ is considered to be a realization of the stationary fuzzy random interference process $(\tilde{\mathrm{R}}_\tau)_{\tau \in \mathbf{T}}$ at time τ. Methods for matching fuzzy random processes are presented in Sect. 3.5.

3.3 Stationary Fuzzy Time Series

If fuzzy time series do not exhibit systematic variations (trends, cycles), they are usually postulated as stationary and ergodic series. The stationary condition presupposes time-invariant characteristics of the fuzzy time series while ergodicity permits a determination of these time-invariant characteristics by statistical evaluation of the fuzzy time series over time.

Essential characteristics of stationary and ergodic fuzzy time series are the fuzzy time-average and the empirical $l_\alpha r_\alpha$-variance.

Definition 3.9. *The fuzzy time-average $\tilde{\overline{x}}$ of a fuzzy time series $(\tilde{x}_\tau)_{\tau \in \mathbf{T}}$ is defined according to Eq. (3.16), whereby N is the number of observed fuzzy variables $\tilde{x}_\tau$.*

$$\tilde{\overline{x}} = \frac{1}{N} \bigoplus_{\tau=1}^{N} \tilde{x}_\tau \tag{3.16}$$

◆

The fuzzy time-average $\tilde{\overline{x}}$ is the central fuzzy variable about which the values of the time series vary.

The empirical $l_\alpha r_\alpha$-variance ${}_{lr}\underline{s}^2_{\tilde{x}_\tau}$ of a fuzzy time series $(\tilde{x}_\tau)_{\tau \in \mathbf{T}}$ is a measure for assessing the scatter of the series. This is represented in the form of a vector of size $[2n]$.

Definition 3.10. *The elements ${}_{lr}s_i^2$ of the vector of the empirical $l_\alpha r_\alpha$-variance ${}_{lr}\underline{s}^2_{\tilde{x}_\tau}$ are defined according to Eq. (3.17). The terms $\Delta\overline{x}_i$ are hereby the $l_\alpha r_\alpha$-increments of the fuzzy time-average $\tilde{\overline{x}}$ and the terms $\Delta x_i(\tau)$ are the $l_\alpha r_\alpha$-increments of the fuzzy variables $\tilde{x}_\tau$ at time τ.*

$${}_{lr}s_i^2 = \frac{1}{N-1} \sum_{\tau=1}^{N} \left(\Delta x_i(\tau) - \Delta\overline{x}_i\right)^2 \quad \text{for} \quad i = 1, 2, ..., 2n \tag{3.17}$$

◆

Remark 3.11. The size of the vector ${}_{lr}\underline{s}^2_{\tilde{x}_\tau}$, i.e. the $l_\alpha r_\alpha$-variance, depends on the number n of the chosen α-levels. The $l_\alpha r_\alpha$-variance ${}_{lr}\underline{s}^2_{\tilde{x}_\tau}(n_1)$ based on an $l_\alpha r_\alpha$-discretization with n_1 α-levels cannot generally be directly converted into the $l_\alpha r_\alpha$-variance ${}_{lr}\underline{s}^2_{\tilde{x}_\tau}(n_2)$ with n_2 discrete α-levels (see example 3.12).

◆

Example 3.12. For a particular fuzzy time series the empirical $l_\alpha r_\alpha$-variance ${}_{lr}\underline{s}^2_{\tilde{x}_\tau}$ according to Eq. (3.17) is determined for an $l_\alpha r_\alpha$-discretization with $n_1 = 2$ α-levels and for an $l_\alpha r_\alpha$-discretization with $n_2 = 3$ α-levels. The result is indicated by Eq. (3.18). A cutout of the fuzzy time series is presented in Fig. 3.8.

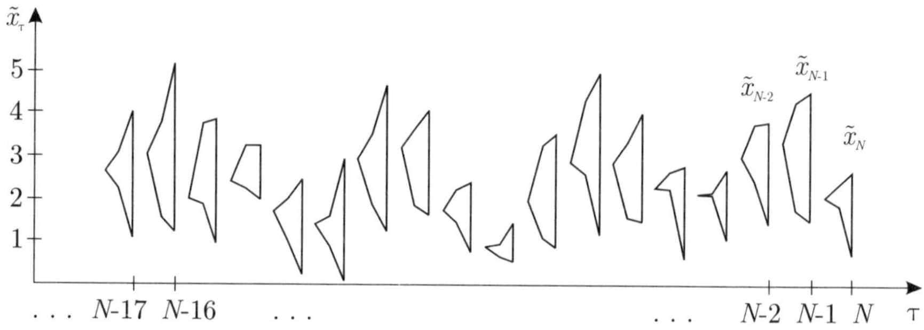

Fig. 3.8. Cutout of a fuzzy time series

$$ {}_{lr}\underline{s}^2_{\tilde{x}_\tau}(n_1) = \begin{bmatrix} 0.185 \\ 0.546 \\ 0 \\ 0.239 \end{bmatrix}, \quad {}_{lr}\underline{s}^2_{\tilde{x}_\tau}(n_2) = \begin{bmatrix} 0.211 \\ 0.197 \\ 0.546 \\ 0 \\ 0.254 \\ 0.238 \end{bmatrix} \tag{3.18} $$

No direct relationship exists between the $l_\alpha r_\alpha$-variance ${}_{lr}\underline{s}^2_{\tilde{x}_\tau}(n_1)$ and the $l_\alpha r_\alpha$-variance ${}_{lr}\underline{s}^2_{\tilde{x}_\tau}(n_2)$. ◆

The empirical $l_\alpha r_\alpha$-standard deviation ${}_{lr}\underline{s}_{\tilde{x}_\tau}$ is obtained by extracting the positive square root from each element of the vector of the empirical $l_\alpha r_\alpha$-variance ${}_{lr}\underline{s}^2_{\tilde{x}_\tau}$ according to Eq. (3.19).

$$ {}_{lr}s_i = +\sqrt{{}_{lr}s_i^2} \quad \text{for} \quad i = 1, 2, ..., 2n \tag{3.19} $$

Remark 3.13. With regard to the relationship that exists in special cases between the $l_\alpha r_\alpha$-standard deviation ${}_{lr}\underline{s}_{\tilde{x}_\tau}(n_1)$ based on an $l_\alpha r_\alpha$-discretization with n_1 α-levels and the $l_\alpha r_\alpha$-standard deviation ${}_{lr}\underline{s}_{\tilde{x}_\tau}(n_2)$ with n_2 discrete α-levels the same conditions apply as already stated for the $l_\alpha r_\alpha$-standard deviation of fuzzy random variables (see Remark 2.55). ◆

Definition 3.14. *Linear dependencies between the $l_\alpha r_\alpha$-increments of the fuzzy values $\tilde{x}_\tau$ and $\tilde{x}_{\tau+\Delta\tau}$ ($\tau = 1, 2, ..., N - \Delta\tau$) of a stationary and ergodic fuzzy time series are given in the form of an empirical $l_\alpha r_\alpha$-covariance matrix ${}_{lr}\underline{\hat{K}}_{\tilde{x}_\tau}(\Delta\tau)$ of size $[2n, 2n]$ according to Eq. (3.20). The term $\Delta\tau$ denotes the relative time lag between two discrete time points and is given as a natural number (e.g. $\Delta\tau = 3$ for $\tilde{x}_2$ and $\tilde{x}_5$). Considering arbitrary time lags $\Delta\tau$, the function ${}_{lr}\underline{\hat{K}}_{\tilde{x}_\tau}(\Delta\tau)$ is represented in the form of an empirical $l_\alpha r_\alpha$-covariance function.*

$$
{}_{lr}\underline{\hat{K}}_{\tilde{x}_\tau}(\Delta\tau) = \begin{bmatrix} \hat{k}^{\alpha_1 l}_{\alpha_1 l}(\Delta\tau) & \hat{k}^{\alpha_2 l}_{\alpha_1 l}(\Delta\tau) & \cdots & \hat{k}^{\alpha_2 r}_{\alpha_1 l}(\Delta\tau) & \hat{k}^{\alpha_1 r}_{\alpha_1 l}(\Delta\tau) \\ \hat{k}^{\alpha_1 l}_{\alpha_2 l}(\Delta\tau) & \hat{k}^{\alpha_2 l}_{\alpha_2 l}(\Delta\tau) & \cdots & \hat{k}^{\alpha_2 r}_{\alpha_2 l}(\Delta\tau) & \hat{k}^{\alpha_1 r}_{\alpha_2 l}(\Delta\tau) \\ \vdots & \vdots & \ddots & \vdots & \vdots \\ \hat{k}^{\alpha_1 l}_{\alpha_2 r}(\Delta\tau) & \hat{k}^{\alpha_2 l}_{\alpha_2 r}(\Delta\tau) & \cdots & \hat{k}^{\alpha_2 r}_{\alpha_2 r}(\Delta\tau) & \hat{k}^{\alpha_1 r}_{\alpha_2 r}(\Delta\tau) \\ \hat{k}^{\alpha_1 l}_{\alpha_1 r}(\Delta\tau) & \hat{k}^{\alpha_2 l}_{\alpha_1 r}(\Delta\tau) & \cdots & \hat{k}^{\alpha_2 r}_{\alpha_1 r}(\Delta\tau) & \hat{k}^{\alpha_1 r}_{\alpha_1 r}(\Delta\tau) \end{bmatrix} \tag{3.20}
$$

◆

The elements of the $l_\alpha r_\alpha$-covariance function ${}_{lr}\underline{\hat{K}}_{\tilde{x}_\tau}(\Delta\tau)$ are computed for each time lag $\Delta\tau$ according to Eq. (3.21).

$$
\hat{k}^{\alpha_i l*}_{\alpha_j r*}(\Delta\tau) = \frac{1}{N - \Delta\tau - 1} \sum_{\tau=1}^{N-\Delta\tau} [(\Delta x_{\alpha_i l*}(\tau) - \Delta\overline{x}_{\alpha_i l*}) \ldots \tag{3.21}
$$

$$
\ldots (\Delta x_{\alpha_j r*}(\tau + \Delta\tau) - \Delta\overline{x}_{\alpha_j r*})]
$$

The indices α_i and α_j with $i, j = 1, 2, ..., n$ denote the α-levels to be analyzed, and $l* = l, r$ and $r* = l, r$ denote the left and right branches of the membership function, respectively. The elements of the leading diagonal of the $l_\alpha r_\alpha$-covariance function ${}_{lr}\underline{\hat{K}}_{\tilde{x}_\tau}(\Delta\tau)$ correspond to the auto-covariances, whereas the remaining elements correspond to the cross-covariances of the $l_\alpha r_\alpha$-increments.

Definition 3.15. *The empirical $l_\alpha r_\alpha$-correlation function ${}_{lr}\underline{\hat{R}}_{\tilde{x}_\tau}(\Delta\tau)$ according to Eq. (3.22) yields a scale-invariant representation of the linear dependencies. This is obtained according to Eq. (3.23) as the result of element-by-element division of the empirical $l_\alpha r_\alpha$-covariance function ${}_{lr}\underline{\hat{K}}_{\tilde{x}_\tau}(\Delta\tau)$ by the corresponding elements of the leading diagonal.*

$$
{}_{lr}\underline{\hat{R}}_{\tilde{x}_\tau}(\Delta\tau) = \begin{bmatrix} \hat{r}^{\alpha_1 l}_{\alpha_1 l}(\Delta\tau) & \hat{r}^{\alpha_2 l}_{\alpha_1 l}(\Delta\tau) & \cdots & \hat{r}^{\alpha_2 r}_{\alpha_1 l}(\Delta\tau) & \hat{r}^{\alpha_1 r}_{\alpha_1 l}(\Delta\tau) \\ \hat{r}^{\alpha_1 l}_{\alpha_2 l}(\Delta\tau) & \hat{r}^{\alpha_2 l}_{\alpha_2 l}(\Delta\tau) & \cdots & \hat{r}^{\alpha_2 r}_{\alpha_2 l}(\Delta\tau) & \hat{r}^{\alpha_1 r}_{\alpha_2 l}(\Delta\tau) \\ \vdots & \vdots & \ddots & \vdots & \vdots \\ \hat{r}^{\alpha_1 l}_{\alpha_2 r}(\Delta\tau) & \hat{r}^{\alpha_2 l}_{\alpha_2 r}(\Delta\tau) & \cdots & \hat{r}^{\alpha_2 r}_{\alpha_2 r}(\Delta\tau) & \hat{r}^{\alpha_1 r}_{\alpha_{n-1} r}(\Delta\tau) \\ \hat{r}^{\alpha_1 l}_{\alpha_1 r}(\Delta\tau) & \hat{r}^{\alpha_2 l}_{\alpha_1 r}(\Delta\tau) & \cdots & \hat{r}^{\alpha_2 r}_{\alpha_1 r}(\Delta\tau) & \hat{r}^{\alpha_1 r}_{\alpha_1 r}(\Delta\tau) \end{bmatrix} \tag{3.22}
$$

$$
\hat{r}^{\alpha_i l*}_{\alpha_j r*}(\Delta\tau) = \frac{\hat{k}^{\alpha_i l*}_{\alpha_j r*}(\Delta\tau)}{\sqrt{\hat{k}^{\alpha_i l*}_{\alpha_i r*}(\Delta\tau) \cdot \hat{k}^{\alpha_j l*}_{\alpha_j r*}(\Delta\tau)}} \tag{3.23}
$$

◆

Example 3.16. The empirical $l_\alpha r_\alpha$-correlation function ${}_{lr}\underline{\hat{R}}_{\tilde{x}_\tau}(\Delta\tau)$ given by Eq. (3.22) is determined for the fuzzy time series shown in Fig. 3.8 for an $l_\alpha r_\alpha$-discretization with $n = 3$ α-levels. The results for $\Delta\tau = 0$, $\Delta\tau = 1$ and $\Delta\tau = 2$ are given by Eqs. (3.24) to (3.26), respectively.

$$ {}_{lr}\underline{\hat{R}}_{\tilde{x}_\tau}(\Delta\tau = 0) = \begin{bmatrix} 0.999 & 0.459 & 0 & 0.707 \\ 0.459 & 0.997 & 0 & 0.402 \\ 0 & 0 & 0 & 0 \\ 0.707 & 0.402 & 0 & 1.000 \end{bmatrix} \tag{3.24} $$

$$ {}_{lr}\underline{\hat{R}}_{\tilde{x}_\tau}(\Delta\tau = 1) = \begin{bmatrix} 0.297 & 0.568 & 0 & 0.875 \\ 0.568 & 0.496 & 0 & 0.497 \\ 0 & 0 & 0 & 0 \\ 0.875 & 0.497 & 0 & 0.298 \end{bmatrix} \tag{3.25} $$

$$ {}_{lr}\underline{\hat{R}}_{\tilde{x}_\tau}(\Delta\tau = 2) = \begin{bmatrix} 0.064 & 0.090 & 0 & 0.140 \\ 0.090 & 0.038 & 0 & 0.080 \\ 0 & 0 & 0 & 0 \\ 0.140 & 0.080 & 0 & 0.042 \end{bmatrix} \tag{3.26} $$

These results indicate a reduction in the linear dependencies between the $l_\alpha r_\alpha$-increments of the fuzzy values $\tilde{x}_\tau$ and $\tilde{x}_{\tau+\Delta\tau}$ with increasing time lag. For $\Delta\tau \geqslant 2$ the elements of the empirical correlation function are negligibly small. ◆

For the modeling and forecasting of fuzzy time series it is also advisable to determine the so-called empirical partial $l_\alpha r_\alpha$-correlation function ${}_{lr}\underline{\hat{P}}_{\tilde{x}_\tau}(\Delta\tau)$.

Definition 3.17. *The empirical partial $l_\alpha r_\alpha$-correlation function ${}_{lr}\underline{\hat{P}}_{\tilde{x}_\tau}(\Delta\tau)$ is defined by Eq. (3.27) as the empirical correlation between the fuzzy values $\tilde{x}_\tau$ and $\tilde{x}_{\tau+\Delta\tau}$ of a stationary fuzzy time series with the exclusion of the influence of the intermediate fuzzy values $\tilde{x}_{\tau+1}$, $\tilde{x}_{\tau+2}$, ..., $\tilde{x}_{\tau+\Delta\tau-1}$.*

$$ {}_{lr}\underline{\hat{P}}_{\tilde{x}_\tau}(\Delta\tau) = \begin{bmatrix} \hat{p}^{\alpha_1 l}_{\alpha_1 l}(\Delta\tau) & \hat{p}^{\alpha_2 l}_{\alpha_1 l}(\Delta\tau) & \cdots & \hat{p}^{\alpha_2 r}_{\alpha_1 l}(\Delta\tau) & \hat{p}^{\alpha_1 r}_{\alpha_1 l}(\Delta\tau) \\ \hat{p}^{\alpha_1 l}_{\alpha_2 l}(\Delta\tau) & \hat{p}^{\alpha_2 l}_{\alpha_2 l}(\Delta\tau) & \cdots & \hat{p}^{\alpha_2 r}_{\alpha_2 l}(\Delta\tau) & \hat{p}^{\alpha_1 r}_{\alpha_2 l}(\Delta\tau) \\ \vdots & \vdots & \ddots & \vdots & \vdots \\ \hat{p}^{\alpha_1 l}_{\alpha_2 r}(\Delta\tau) & \hat{p}^{\alpha_2 l}_{\alpha_2 r}(\Delta\tau) & \cdots & \hat{p}^{\alpha_2 r}_{\alpha_2 r}(\Delta\tau) & \hat{p}^{\alpha_1 r}_{\alpha_{n-1} r}(\Delta\tau) \\ \hat{p}^{\alpha_1 l}_{\alpha_1 r}(\Delta\tau) & \hat{p}^{\alpha_2 l}_{\alpha_1 r}(\Delta\tau) & \cdots & \hat{p}^{\alpha_2 r}_{\alpha_1 r}(\Delta\tau) & \hat{p}^{\alpha_1 r}_{\alpha_1 r}(\Delta\tau) \end{bmatrix} \tag{3.27} $$

◆

The determination of the empirical partial $l_\alpha r_\alpha$-correlation function ${}_{lr}\underline{\hat{P}}_{\tilde{x}_\tau}(\Delta\tau)$ is equivalent to the determination of the empirical $l_\alpha r_\alpha$-correlation function of the fuzzy time series suitably adjusted to exclude the influence of the fuzzy values $\tilde{x}_{\tau+1}$, $\tilde{x}_{\tau+2}$, ..., $\tilde{x}_{\tau+\Delta\tau-1}$. The elements of the empirical partial $l_\alpha r_\alpha$-correlation function ${}_{lr}\underline{\hat{P}}_{\tilde{x}_\tau}(\Delta\tau)$ are computed with the aid of Eqs. (3.28) and (3.29). The indices α_i and α_j with $i, j = 1, 2, ..., n$ again denote the α-levels to be analyzed, while $l^* = l, r$ and $r^* = l, r$ indicate the left and right branches of the membership function, respectively.

$$\hat{p}^{\alpha_i l*}_{\alpha_j r*}(\Delta\tau) = \frac{\hat{h}^{\alpha_i l*}_{\alpha_j r*}(\Delta\tau)}{\sqrt{\hat{h}^{\alpha_i l*}_{\alpha_i r*}(\Delta\tau)\hat{h}^{\alpha_j l*}_{\alpha_j r*}(\Delta\tau)}} \tag{3.28}$$

$$\text{with} \quad \hat{h}^{\alpha_i l*}_{\alpha_j r*}(\Delta\tau) = \frac{1}{N-\Delta\tau}\sum_{\tau=1}^{N-\Delta\tau}\left[\left(\Delta z_{\alpha_i l*}(\tau) - \Delta\overline{z}_{\alpha_i l*}(\tau)\right)\ldots\right. \tag{3.29}$$
$$\left.\ldots\left(\Delta z_{\alpha_j r*}(\tau+\Delta\tau) - \Delta\overline{z}_{\alpha_j r*}(\tau+\Delta\tau)\right)\right]$$

The $l_\alpha r_\alpha$-increments $\Delta z_{\alpha_i l*}(\tau)$ and $\Delta z_{\alpha_j r*}(\tau+\Delta\tau)$ are computed with the aid of Eqs. (3.30) and (3.31). In both equations the fuzzy values $\tilde{x}_\tau$ and $\tilde{x}_{\tau+\Delta\tau}$ are adjusted to exclude the influence of the fuzzy values $\tilde{x}_{\tau+1}$, $\tilde{x}_{\tau+2}$, ..., $\tilde{x}_{\tau+\Delta\tau-1}$ lying between τ and $\tau + \Delta\tau$.

$$\tilde{z}_\tau = \tilde{x}_\tau - \check{\tilde{x}}_\tau \tag{3.30}$$

$$\tilde{z}_{\tau+\Delta\tau} = \tilde{x}_{\tau+\Delta\tau} - \check{\tilde{x}}_{\tau+\Delta\tau} \tag{3.31}$$

The fuzzy values $\check{\tilde{x}}_\tau$ and $\check{\tilde{x}}_{\tau+\Delta\tau}$ which yield the best linear approximations of the fuzzy values $\tilde{x}_\tau$ and $\tilde{x}_{\tau+\Delta\tau}$ are determined according to Eqs. (3.32) and (3.33).

$$\check{\tilde{x}}_\tau = \underline{A}_1 \odot \tilde{x}_{\tau+1} \oplus \underline{A}_2 \odot \tilde{x}_{\tau+2} \oplus \ldots \oplus \underline{A}_{h-1} \odot \tilde{x}_{\tau+\Delta\tau-1} \tag{3.32}$$

$$\check{\tilde{x}}_{\tau+\Delta\tau} = \underline{B}_1 \odot \tilde{x}_{\tau+1} \oplus \underline{B}_2 \odot \tilde{x}_{\tau+2} \oplus \ldots \oplus \underline{B}_{h-1} \odot \tilde{x}_{\tau+\Delta\tau-1} \tag{3.33}$$

The matrices $\underline{A}_k$ and $\underline{B}_k$ are the coefficients of the best linear approximations. The elements $a_k[i,j]$ and $b_k[i,j]$ of the $[2n, 2n]$ coefficient matrices $\underline{A}_k$ and $\underline{B}_k$ (k = 1, 2, ..., $h-1$) are determined according to the minimization requirements given by Eqs. (3.34) and (3.35).

$$\sum_{g=1}^{N-h}\sum_{i=1}^{2n}\left[\Delta\mathrm{x}_i(g) - \sum_{j=1}^{2n}\sum_{k=1}^{h-1} a_k[i,j]\,\Delta\mathrm{x}_j(g+k)\right]^2 \stackrel{!}{=} \min \tag{3.34}$$

$$\sum_{g=1}^{N-h}\sum_{i=1}^{2n}\left[\Delta\mathrm{x}_i(g+h) - \sum_{j=1}^{2n}\sum_{k=1}^{h-1} b_k[i,j]\,\Delta\mathrm{x}_j(g+k)\right]^2 \stackrel{!}{=} \min \tag{3.35}$$

The $l_\alpha r_\alpha$-increments $\Delta\overline{z}_{\alpha_i l*}(\tau)$ and $\Delta\overline{z}_{\alpha_j r*}(\tau+\Delta\tau)$ also required in Eq. (3.29) are computed with the aid of Eqs. (3.36) and (3.37).

$$\tilde{\bar{z}}_{\tau} = \frac{1}{N - \Delta\tau} \bigoplus_{g=1}^{N-\Delta\tau} \tilde{z}_g \tag{3.36}$$

$$\tilde{\bar{z}}_{\tau+\Delta\tau} = \frac{1}{N - \Delta\tau} \bigoplus_{g=1}^{N-\Delta\tau} \tilde{z}_{g+\Delta\tau} \tag{3.37}$$

Numerical Tests for Stationarity

Reverse argument is applied to verify the stationarity of fuzzy time series. The non-existence of non-stationarity such as fuzzy trends or fuzzy cycles is verified numerically. Although the following methods do not directly verify stationarity, they indicate the extent to which non-stationarity exists. In particular, very short fuzzy time series or segments of a fuzzy time series may erroneously give the expression that stationarity or non-stationarity is present.

- In accordance with Sect. 3.1 the plot should always serve as the first step in the analysis of a time series. On the basis of the plot it is possible to ascertain the non-existence of a fuzzy trend or fuzzy cycle. The $l_\alpha r_\alpha$-increment plot is especially suitable for this purpose. An obvious non-stationarity should not exist for each individual $l_\alpha r_\alpha$-increment.
- Sufficiently long fuzzy time series may be subdivided into several segments. The empirical moments according to Sect. 3.3 are computed separately for each segment. In the case of stationarity the empirical moments of individual segments are equal. If the empirical moments of the individual segments show significant differences, non-stationarity must be assumed. A precondition for this approach is that stationarity is assumed a priori for each individual segment. This assumption may be checked a posteriori by modifying the segmental subdivision.
- In order to check stationarity the fuzzy component model according to Sect. 3.2 may also be applied. The matching of a fuzzy trend function $\tilde{t}(\tau)$ and a fuzzy cycle function $\tilde{z}(\tau)$ may be also be applied to verify stationarity or non-stationarity. If the matching of a fuzzy trend function with a non-negligible ascent and/or a fuzzy cycle function with a non-negligible amplitude is already possible for an individual $l_\alpha r_\alpha$-increment, non-stationarity must be assumed.

3.4 Transformation of Fuzzy Time Series Using Filters

The transformation of a given fuzzy time series $(\tilde{x}_\tau)_{\tau\in\mathbf{T}}$ into a fuzzy time series $(\tilde{z}_\tau)_{\tau\in\mathbf{T}*}$ is carried out by means of so-called filters. The aim of the transformation is to smooth irregular local and global fluctuations of the given fuzzy time series $(\tilde{x}_\tau)_{\tau\in\mathbf{T}}$. Smoothed time series facilitate the modeling of trends and cycles.

3.4.1 Smoothing of Fuzzy Time Series

Definition 3.18. *Analogous to classical time series analysis, the transformation L according to Eq. (3.38) is defined as a linear filter for fuzzy time series.*

$$\tilde{z}_\tau = L\tilde{x}_\tau = \bigoplus_{i=-a}^{b} c_i \tilde{x}_{\tau+i} \tag{3.38}$$

In the above equation, $(\tilde{x}_\tau)_{\tau\in\mathbf{T}}$ *is the fuzzy time series to be transformed,* $(\tilde{z}_\tau)_{\tau\in\mathbf{T}^*}$ *is the smoothed fuzzy time series with* $\mathbf{T}^* = \{a+1, ..., N-b\}$*, and* c_i *are the filter coefficients.* ◆

An affine transformation is performed for all n $l_\alpha r_\alpha$-increments of the fuzzy values $\tilde{x}_\tau$ using the real-valued filter coefficients c_i. An extension of the latter to non-affine transformations is discussed in Sect. 3.4.3.

Remark 3.19. The filter coefficients must be chosen in such a way that the smoothed fuzzy time series only includes fuzzy values in the proper sense. ◆

The filter given by Eq. (3.38) defines a linear combination of fuzzy values. The linearity of the filter given by Eq. (3.38) may be demonstrated as follows:

Applying the incremental fuzzy arithmetic introduced in Sect. 2.1.2, the following holds for the sum of two filters:

$$L\tilde{x}_\tau \oplus L\tilde{y}_\tau = \bigoplus_{i=-a}^{b} c_i \tilde{x}_{\tau+i} \oplus \bigoplus_{i=-a}^{b} c_i \tilde{y}_{\tau+i} \tag{3.39}$$

$$= \bigoplus_{i=-a}^{b} \left(c_i \tilde{x}_{\tau+i} \oplus c_i \tilde{y}_{\tau+i}\right) \tag{3.40}$$

$$= \bigoplus_{i=-a}^{b} c_i \left(\tilde{x}_{\tau+i} \oplus \tilde{y}_{\tau+i}\right) \tag{3.41}$$

$$L\tilde{x}_\tau \oplus L\tilde{y}_\tau = L\left(\tilde{x}_\tau \oplus \tilde{y}_\tau\right) . \tag{3.42}$$

In contrast to the application of the extension principle, linearity is only ensured by applying incremental fuzzy arithmetic.

Definition 3.20. *If* $\sum c_i = 1$ *holds for the coefficients, the linear filter is referred to as a moving fuzzy average. Considering the simplest case of a linear filter, the simple moving fuzzy average for fuzzy time series is defined according to Eq. (3.43).*

$$\tilde{z}_\tau = \frac{1}{2a+1} \bigoplus_{i=-a}^{a} \tilde{x}_{\tau+i} \quad \text{mit} \quad \tau = a+1, ..., N-a \tag{3.43}$$

◆

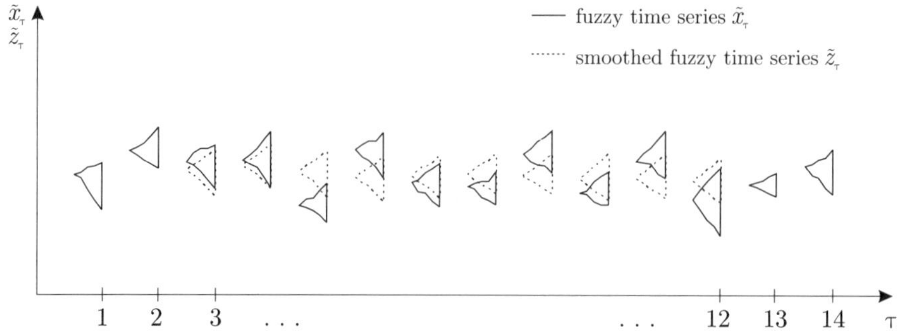

Fig. 3.9. Moving fuzzy average of a fuzzy time series $(\tilde{x}_\tau)_{\tau\in\mathbf{T}}$

Example 3.21. The simple moving fuzzy average according to Eq. (3.43) with $a = 2$ is shown by way of example for the fuzzy time series presented in Fig. 3.9. ◆

By means of the linear filter given by Eq. (3.38) it is possible to transform fuzzy time series arbitrarily. For example, a local polynomial approximation of a fuzzy time series may be obtained by a suitable choice of the filter coefficients c_i. The filter coefficients c_i corresponding to the desired polynomial degree are thereby determined by the following condition. A nontrivial linear filter L results in a local polynomial approximation of a fuzzy time series provided it reproduces the underlying series when applied to a fuzzy time series which follows a polynomial curve.

Example 3.22. This postulation, which is valid for general polynomial curves, is demonstrated by the example of a fuzzy time series $\tilde{x}_\tau = \tilde{p}_1 \oplus \tau\tilde{p}_2 \oplus \tau^2\tilde{p}_3$ following a quadratic curve (the coefficients $\tilde{p}_1$, $\tilde{p}_2$ and $\tilde{p}_3$ are arbitrary fuzzy values). The application of the filter L to $\tilde{x}_\tau$ according to Eq. (3.44) again yields the original series $\tilde{x}_\tau$.

$$\begin{aligned}\tilde{z}_\tau &= L\tilde{x}_\tau = L\left(\tilde{p}_1 \oplus \tau\tilde{p}_2 \oplus \tau^2\tilde{p}_3\right) = \bigoplus_{i=-2}^{2} c_i\tilde{x}_{\tau+i} \\ &= \frac{1}{35}\left(-3\tilde{x}_{\tau-2} \oplus 12\tilde{x}_{\tau-1} \oplus 17\tilde{x}_\tau \oplus 12\tilde{x}_{\tau+1} \ominus 3\tilde{x}_{\tau+2}\right) \\ &= \tilde{p}_1 \oplus \tau\tilde{p}_2 \oplus \tau^2\tilde{p}_3 = \tilde{x}_\tau\end{aligned} \tag{3.44}$$

◆

Remark 3.23. The linear filter defined by Eq. (3.38) possesses the following numerical property. If the smoothed time series values $\tilde{z}_\tau$ are computed according to the extension principle (see Sect. 2.1.3), the uncertainty of the $\tilde{z}_\tau$ values increases for negative coefficients c_i. If, on the other hand, $l_\alpha r_\alpha$-addition (see Def. 2.10) is applied, an increase in the uncertainty of the $\tilde{z}_\tau$ values due to numerical effects does not occur. The moving fuzzy averages computed for a time series with constant fuzzy values according to Eq. (3.43) using incremental fuzzy arithmetic and the extension principle are compared in Fig. 3.10. ◆

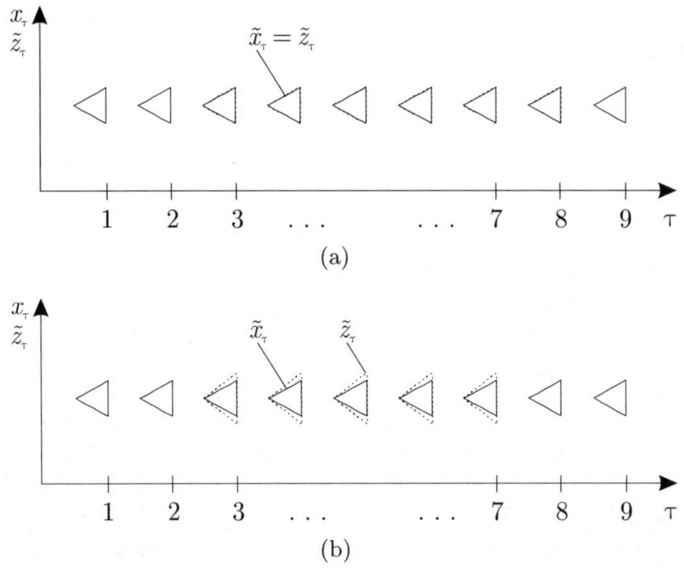

Fig. 3.10. Moving fuzzy average according to Eq. (3.43) using (a) incremental fuzzy arithmetic and (b) the extension principle

3.4.2 Fuzzy Difference Filter

For the local polynomial approximation of a fuzzy time series or the matching of a polynomial to a fuzzy time series it is necessary to select the degree of the polynomial to be used. For this purpose the difference filter applied in classical time series analysis is extended to deal with fuzzy time series.

Definition 3.24. *The fuzzy difference filter D^p of pth order is defined recursively by Eq. (3.45) for $p > 1$.*

$$D^p \tilde{x}_\tau = D^{p-1} \tilde{x}_\tau \ominus D^{p-1} \tilde{x}_{\tau-1} \oplus \tilde{d}^p \quad \text{with} \quad \tau = p+1, ..., N \tag{3.45}$$

◆

By means of the fuzzy difference filter D^p the degree of a polynomial of order p^* is reduced to the degree of order $p^* - p$.

As a 1st order fuzzy difference filter D^1 the transformation according to Eq. (3.46) follows from Eq. (3.45).

$$D^1\tilde{x}_\tau = \tilde{x}_\tau \ominus \tilde{x}_{\tau-1} \oplus \tilde{d}^1 \quad \text{with} \quad \tau = 2, 3, ..., N \tag{3.46}$$

The fuzzy correction factor $\tilde{d}^p$ ensures that Eq. (2.18) is complied with for the fuzzy result. The determination of the $l_\alpha r_\alpha$-increments of $\tilde{d}^p$ is shown by way of example for $p = 1$ in Eqs. (3.47) to (3.49). The computations for $p > 1$ are performed in a similar manner.

$$\Delta d^1_{\alpha_i l} = \begin{cases} -h_{\alpha_i l} & \text{for } h_{\alpha_i l} < 0 \\ 0 & \text{for } h_{\alpha_i l} \geqslant 0 \end{cases} \quad \text{for} \quad i = 1, 2, ..., n-1 \tag{3.47}$$

$$\Delta d^1_{\alpha_n l} = 0 \tag{3.48}$$

$$\Delta d^1_{\alpha_i r} = \begin{cases} -h_{\alpha_i r} & \text{for } h_{\alpha_i r} < 0 \\ 0 & \text{for } h_{\alpha_i r} \geqslant 0 \end{cases} \quad \text{for} \quad i = 1, 2, ..., n \tag{3.49}$$

$$\text{with} \quad h_{\alpha_i l/r} = \min_{\tau=2,3,...,N} \left[\Delta x_{\alpha_i l/r}(\tau) - \Delta x_{\alpha_i l/r}(\tau-1)\right]$$

A non-stationary fuzzy time series with a linear trend is transformed into a stationary fuzzy time series by applying a 1st order fuzzy difference filter. A non-stationary fuzzy time series whose trend is described by a pth order polynomial reduces to a stationary fuzzy time series when a fuzzy difference filter D^p is applied. By this means it is thus possible to determine the degree of the polynomial to be chosen for a polynomial of unknown order p. This procedure is also suitable for the polynomial approximation of local sections of a fuzzy time series.

Cyclic fluctuations of a fuzzy time series may also be eliminated with the aid of a cyclic fuzzy difference filter D_z according to Eq. (3.50). The fuzzy correction factor $\tilde{d}_z$ is determined analogously according to Eqs. (3.47) to (3.49).

$$D_z\tilde{x}_\tau = \tilde{x}_\tau \ominus \tilde{x}_{\tau-z} \oplus \tilde{d}_z \quad \text{with} \quad \tau = z, z+1, ..., N \tag{3.50}$$

3.4.3 Extended Smoothing and Extended Fuzzy Difference Filter

The filters introduced in Sects. 3.4.1 and 3.4.2 for fuzzy time series are characterized by real-valued filter coefficients. Affine transformations of the $l_\alpha r_\alpha$-increments are performed by means of these filters. In some cases, however, a non-affine transformation, i.e. the different transformation of individual $l_\alpha r_\alpha$-increments, is often advantageous. For this purpose the extended linear fuzzy filter L_e is introduced.

Definition 3.25. *The extended linear fuzzy filter L_e is defined according to Eq. (3.51).*

$$\tilde{z}_\tau = L_e\tilde{x}_\tau = \bigoplus_{i=-a}^{b} \underline{C}_i \odot \tilde{x}_{\tau+i} \tag{3.51}$$

◆

The $\underline{C}_i$ diagonal matrices of size $[2n, 2n]$ hereby contain real-valued elements. By this means a different transformation is possible on each α-level. If the elements of the coefficient matrices $\underline{C}_i$ are identical, Eq. (3.51) describes an affine transformation just as Eq. (3.38) as a special case.

Analogous to the extended linear filter, it is also possible to define an extended fuzzy difference filter. The extended fuzzy difference filter D_e^1 of 1st order is presented by way of example in Eq. (3.52).

$$D_e^1 \tilde{x}_\tau = \tilde{x}_\tau \ominus \underline{D}^1 \odot \tilde{x}_{\tau-1} \oplus \tilde{d}^1 \quad \text{with} \quad \tau = 2, 3, ..., N \tag{3.52}$$

$\underline{D}^1$ is hereby a diagonal matrix whose elements may be assigned the values 0 or 1.

The following holds for the extended fuzzy difference filter D_e^p of pth order:

$$D_e^p \tilde{x}_\tau = D_e^{p-1} \odot \tilde{x}_\tau \ominus D_e^{p-1} \odot \tilde{x}_{\tau-1} \oplus \tilde{d}^p \quad \text{with} \quad \tau = p+1, ..., N\,. \tag{3.53}$$

By means of the extended difference filter it is possible to take account of different trends in the $l_\alpha r_\alpha$-increments on different α-levels. By repeating the extended difference computation p-times it is possible to transform the trends on an α-level to α-level basis.

3.5 Modeling on the Basis of Specific Fuzzy Random Processes

As already introduced in Sect. 2.3, a fuzzy time series $(\tilde{x}_\tau)_{\tau \in \mathbf{T}}$ may be interpreted as the realization of a fuzzy random process $(\tilde{\mathrm{X}}_\tau)_{\tau \in \mathbf{T}}$. In the following section, specific fuzzy random processes and specific methods are presented which permit the identification of the hypothetical underlying fuzzy random process of a given time series.

Remark 3.26. The fuzzy random processes $(\tilde{\mathrm{X}}_\tau)_{\tau \in \mathbf{T}}$ are exclusively evaluated at equidistant discrete time points τ in the following, i.e. only discrete processes are considered. In order to describe the random $l_\alpha r_\alpha$- increments of a fuzzy random variable $\tilde{\mathrm{X}}_\tau$ of the fuzzy random process $(\tilde{\mathrm{X}}_\tau)_{\tau \in \mathbf{T}}$ at time point τ the bracket notation $\Delta \mathrm{X}_{\alpha_i l}(\tau)$ and $\Delta \mathrm{X}_{\alpha_i r}(\tau)$ is used. ◆

3.5.1 Fuzzy White-Noise Processes

Definition 3.27. *A fuzzy white-noise process $(\tilde{\mathcal{E}}_\tau)_{\tau \in \mathbf{T}}$ is a stationary and ergodic fuzzy random process. A fuzzy white-noise variable $\tilde{\mathcal{E}}_\tau$ with the fuzzy realizations $\tilde{\varepsilon}_\tau$ is assigned to each time point τ. The random $l_\alpha r_\alpha$-increments $\Delta\mathcal{E}_j(\tau)$ and $\Delta\mathcal{E}_j(\tau + \Delta\tau)$ of the fuzzy white-noise variables $\tilde{\mathcal{E}}_\tau$ and $\tilde{\mathcal{E}}_{\tau+\Delta\tau}$ are fully independent real-valued random variables with a constant expected value $E[\Delta\mathcal{E}_j(\tau)] = E[\Delta\mathcal{E}_j(\tau + \Delta\tau)]$ and a constant variance $Var[\Delta\mathcal{E}_j(\tau)] = Var[\Delta\mathcal{E}_j(\tau + \Delta\tau)]$. The independency postulation exclusively holds for time* *differences of $\Delta\tau \neq 0$. In accordance with Eq. (3.56) the $l_\alpha r_\alpha$-covariance function is permitted for $\Delta\tau = 0$. The following holds for a white-noise process:*

$$E[\tilde{\mathcal{E}}_\tau] = \tilde{m}_{\tilde{\mathcal{E}}_\tau} \quad = \text{constant} \quad \forall\ \tau \in \mathbf{T} \tag{3.54}$$

$$_{lr}Var[\tilde{\mathcal{E}}_\tau] = {}_{lr}\underline{\sigma}^2_{\tilde{\mathcal{E}}_\tau} \quad = \text{constant} \quad \forall\ \tau \in \mathbf{T} \tag{3.55}$$

$$_{lr}\underline{K}_{\tilde{\mathcal{E}}_\tau}(\Delta\tau) = \begin{cases} {}_{lr}\underline{K}_{\tilde{\mathcal{E}}_\tau}(0) & \text{for} \quad \Delta\tau = 0 \\ \underline{0} & \text{for} \quad \Delta\tau \neq 0 \end{cases} . \tag{3.56}$$

◆

Each of the random $l_\alpha r_\alpha$-increments $\Delta\mathcal{E}_j(\tau)$ is characterized by a corresponding probability density function $f_{\Delta\mathcal{E}_j}(\Delta\varepsilon_j)$. All realizations $\Delta\varepsilon_j(\tau)$ of the random $l_\alpha r_\alpha$- increments $\Delta\mathcal{E}_j(\tau)$ must satisfy the requirements of Eqs. (3.57) to (3.59).

$$\Delta\varepsilon_j(\tau) \geqslant 0 \qquad \text{for} \quad j = 1, 2, ..., n-1 \tag{3.57}$$

$$\Delta\varepsilon_j(\tau) \text{ arbitrary} \quad \text{for} \quad j = n \tag{3.58}$$

$$\Delta\varepsilon_j(\tau) \geqslant 0 \qquad \text{for} \quad j = n+1, n+2, ..., 2n \tag{3.59}$$

The non-negativity requirement of Eqs. (3.57) and (3.59) does not apply to the mean value. The realizations $\Delta\varepsilon_n(\tau)$ of the random $l_\alpha r_\alpha$-increment $\Delta\mathcal{E}_n(\tau)$ of the mean value may also take on negative values.

Example 3.28. A realization of typical fuzzy white-noise process is shown in Fig. 3.11.

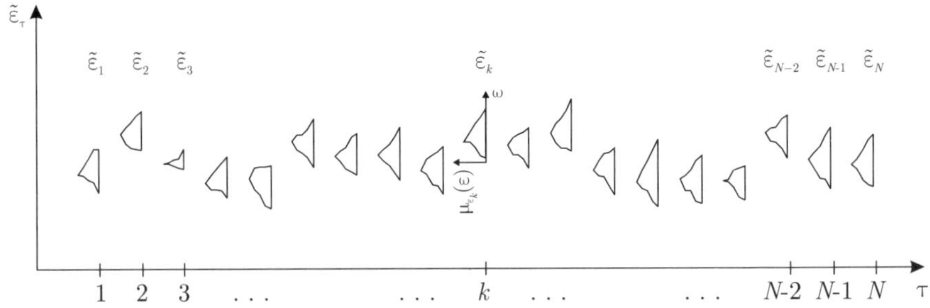

Fig. 3.11. Realization of a typical fuzzy white-noise process

◆

By means of Eq. (3.57) and Eq. (3.59) lower bounds are specified for the realizations $\Delta\varepsilon_j(\tau)$. Moreover, upper bounds may also exist.

It is advantageous to adopt a stepped function according to Eq. (3.60) as the probability density function $f_{\Delta\mathcal{E}_j}(\Delta\varepsilon_j)$ for the random $l_\alpha r_\alpha$-increments $\Delta\mathcal{E}_j(\tau)$. The basic form of such a probability density function is shown in Fig. 3.12.

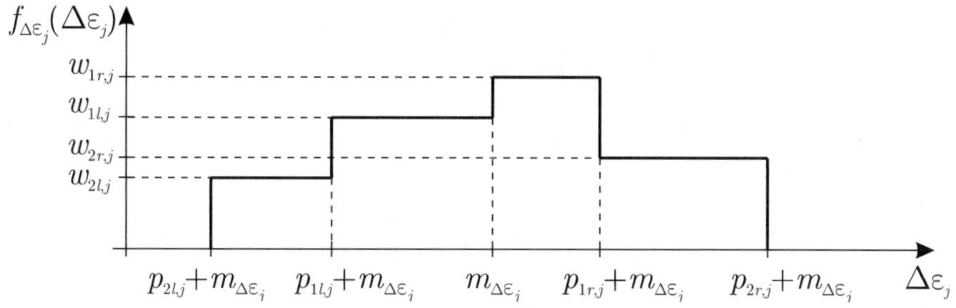

Fig. 3.12. Probability density function $f_{\Delta\mathcal{E}_j}(\Delta\varepsilon_j(\tau))$ for a random $l_\alpha r_\alpha$-increment $\Delta\mathcal{E}_j(\tau)$ of a fuzzy white-noise process

The following holds for each random $l_\alpha r_\alpha$-increment $\Delta\mathcal{E}_j(\tau)$:

$$f_{\Delta\mathcal{E}_j}(\Delta\varepsilon_j,\tau) = \begin{cases} w_{2l} & \text{for } p_{2l} \leqslant (\Delta\varepsilon_j - m_{\Delta\mathcal{E}_j}) \leqslant p_{1l} \\ w_{1l} & \text{for } p_{1l} \leqslant (\Delta\varepsilon_j - m_{\Delta\mathcal{E}_j}) \leqslant 0 \\ w_{1r} & \text{for } 0 < (\Delta\varepsilon_j - m_{\Delta\mathcal{E}_j}) \leqslant p_{1r} \\ w_{2r} & \text{for } p_{1r} \leqslant (\Delta\varepsilon_j - m_{\Delta\mathcal{E}_j}) \leqslant p_{2r} \\ 0 & \text{else} \end{cases} \tag{3.60}$$

$$\text{with} \quad -p_{1l}w_{1l} + (p_{1l} - p_{2l})w_{2l} = 0.5$$
$$p_{1r}w_{1r} + (p_{2r} - p_{1r})w_{2r} = 0.5\,.$$

In order to simplify the notation in Eq. (3.60) the subscript j is omitted for the parameters.

By means of this function it is possible to arbitrarily specify the upper and lower limits for a given variance. The upper and lower limits are defined by the parameters p_{2l} and p_{2r}. The prescribed mean values $E[\Delta\mathcal{E}_j(\tau)] = m_{\Delta\mathcal{E}_j}$ and variances $Var[\Delta\mathcal{E}_j(\tau)] = \sigma^2_{\Delta\mathcal{E}_j}$ lead in each case to Eqs. (3.61) and (3.62) as additional constraints for the parameters p_{1r}, p_{1l}, w_{2r}, w_{2l}, w_{1r} and w_{1l}.

$$\sigma^2_{\Delta\mathcal{E}_j} = \frac{w_{2l}}{3}\left(p_{1l}^3 - p_{2l}^3\right) - \frac{w_{1l}}{3}\left(p_{1l}^3\right) + \frac{w_{1r}}{3}\left(p_{1r}^3\right) + \frac{w_{2r}}{3}\left(p_{2r}^3 - p_{1r}^3\right) \tag{3.61}$$

$$0 = \frac{w_{2l}}{2}\left(p_{1l}^2 - p_{2l}^2\right) - \frac{w_{1l}}{2}\left(p_{1l}^2\right) + \frac{w_{1r}}{2}\left(p_{1r}^2\right) + \frac{w_{2r}}{2}\left(p_{2r}^2 - p_{1r}^2\right) \tag{3.62}$$

The functional values w_{2r}, w_{2l}, w_{1r} and w_{1l} are computed from Eqs. (3.63) to (3.66), respectively.

$$w_{2r} = \frac{6\sigma^2_{\Delta\mathcal{E}_j} + p_{1l}p_{1r} + p_{2l}p_{1l} - p^2_{1r} + p_{2l}p_{1r}}{2p_{2r}\left(p_{1l}p_{1r} - p_{2r}p_{1l} + p^2_{2r} - p^2_{1r} + p_{2l}p_{1r} - p_{2l}p_{2r}\right)} \tag{3.63}$$

$$w_{2l} = \frac{p_{1l} - 2p_{1r}w_{2r}p_{2r} + p_{1r} + 2w_{2r}p^2_{2r}}{2p_{2l}\left(p_{2l} - p_{1l}\right)} \tag{3.64}$$

$$w_{1r} = \frac{1 - 2w_{2r}p_{2r} + w_{2r}p_{1r}}{2p_{1r}} \tag{3.65}$$

$$w_{1l} = \frac{2w_{2l}p_{1l} - 2p_{2l}w_{2l} - 1}{2p_{1l}} \tag{3.66}$$

The parameters $0 < p_{1r} < p_{2r}$ and $p_{2l} < p_{1l} < 0$ must be chosen in such a way that the following holds:

$$w_{2r},\ w_{2l},\ w_{1r},\ w_{1l} \geqslant 0\,. \tag{3.67}$$

The corresponding probability distribution function $F_{\Delta\mathcal{E}_j}\left(\Delta\varepsilon_j\right)$ is obtained by integrating $f_{\Delta\mathcal{E}_j}\left(\Delta\varepsilon_j\right)$.

Numerical realization. The simulation of a fuzzy white-noise process $(\tilde{\mathcal{E}}_\tau)_{\tau\in\mathbf{T}}$, i.e. the determination of the realizations $(\tilde{\varepsilon}_\tau)_{\tau\in\mathbf{T}}$, is based on the Monte Carlo simulation of continuous fuzzy random variables. Use is thereby made of the characteristic properties of the fuzzy white-noise process. In accordance with Eqs. (3.54) and (3.55) the fuzzy white-noise variables $\tilde{\mathcal{E}}_\tau$ for different values of τ are fully-independent fuzzy random variables with a constant fuzzy expected value $E[\tilde{\mathcal{E}}_\tau] = \tilde{m}_{\tilde{\mathcal{E}}_\tau}$ and a constant $l_\alpha r_\alpha$-variance ${}_{lr}Var[\tilde{\mathcal{E}}_\tau] = {}_{lr}\underline{\sigma}^2_{\tilde{\mathcal{E}}_\tau}$. According to Eq. (3.56) the $l_\alpha r_\alpha$-covariance function ${}_{lr}\underline{K}_{\tilde{\mathcal{E}}_\tau}(\Delta\tau)$ only exists for $\Delta\tau = 0$. From this it follows that the simulation of the fuzzy white-noise process $(\tilde{\mathcal{E}}_\tau)_{\tau\in\mathbf{T}}$ at each time point τ may be reduced to the Monte Carlo simulation (see Sect. 2.2.4) of a fuzzy white-noise variable $\tilde{\mathcal{E}}$ with the properties $E[\tilde{\mathcal{E}}] = E[\tilde{\mathcal{E}}_\tau]$, ${}_{lr}Var[\tilde{\mathcal{E}}] = {}_{lr}Var[\tilde{\mathcal{E}}_\tau]$ and ${}_{lr}\underline{K}_{\tilde{\mathcal{E}}} = {}_{lr}\underline{K}_{\tilde{\mathcal{E}}_\tau}(\Delta\tau = 0)$. Analogous to Sect. 2.2.4, the continuous fuzzy random variable $\tilde{\mathcal{E}}$ is thus computed as a one-to-one mapping of the uniformly distributed fuzzy random variable $\tilde{Y}$ according to Eq. (3.68).

$$\tilde{\mathcal{E}} = f_c(\tilde{Y}) \tag{3.68}$$

In the case of a discrete fuzzy random variable $\tilde{\mathcal{E}}$ the simulation is computed as a one-to-one mapping of the uniformly distributed random variable Y according to Eq. (3.69) (see also Sect. 2.2.4).

$$\tilde{\mathcal{E}} = f_d(Y) \tag{3.69}$$

3.5.2 Fuzzy Moving Average Processes

Definition 3.29. *A fuzzy random process* $(\tilde{X}_\tau)_{\tau \in \mathbf{T}}$ *is referred to as a fuzzy moving average process of order q (fuzzy MA[q] process for short) if it can be described by Eq. (3.70) at each time point* τ.

$$\tilde{X}_\tau = \tilde{\mathcal{E}}_\tau \ominus \underline{B}_1 \odot \tilde{\mathcal{E}}_{\tau-1} \ominus ... \ominus \underline{B}_q \odot \tilde{\mathcal{E}}_{\tau-q} \tag{3.70}$$

◆

The parameters $\underline{B}_j$ $(j = 1, 2, ..., q)$ are real-valued $[2n, 2n]$ matrices, whereby n denotes the number of chosen α-levels. The variables $\tilde{\mathcal{E}}_\tau$, $\tilde{\mathcal{E}}_{\tau-1}$, ..., $\tilde{\mathcal{E}}_{\tau-q}$ are the fuzzy random variables of a fuzzy white-noise process $(\tilde{\mathcal{E}}_\tau)_{\tau \in \mathbf{T}}$ at time points τ, $\tau - 1$, ..., $\tau - q$ (see Sect. 3.5.1).

Fuzzy moving average processes $(\tilde{X}_\tau)_{\tau \in \mathbf{T}}$ are generally stationary fuzzy random processes. The fuzzy expected values and the $l_\alpha r_\alpha$-variance of a fuzzy MA process may be determined by means of Eqs. (3.71) and (3.73).

The fuzzy expected value $E[\tilde{X}_\tau]$ is computed using the fuzzy expected value $E[\tilde{\mathcal{E}}_\tau] = \tilde{m}_{\tilde{\mathcal{E}}_\tau}$ of the fuzzy white-noise process $(\tilde{\mathcal{E}}_\tau)_{\tau \in \mathbf{T}}$ as follows:

$$E[\tilde{X}_\tau] = \tilde{m}_{\tilde{X}_\tau} = \left[\sum_{j=0}^{q} (-\underline{B}_j) \right] \odot \tilde{m}_{\tilde{\mathcal{E}}_\tau} . \tag{3.71}$$

The parameter matrices $\underline{B}_1$, $\underline{B}_2$, ..., $\underline{B}_q$ are obtained from Eq. (3.70). The negative unit matrix according to Eq. (3.72) is applied to compute $\underline{B}_0$.

$$\underline{B}_0 = \begin{bmatrix} -1 & \dots & 0 \\ \vdots & \ddots & \vdots \\ 0 & \cdots & -1 \end{bmatrix} \tag{3.72}$$

The $l_\alpha r_\alpha$-variance ${}_{lr}Var[\tilde{X}_\tau]$ is computed using the $l_\alpha r_\alpha$-variance ${}_{lr}Var[\tilde{\mathcal{E}}_\tau] = {}_{lr}\underline{\sigma}^2_{\tilde{\mathcal{E}}_\tau}$ of the fuzzy white-noise process.

$${}_{lr}Var[\tilde{X}_\tau] = {}_{lr}\underline{\sigma}^2_{\tilde{X}_\tau} = \left[\sum_{j=0}^{q} (\underline{B}_j \bullet \underline{B}_j) \right] \cdot {}_{lr}\underline{\sigma}^2_{\tilde{\mathcal{E}}_\tau} \tag{3.73}$$

The operator • hereby represents the naive element-by-element multiplication of the parameter matrices.

The $l_\alpha r_\alpha$-covariance function ${}_{lr}\underline{K}_{\tilde{X}_\tau}(\Delta\tau)$ of a fuzzy moving average process $(\tilde{X}_\tau)_{\tau \in \mathbf{T}}$ may be determined for $\Delta\tau = 0, 1, ..., q$ according to Eq. (3.74). The term ${}_{lr}\underline{K}_{\tilde{\mathcal{E}}_\tau}(\Delta\tau = 0)$ hereby represents the $l_\alpha r_\alpha$-covariance function of the corresponding fuzzy white-noise process $(\tilde{\mathcal{E}}_\tau)_{\tau \in \mathbf{T}}$ for $\Delta\tau = 0$.

$${}_{lr}\underline{K}_{\tilde{X}_\tau}(\Delta\tau) = \sum_{c=0}^{q-\Delta\tau} \underline{B}_{c+\Delta\tau} \; {}_{lr}\underline{K}_{\tilde{\mathcal{E}}_\tau}(\Delta\tau = 0) \, \underline{B}_c^T \tag{3.74}$$

The elements $k_{\tilde{X}_\tau}[i,j](\Delta\tau)$ of the $l_\alpha r_\alpha$-covariance function ${}_{lr}\underline{K}_{\tilde{X}_\tau}(\Delta\tau)$ may be determined for $i,j = 1,2,...,2n$ using Eq. (3.75). The variables $b_c[j,b]$ and $b_{c+\Delta\tau}[i,a]$ are hereby the elements of the parameter matrices $\underline{B}_c$ and $\underline{B}_{c+\Delta\tau}$, respectively, and $k_{\tilde{\mathcal{E}}_\tau}[a,b](\Delta\tau = 0)$ are the elements of the $l_\alpha r_\alpha$-covariance function ${}_{lr}\underline{K}_{\tilde{\mathcal{E}}_\tau}(\Delta\tau = 0)$.

$$k_{\tilde{X}_\tau}[i,j](\Delta\tau) = \sum_{c=0}^{q-\Delta\tau} \sum_{b=1}^{2n} \sum_{a=1}^{2n} b_c[j,b] \; k_{\tilde{\mathcal{E}}_\tau}[a,b](\Delta\tau = 0) \; b_{c+\Delta\tau}[i,a] \quad (3.75)$$

Numerical realization. The numerical simulation of a fuzzy MA process $(\tilde{X}_\tau)_{\tau\in\mathbf{T}}$ is a special case of the numerical simulation of a fuzzy ARMA process (see Sect. 3.5.4).

Example 3.30. For a fuzzy MA[1] process it follows from Eq. (3.70) that

$$\tilde{X}_\tau = \tilde{\mathcal{E}}_\tau \ominus \underline{B}_1 \odot \tilde{\mathcal{E}}_{\tau-1} \ominus ... \ominus \underline{B}_q \odot \tilde{\mathcal{E}}_{\tau-q} \,. \quad (3.76)$$

For an $l_\alpha r_\alpha$-discretization with $n = 3$ α-levels the following is obtained according to the notation used in Eq. (2.89).

$$\tilde{X}_\tau = \begin{bmatrix} \Delta X_1(\tau) \\ \Delta X_2(\tau) \\ \Delta X_3(\tau) \\ \Delta X_4(\tau) \\ \Delta X_5(\tau) \\ \Delta X_6(\tau) \end{bmatrix} = \begin{bmatrix} \Delta\mathcal{E}_1(\tau) \\ \Delta\mathcal{E}_2(\tau) \\ \Delta\mathcal{E}_3(\tau) \\ \Delta\mathcal{E}_4(\tau) \\ \Delta\mathcal{E}_5(\tau) \\ \Delta\mathcal{E}_6(\tau) \end{bmatrix} + \begin{bmatrix} 0.5 & 0.1 & 0.5 & 0 & 0.2 & 0.3 \\ 0.1 & 0.5 & 0.8 & 0 & 0.4 & 0.2 \\ 0.7 & 0.9 & 0.8 & 0 & 0.9 & 0.8 \\ 0 & 0 & 0 & 0 & 0 & 0 \\ 0.1 & 0.3 & 0.2 & 0 & 0.6 & 0.2 \\ 0.4 & 0.2 & 0.1 & 0 & 0 & 0.6 \end{bmatrix} \begin{bmatrix} \Delta\mathcal{E}_1(\tau-1) \\ \Delta\mathcal{E}_2(\tau-1) \\ \Delta\mathcal{E}_3(\tau-1) \\ \Delta\mathcal{E}_4(\tau-1) \\ \Delta\mathcal{E}_5(\tau-1) \\ \Delta\mathcal{E}_6(\tau-1) \end{bmatrix} \quad (3.77)$$

The random $l_\alpha r_\alpha$-increments $\Delta\mathcal{E}_1(\tau), ..., \Delta\mathcal{E}_6(\tau)$ of the fuzzy white-noise variables $\tilde{\mathcal{E}}_\tau$ for this example are equally-distributed random variables in the interval [0,1]. The linear dependency between the $\Delta\mathcal{E}_1(\tau), ..., \Delta\mathcal{E}_6(\tau)$ is expressed by means of the $l_\alpha r_\alpha$-correlation function

$${}_{lr}\underline{R}_{\tilde{\mathcal{E}}_\tau}(\Delta\tau = 0) = \begin{bmatrix} 1 & 0.3 & 0.3 & - & 0.4 & 0.5 \\ 0.3 & 1 & 0.4 & - & 0.4 & 0.4 \\ 0.3 & 0.4 & 1 & - & 0.3 & 0.4 \\ - & - & - & - & - & - \\ 0.4 & 0.4 & 0.3 & - & 1 & 0.5 \\ 0.5 & 0.4 & 0.4 & - & 0.5 & 1 \end{bmatrix} . \quad (3.78)$$

A segment of a typical realization is shown in Fig. 3.13.

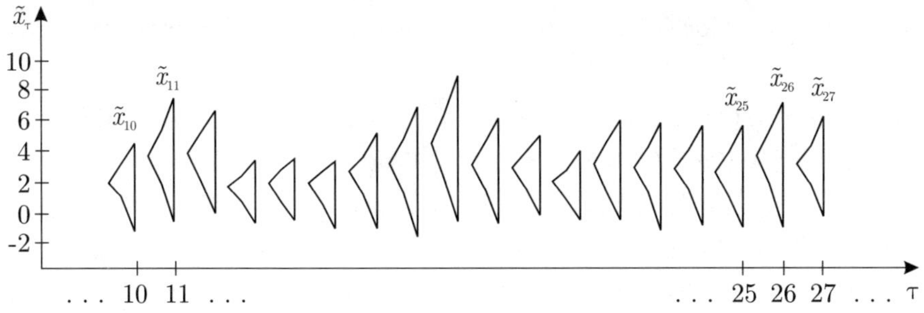

Fig. 3.13. Segment of a realization of the fuzzy MA[1] process according to Eq. (3.77)

◆

3.5.3 Fuzzy Autoregressive Processes

Definition 3.31. *A fuzzy random process* $(\tilde{\mathrm{X}}_\tau)_{\tau\in\mathbf{T}}$ *is referred to as a fuzzy autogressive process of order p (fuzzy AR[p] process for short) if it may be represented by Eq. (3.79) at each time point* τ.

$$\tilde{\mathrm{X}}_\tau = \underline{A}_1 \odot \tilde{\mathrm{X}}_{\tau-1} \oplus ... \oplus \underline{A}_p \odot \tilde{\mathrm{X}}_{\tau-p} \oplus \tilde{\mathcal{E}}_\tau \tag{3.79}$$

◆

The parameters $\underline{A}_j$ are real-valued $[2n, 2n]$ matrices, where n denotes the number of chosen α-levels. The term $\tilde{\mathcal{E}}_\tau$ is the fuzzy random variable of a fuzzy white-noise process $(\tilde{\mathcal{E}}_\tau)_{\tau\in\mathbf{T}}$ at time point τ (see Sect. 3.5.1).

By means of fuzzy AR processes it is possible to model stationary as well as non-stationary fuzzy time series. The characteristic moments of a fuzzy AR process may be computed by means of the numerical methods described in Sect. 3.5.4 for fuzzy ARMA processes.

The fuzzy random variable $\tilde{\mathrm{X}}_\tau$ and its random $l_\alpha r_\alpha$-increments $\Delta\mathrm{X}_i(\tau)$ with $i = 1, 2, ..., 2n$ depend on the previous values $\tilde{\mathrm{X}}_{\tau-1}$, $\tilde{\mathrm{X}}_{\tau-2}$, ..., $\tilde{\mathrm{X}}_{\tau-p}$ and are thus dependent on the random $l_\alpha r_\alpha$-increments $\Delta\mathrm{X}_k(\tau - l)$ with $k = 1, 2, ..., 2n$ and $l = 1, 2, ..., p$. This causal dependency is weighted by the parameter matrices $\underline{A}_j$ and may be formalized with the aid of the GRANGER causality concept [17].

The GRANGER causality concept demands the fulfillment of two requirements: firstly, causality is only defined for variables with stochastic characteristics. The second requirement is that future realizations are solely influenced by past values. According to Sect. 2.2, the random $l_\alpha r_\alpha$-increments of a fuzzy random variable $\tilde{\mathrm{X}}_\tau$ at time τ are real-valued random variables $\Delta\mathrm{X}_i(\tau)$ with $i = 1, 2, ..., 2n$. For time series comprised of fuzzy data, as considered here, it follows from both requirements that a random $l_\alpha r_\alpha$-increment $\Delta\mathrm{X}_k$ of a fuzzy random variable $\tilde{\mathrm{X}}_\tau$ is GRANGER-causal for a different random $l_\alpha r_\alpha$-increment $\Delta\mathrm{X}_i$ of a fuzzy random variable $\tilde{\mathrm{X}}_{\tau+\Delta\tau}$ for $\Delta\tau > 0$, provided improved predictions of ΔX_imay be obtained with the aid of previous values of ΔX_k.

Definition 3.32. *In a fuzzy autoregressive process, $\Delta X_k$is* GRANGER-*causal in relation to ΔX_i if Eq. (3.80) is not satisfied.*

$$a_1[i,k] = a_2[i,k] = ... = a_p[i,k] = 0 \tag{3.80}$$

◆

If all parameter matrices of a fuzzy autoregressive process are diagonal matrices $\underline{A}_j$, GRANGER causality does not exist between the different α-levels. For a general configuration of the $\underline{A}_j$ matrices the fuzzy AR process is considered to be GRANGER-causal with regard to fuzzy random process modeling.

The term causality must not be confused with the term correlation. If a statistical relationship exists between two random $l_\alpha r_\alpha$-increments ΔX_i and ΔX_k, it is not possible make any statements regarding causal relationships. Conversely, the fact that previous values of ΔX_k may lead to GRANGER-causal realizations of ΔX_i at time τ , they do not necessarily lead to correlation.

From GRANGER causality it follows that it is not possible to deduce the occupancy of the parameter matrices $\underline{A}_j$ from the $l_\alpha r_\alpha$-correlation function. The $\underline{A}_j$ must be determined using the methods presented in Sect. 3.5.5.

In contrast to fuzzy MA processes, fuzzy AR processes permit the modeling of stationary as well as non-stationary fuzzy time series. Whether or not a specific fuzzy AR process is stationary or non-stationary may be determined numerically. It is possible to verify stationarity from a numerical simulation of the fuzzy AR process and a statistical evaluation of a sufficiently long series of realizations according to the methods described in Sect. 3.3.

Numerical realization. The numerical simulation of a fuzzy AR process $(\tilde{X}_\tau)_{\tau\in\mathbf{T}}$ is a special case of the numerical simulation of a fuzzy ARMA process, as described in the following section.

3.5.4 Fuzzy Autoregressive Moving Average Processes

Definition 3.33. *A fuzzy random process $(\tilde{X}_\tau)_{\tau\in\mathbf{T}}$ is referred to as a fuzzy autoregressive moving average process of order $[p,q]$ (fuzzy ARMA$[p,q]$ process for short) if it may be described by Eq. (3.81) at each time point τ.*

$$\tilde{X}_\tau = \underbrace{\underline{A}_1 \odot \tilde{X}_{\tau-1} \oplus ... \oplus \underline{A}_p \odot \tilde{X}_{\tau-p}}_{\textit{fuzzy AR component}} \oplus \underbrace{\tilde{\mathcal{E}}_\tau \ominus \underline{B}_1 \odot \tilde{\mathcal{E}}_{\tau-1} \ominus ... \ominus \underline{B}_q \odot \tilde{\mathcal{E}}_{\tau-q}}_{\textit{fuzzy MA component}} \tag{3.81}$$

◆

The parameters $\underline{A}_j$ and $\underline{B}_j$ are real-valued $[2n, 2n]$ matrices, whereby n denotes the number of chosen α-levels. The variables $\tilde{\mathcal{E}}_\tau$, $\tilde{\mathcal{E}}_{\tau-1}$, ..., $\tilde{\mathcal{E}}_{\tau-q}$ are the fuzzy random variables of a fuzzy white-noise process $(\tilde{\mathcal{E}}_\tau)_{\tau\in\mathbf{T}}$ at times τ, $\tau-1$, ..., $\tau-q$ (see Sect. 3.5.1). Fuzzy autoregressive moving average processes are hence a combination of the processes presented in the previous two sections.

By means of fuzzy ARMA processes it is possible to model stationary as well as non-stationary fuzzy time series. Whether or not a specific fuzzy ARMA process is stationary or non-stationary may be tested numerically (see also Sect. 3.5.3). It is possible to verify stationarity from a numerical simulation of a sufficiently long realization of the fuzzy ARMA process and a statistical evaluation of the latter using the methods described in Sect. 3.3.

The characteristic moments such as the fuzzy expected value function, the $l_\alpha r_\alpha$-variance function, and the $l_\alpha r_\alpha$-covariance function may be determined from a numerical simulation followed by a statistical evaluation of the realizations. In the case of a stationary fuzzy ARMA process a sufficiently long realization of the process is simulated in order to determine the characteristic moments. For these fuzzy time series the characteristic moments are computed according to Sect. 3.3 and used as estimators for the characteristic moments of the fuzzy ARMA process.

In the case of a non-stationary fuzzy ARMA process, s realizations $(\tilde{x}_\tau^k)_{\tau\in\mathbf{T}}$ $(k = 1, 2, ..., s)$ are simulated in order to determine the characteristic moments. The local fuzzy mean value $\tilde{\overline{x}}(\tau)$ according to Eq. (3.82) then serves as an estimator for the fuzzy expected value function $E[\tilde{\mathrm{X}}_\tau] = \tilde{m}_{\tilde{\mathrm{X}}_\tau}$.

$$\hat{E}[\tilde{\mathrm{X}}_\tau] = \hat{\tilde{m}}_{\tilde{\mathrm{X}}_\tau} = \tilde{\overline{x}}(\tau) = \frac{1}{s} \bigoplus_{k=1}^{s} \tilde{x}_\tau^k \tag{3.82}$$

The elements of the $l_\alpha r_\alpha$-covariance function ${}_{lr}\underline{K}_{\tilde{\mathrm{X}}_\tau}(\tau_a, \tau_b)$ are estimated according to Eq. (3.83).

$$\hat{k}_{\alpha_j r*}^{\alpha_i l*}(\tau_a, \tau_b) = \frac{1}{s-1} \sum_{k=1}^{s} [(\Delta x_{\alpha_i l*}^k(\tau_a) - \Delta\overline{x}_{\alpha_i l*}(\tau_a))\ldots \\ \ldots(\Delta x_{\alpha_j r*}^k(\tau_b) - \Delta\overline{x}_{\alpha_j r*}(\tau_b))] \tag{3.83}$$

The subscripts α_i and α_j denote the α-levels under consideration whereas $l* = l, r$ and $r* = l, r$ denote the left and right branches of the membership function, respectively. The leading diagonal elements of ${}_{lr}\hat{\underline{K}}_{\tilde{\mathrm{X}}_\tau}(\tau_a, \tau_b)$ for $\tau_a = \tau_b = \tau$ are estimators for the elements of the $l_\alpha r_\alpha$-variance function ${}_{lr}Var[\tilde{\mathrm{X}}_\tau] = {}_{lr}\underline{\sigma}^2_{\tilde{\mathrm{X}}_\tau}$ at each time point τ.

The superposition of deterministic ARMA processes, as investigated in [18] within the framework of classical time series analysis, may also be applied to fuzzy autoregressive moving average processes. The fuzzy sum of two independent fuzzy ARMA processes $(\tilde{\mathrm{X}}_\tau)_{\tau\in\mathbf{T}}$ and $(\tilde{\mathrm{Y}}_\tau)_{\tau\in\mathbf{T}}$ of orders $[p_{\tilde{\mathrm{X}}_\tau}, q_{\tilde{\mathrm{X}}_\tau}]$ and $[p_{\tilde{\mathrm{Y}}_\tau}, q_{\tilde{\mathrm{Y}}_\tau}]$ according to Eq. (3.84)

$$\tilde{Z}_\tau = \tilde{X}_\tau \oplus \tilde{Y}_\tau \tag{3.84}$$

again yields a fuzzy ARMA process $(\tilde{Z}_\tau)_{\tau\in\mathbf{T}}$ of order $[p_{\tilde{Z}_\tau}, q_{\tilde{Z}_\tau}]$. The upper bounds for $p_{\tilde{Z}_\tau}$ and $q_{\tilde{Z}_\tau}$ are specified by the inequalities (3.85) and (3.86), respectively.

$$p_{\tilde{Z}_\tau} \leqslant p_{\tilde{X}_\tau} + p_{\tilde{Y}_\tau} \tag{3.85}$$

$$q_{\tilde{Z}_\tau} \leqslant \max[p_{\tilde{X}_\tau} + q_{\tilde{Y}_\tau}, p_{\tilde{Y}_\tau} + q_{\tilde{X}_\tau}] \tag{3.86}$$

Numerical realization. The simulation of a fuzzy ARMA$[p, q]$ process $(\tilde{X}_\tau)_{\tau\in\mathbf{T}}$, i.e. the determination of realizations $(\tilde{x}_\tau)_{\tau\in\mathbf{T}}$, is based on a Monte Carlo simulation of the included fuzzy white-noise process $(\tilde{\mathcal{E}}_\tau)_{\tau\in\mathbf{T}}$. This follows the recursive procedure according to Eq. (3.87). The simulation of fuzzy AR and fuzzy MA processes is included as a special case ($q = 0$ and $p = 0$).

$$\begin{aligned}\tilde{X}_\tau = {} & \underline{A}_1 \odot \tilde{x}_{\tau-1} \oplus \ldots \oplus \underline{A}_p \odot \tilde{x}_{\tau-p} \oplus \tilde{\mathcal{E}}_\tau \ominus \\ & \underline{B}_1 \odot \tilde{\varepsilon}_{\tau-1} \ominus \ldots \ominus \underline{B}_q \odot \tilde{\varepsilon}_{\tau-q}\end{aligned} \tag{3.87}$$

$$\text{with} \quad \tilde{x}_{\tau-u} = \begin{cases} 0 & \text{for } \tau - u < 1 \\ \tilde{x}_{\tau-u} & \text{for } \tau - u \geqslant 1 \end{cases}, \quad u = 1, 2, ..., p \tag{3.88}$$

$$\text{and} \quad \tilde{\varepsilon}_{\tau-v} = \begin{cases} E[\tilde{\mathcal{E}}_\tau] & \text{for } \tau - v < 1 \\ \tilde{\varepsilon}_{\tau-v} & \text{for } \tau - v \geqslant 1 \end{cases}, \quad v = 1, 2, ..., q \tag{3.89}$$

A precondition for the application of Eq. (3.87) is that the order $[p, q]$ of the process as well as the parameter matrices $\underline{A}_1$, ... , $\underline{A}_p$ and $\underline{B}_1$, ... , $\underline{B}_q$ are known. Methods for specifying the order $[p, q]$ are presented in Sect. 3.5.5. Methods for estimating the parameter matrices are developed in Sect. 3.5.6.

In the first step, a realization $\tilde{x}_1$ of the fuzzy ARMA process $(\tilde{X}_\tau)_{\tau\in\mathbf{T}}$ at time $\tau = 1$ is determined. In accordance with Eq. (3.87), the realization $\tilde{x}_1$ of the fuzzy random variable $\tilde{X}_1$ is dependent on the realization $\tilde{\varepsilon}_1$ of the fuzzy white-noise variable $\tilde{\mathcal{E}}_1$. The fuzzy variables $\tilde{x}_\tau$ and $\tilde{\varepsilon}_\tau$ at times $\tau < 1$ are hereby given as zero according to Eq. (3.88) or as the fuzzy expected value of the fuzzy white-noise process according to Eq. (3.89). Following the Monte Carlo simulation of a realization $\tilde{\varepsilon}_1$ (see Sect. 3.5.1), the corresponding realization $\tilde{x}_1$ may be computed.

Using the simulated fuzzy variable $\tilde{x}_1$, the realization $\tilde{x}_2$ is obtained in the next step from a repeated Monte Carlo simulation of a fuzzy white-noise variable $\tilde{\varepsilon}_2$. The successive repetition of this procedure for time points $\tau =$ 1, 2, ... yields a realization of the fuzzy ARMA process $(\tilde{X}_\tau)_{\tau\in\mathbf{T}}$. It should be noted here that the initial time points are required for the settling time of the fuzzy ARMA process, and that the corresponding fuzzy variables are rejected. By repeating this procedure s-times it is possible to simulate s sequences of realizations.

Remark 3.34. For fuzzy ARMA processes as well as for the special cases of fuzzy AR and fuzzy MA processes, the inclusion condition given by Eq. (2.80) must be complied with for the fuzzy random variables $\tilde{X}_\tau$ with $\tau = 1, 2, ...$. It must therefore be ensured that the $l_\alpha r_\alpha$-increments $\Delta x_j(\tau)$ $(j = 1, 2, ..., n-1, n+1, ..., 2n)$ of all realizations $(\tilde{x}_\tau)_{\tau \in \mathbf{T}}$ satisfy the requirement of non-negativity. For this purpose, s sufficiently long sequences of realizations are simulated. The $l_\alpha r_\alpha$-increments $\Delta x_j(\tau)$ of the realizations in each sequence must be non-negative. If the $l_\alpha r_\alpha$-increments for all s simulated realizations satisfy the condition $\Delta x_j(\tau) \geqslant 0$, the requirement of non-negativity is fulfilled with an estimated probability of error of $< \frac{1}{s}$. ◆

Remark 3.35. Eq. (3.81) describing the fuzzy ARMA$[p, q]$ process is rearranged as follows.

$$\underbrace{\tilde{X}_\tau \ominus \underline{A}_1 \odot \tilde{X}_{\tau-1} \ominus ... \ominus \underline{A}_p \odot \tilde{X}_{\tau-p}}_{\text{fuzzy AR component}} = \underbrace{\tilde{\mathcal{E}}_\tau \ominus \underline{B}_1 \odot \tilde{\mathcal{E}}_{\tau-1} \ominus ... \ominus \underline{B}_q \odot \tilde{\mathcal{E}}_{\tau-q}}_{\text{fuzzy MA process}} \quad (3.90)$$

The following therefore also holds for each realization of the process.

$$\tilde{x}_\tau \ominus \underline{A}_1 \odot \tilde{x}_{\tau-1} \ominus ... \ominus \underline{A}_p \odot \tilde{x}_{\tau-p} = \tilde{\varepsilon}_\tau \ominus \underline{B}_1 \odot \tilde{\varepsilon}_{\tau-1} \ominus ... \ominus \underline{B}_q \odot \tilde{\varepsilon}_{\tau-q} \quad (3.91)$$
$$\text{with} \quad \tau = 1, 2, ..., N$$

From Eqs. (3.90) and (3.92) it follows that the left-hand side of Eq. (3.92) is the realization of a fuzzy MA$[q]$ process.

A given time series $\tilde{x}_1, \tilde{x}_2, ..., \tilde{x}_N$, which is a specific realization of the fuzzy ARMA$[p, q]$ process, may be transformed into the realization of a fuzzy MA$[q]$ process by means of

$$_{MA}\tilde{x}_\tau = \tilde{x}_\tau \ominus \underline{A}_1 \odot \tilde{x}_{\tau-1} \ominus ... \ominus \underline{A}_p \odot \tilde{x}_{\tau-p} \quad (3.92)$$
$$\text{with} \quad \tau = p+1, p+2, ..., N\,.$$

The realization $_{MA}\tilde{x}_\tau$ is a fuzzy time series truncated to $N-p$ elements which is always stationary. ◆

3.5.5 Specification of Model Order

In the following section, methods are presented for specifying the model order $[p, q]$, i.e. methods for determining the parameters p and q of the fuzzy ARMA$[p, q]$ process.

Specification of Model Order by the BOX-JENKINS Method

A precondition for the application of the classical BOX-JENKINS method to time series comprised of crisp (i.e. real-valued) data is that the considered time series is stationary (or adjusted to remove trends and cycles) [6]. For this reason, non-stationary real-valued time series are converted into stationary time series with the aid of classical difference filters. An extension of the classical BOX-JENKINS method to deal with time series containing fuzzy data is presented in the following. The application presupposes stationary fuzzy time series. Non-stationary fuzzy time series may be converted into stationary fuzzy time series with the aid of the adjustment procedures described in Sect. 3.2 for removing fuzzy trends and cycles or by means of suitable fuzzy difference filters according to Sect. 3.4.

The extended BOX-JENKINS method permits a specification of the model order $[p, q]$ of fuzzy ARMA processes. The $l_\alpha r_\alpha$-correlation function ${}_{lr}\underline{R}_{\tilde{X}_\tau}(\Delta\tau)$ and the partial $l_\alpha r_\alpha$-correlation function ${}_{lr}\underline{P}_{\tilde{X}_\tau}(\Delta\tau)$ of a stationary fuzzy ARMA process exhibit characteristic properties which are dependent on the model parameters p and q. These properties may be recognized from the structure of the empirical $l_\alpha r_\alpha$-correlation function ${}_{lr}\underline{\hat{R}}_{\tilde{x}_\tau}(\Delta\tau)$ and the empirical partial $l_\alpha r_\alpha$-correlation function ${}_{lr}\underline{\hat{P}}_{\tilde{x}_\tau}(\Delta\tau)$ of the stationary fuzzy time series in question (see Sect. 3.3). From the structure it is possible to deduce the model order of the hypothetically underlying fuzzy ARMA$[p, q]$ process. In the following, characteristic properties of fuzzy ARMA$[p, q]$ processes are described and methods for evaluating the empirical parameters of a fuzzy time series are developed.

Fuzzy ARMA$[p, 0]$ process. Eq. (3.93) given below holds for the elements $p_{i,j}(\Delta\tau)$ of the partial $l_\alpha r_\alpha$-correlation function ${}_{lr}\underline{P}_{\tilde{X}_\tau}(\Delta\tau)$ of a fuzzy ARMA$[p, 0]$ process $(\tilde{X}_\tau)_{\tau\in\mathbf{T}}$ or a fuzzy AR process $(\tilde{X}_\tau)_{\tau\in\mathbf{T}}$ of order $[p]$.

$$p_{i,j}(\Delta\tau) = \begin{cases} p_{i,j}(\Delta\tau) & \text{for } \Delta\tau \leqslant p \\ 0 & \text{for } \Delta\tau > p \end{cases} \quad \forall \quad i,j = 1,\, 2,\, ...,\, 2n \tag{3.93}$$

If this characteristic property according to Eq. (3.93) is recognizable in the empirical partial $l_\alpha r_\alpha$-correlation function ${}_{lr}\underline{\hat{P}}_{\tilde{x}_\tau}(\Delta\tau)$ of a given fuzzy time series, a fuzzy AR process $(\tilde{X}_\tau)_{\tau\in\mathbf{T}}$ of order $[p]$ may be assumed. As a rule, ${}_{lr}\underline{\hat{P}}_{\tilde{x}_\tau}(\Delta\tau)$ does not exhibit a pronounced jump according to Eq. (3.93) because only the given fuzzy time series, i.e. only a single realization of length N of the hypothetically underlying fuzzy AR$[p]$ process, is available for determining ${}_{lr}\underline{\hat{P}}_{\tilde{x}_\tau}(\Delta\tau)$.

For specifying the model order a tolerance interval ${}_PI_{i,j}(\Delta\tau)$ is therefore stipulated for the elements $\hat{p}_{i,j}(\Delta\tau)$ of the empirical partial $l_\alpha r_\alpha$-correlation function ${}_{lr}\underline{\hat{P}}_{\tilde{x}_\tau}(\Delta\tau > p)$, within which they must lie for $\Delta\tau > p$. If, according to Eq. (3.94), all empirical partial correlations $\hat{p}_{i,j}(\Delta\tau)$ lie within the corresponding interval ${}_PI_{i,j}(\Delta\tau)$ beyond a time step $\Delta\tau$, the model order $p = \Delta\tau - 1$ may be assumed for the fuzzy AR process underlying the fuzzy time series.

$$\hat{p}_{i,j}(\Delta\tau) \in {}_PI_{i,j}(\Delta\tau) \quad \forall \quad i,j = 1,\, 2,\, ...,\, 2n \tag{3.94}$$

Fuzzy ARMA[0, q] process. For a fuzzy ARMA process $(\tilde{\mathrm{X}}_\tau)_{\tau\in\mathbf{T}}$ of order $[0,q]$, i.e. for a fuzzy MA[q] process $(\tilde{\mathrm{X}}_\tau)_{\tau\in\mathbf{T}}$, the following equation holds for the elements $r_{i,j}(\Delta\tau)$ of the corresponding $l_\alpha r_\alpha$-correlation function ${}_{lr}\underline{R}_{\tilde{\mathrm{X}}_\tau}(\Delta\tau)$.

$$r_{i,j}(\Delta\tau) = \begin{cases} r_{i,j}(\Delta\tau) \text{ for } \Delta\tau \leqslant q \\ \\ 0 \quad \text{ for } \Delta\tau > q \end{cases} \quad \forall \quad i,j = 1,\, 2,\, ...,\, 2 \tag{3.95}$$

If the empirical $l_\alpha r_\alpha$-correlation function ${}_{lr}\hat{\underline{R}}_{\tilde{x}_\tau}(\Delta\tau)$ of a given fuzzy time series exhibits the properties of Eq. (3.95), a fuzzy MA process $(\tilde{\mathrm{X}}_\tau)_{\tau\in\mathbf{T}}$ of order $[q]$ may assumed. Analogous to fuzzy AR processes, a pronounced jump as in Eq. (3.95) is normally not present in the empirical $l_\alpha r_\alpha$-correlation function ${}_{lr}\hat{\underline{R}}_{\tilde{x}_\tau}(\Delta\tau)$ of a given fuzzy time series. Analogous to the procedure adopted for fuzzy AR processes, a tolerance interval ${}_R I_{i,j}(\Delta\tau)$ is thus stipulated for the elements $\hat{r}_{i,j}(\Delta\tau)$ of the empirical $l_\alpha r_\alpha$-correlation function ${}_{lr}\hat{\underline{R}}_{\tilde{x}_\tau}(\Delta\tau > q)$, within which they must lie. The value $q = \Delta\tau - 1$ may be assumed for the model order of the fuzzy time series of the underlying fuzzy MA process, provided all empirical correlations $\hat{r}_{i,j}(\Delta\tau)$ beyond the time step $\Delta\tau$ lie within the corresponding interval ${}_R I_{i,j}(\Delta\tau)$ according to Eq. (3.96).

$$\hat{r}_{i,j}(\Delta\tau) \in {}_R I_{i,j}(\Delta\tau) \quad \forall \quad i,j = 1,\, 2,\, ...,\, 2n \tag{3.96}$$

Remark 3.36. In accordance with Eqs. (3.94) and (3.96), the model orders p and q for fuzzy AR and fuzzy MA processes may be specified with the aid of predefined tolerance intervals ${}_P I_{i,j}(\Delta\tau)$ and ${}_R I_{i,j}(\Delta\tau)$. By means of ${}_P I_{i,j}(\Delta\tau)$ and ${}_R I_{i,j}(\Delta\tau)$ it is also possible to specify individual elements of the parameter matrices $\underline{A}_1$, ..., $\underline{A}_p$ and $\underline{B}_1$, ..., $\underline{B}_q$ a priori. If the empirical partial correlations $\hat{p}_{i,j}(\Delta\tau)$ and the empirical correlations $\hat{r}_{i,j}(\Delta\tau)$ for each time step $\Delta\tau > d$ lie within the tolerance intervals ${}_P I_{i,j}(\Delta\tau)$ and ${}_R I_{i,j}(\Delta\tau)$, respectively, the corresponding elements $a_{i,j}(d+1)$, $a_{i,j}(d+2)$, ... and $b_{i,j}(d+1)$, $b_{i,j}(d+2)$, ... of the parameter matrices $\underline{A}_{d+1}$, $\underline{A}_{d+2}$, ... and $\underline{B}_{d+1}$, $\underline{B}_{d+2}$, ..., respectively, may be set to zero according to Eqs. (3.97) and (3.98).

$$a_{i,j}(d+1),\, a_{i,j}(d+2),\, ... = 0 \;\text{ if }\; \hat{p}_{i,j}(\Delta\tau) \in {}_P I_{i,j}(\Delta\tau) \;\forall\; \Delta\tau > d \tag{3.97}$$

$$b_{i,j}(d+1),\, b_{i,j}(d+2),\, ... = 0 \;\text{ if }\; \hat{r}_{i,j}(\Delta\tau) \in {}_R I_{i,j}(\Delta\tau) \;\forall\; \Delta\tau > d \tag{3.98}$$

◆

Fuzzy ARMA[p, q] process. If the empirical partial $l_\alpha r_\alpha$-correlation function ${}_{lr}\hat{\underline{P}}_x(\Delta\tau)$ and the empirical $l_\alpha r_\alpha$-correlation function ${}_{lr}\hat{\underline{R}}_{\tilde{x}_\tau}(\Delta\tau)$ of a given fuzzy time series do not exhibit specific properties according to Eqs. (3.93) and (3.95), a general fuzzy ARMA[p,q] process may be assumed according to the case in question. A specification of the model order $[p,q]$ of a fuzzy

ARMA process by means of the BOX-JENKINS method, i.e. the recognition and evaluation of particular patterns in ${}_{lr}\underline{\hat{P}}_{\tilde{x}_\tau}(\Delta\tau)$ and ${}_{lr}\underline{\hat{R}}_{\tilde{x}_\tau}(\Delta\tau)$, is difficult, however, due to the fact that the $l_\alpha r_\alpha$-correlation functions ${}_{lr}\underline{R}_{\tilde{\mathrm{X}}_\tau}(\Delta\tau)$ and the partial $l_\alpha r_\alpha$-correlation functions ${}_{lr}\underline{P}_{\tilde{\mathrm{X}}_\tau}(\Delta\tau)$ of fuzzy ARMA$[p,q]$ processes are not marked by clearly defined and easily recognizable characteristics. According to the procedures adopted in classical time series analysis (see e.g. [65]), it is feasible to construct diagrams of the $l_\alpha r_\alpha$-correlation functions ${}_{lr}\underline{R}_{\tilde{\mathrm{X}}_\tau}(\Delta\tau)$ and the partial $l_\alpha r_\alpha$-correlation functions ${}_{lr}\underline{P}_{\tilde{\mathrm{X}}_\tau}(\Delta\tau)$ corresponding to all conceivable model orders $[p,q]$ for the purpose of identification. A far more suitable approach for specifying the model order $[p,q]$ of stationary fuzzy ARMA processes, however, is offered by the method presented in the following, which makes use of $l_\alpha r_\alpha$-correlation tables for time series comprised of fuzzy data.

Specification of Model Order by Means of $l_\alpha r_\alpha$-Correlation Tables

In classical time series analysis the underlying (deterministic) ARMA$[p,q]$ process is specified by computing empirical vector correlations $\lambda(p,\,q)$ of the time series concerned. The arrangement of the vector correlations $\lambda(p,\,q)$ in the so-called correlation table and their evaluation yields information regarding the sought model order $[p,q]$ (see e.g. [60]). The developed $l_\alpha r_\alpha$-discretization permits the consistent extension of this approach to the fuzzy vectors $\underline{\tilde{x}}$ and $\underline{\tilde{y}}$ according to Eqs. (3.99) and (3.100), which are segments of the given stationary fuzzy time series. Both vectors are of length $p+1$ and are displaced by $q+1$ towards each other on the time series.

$$\underline{\tilde{x}} = \left(\tilde{x}_\tau,\, \tilde{x}_{\tau+1},\, ...,\, \tilde{x}_{\tau+p}\right)^T \tag{3.99}$$

$$\underline{\tilde{y}} = \left(\tilde{x}_{\tau+q+1},\, \tilde{x}_{\tau+q+2},\, ...,\, \tilde{x}_{\tau+q+p+1}\right)^T \tag{3.100}$$

The empirical $l_\alpha r_\alpha$-vector correlation matrix ${}_{lr}\underline{\hat{\Lambda}}_{\tilde{x}_\tau}(p,q)$ is then given as being dependent on the order $[p,q]$ according to Eq. (3.101). This is a measure of the correlation between the fuzzy vectors $\underline{\tilde{x}}$ and $\underline{\tilde{y}}$.

$${}_{lr}\underline{\hat{\Lambda}}_{\tilde{x}_\tau}(p,q) = \begin{bmatrix} \lambda_{1,1} & \lambda_{1,2} & \cdots & \lambda_{1,(2n)} \\ \lambda_{2,1} & \lambda_{2,2} & \cdots & \lambda_{2,(2n)} \\ \vdots & \vdots & \ddots & \vdots \\ \lambda_{(2n),1} & \lambda_{(2n,)2} & \cdots & \lambda_{(2n),(2n)} \end{bmatrix} \tag{3.101}$$

The elements of the empirical $l_\alpha r_\alpha$-vector correlation matrix ${}_{lr}\underline{\hat{\Lambda}}_{\tilde{x}_\tau}(p,q)$ are determined for $i,j = 1,\, 2,\, ...,\, 2n$ according to Eq. (3.102).

$$\lambda_{i,j}(p,q) = \frac{\det {}_{lr}\underline{\hat{\mathsf{K}}}_{\underline{\tilde{x}}\underline{\tilde{y}}}(i,j)}{\sqrt{\left(\det {}_{lr}\underline{\hat{\mathsf{K}}}_{\underline{\tilde{x}}\underline{\tilde{x}}}(i,i)\right)\left(\det {}_{lr}\underline{\hat{\mathsf{K}}}_{\underline{\tilde{y}}\underline{\tilde{y}}}(j,j)\right)}} \tag{3.102}$$

The terms ${}_{lr}\underline{\hat{\mathsf{K}}}_{\tilde{\underline{x}}\tilde{\underline{x}}}(i,i)$, ${}_{lr}\underline{\hat{\mathsf{K}}}_{\tilde{\underline{y}}\tilde{\underline{y}}}(j,j)$ and ${}_{lr}\underline{\hat{\mathsf{K}}}_{\tilde{\underline{x}}\tilde{\underline{y}}}(i,j)$ are hereby special empirical $l_\alpha r_\alpha$-covariance matrices according to Eqs. (3.103) to (3.105).

$$ {}_{lr}\underline{\hat{\mathsf{K}}}_{\tilde{\underline{x}}\tilde{\underline{x}}}(i,i) = \begin{bmatrix} \hat{k}^{\Delta x_i(\tau)}_{\Delta x_i(\tau)} & \hat{k}^{\Delta x_i(\tau)}_{\Delta x_i(\tau+1)} & \cdots & \hat{k}^{\Delta x_i(\tau)}_{\Delta x_i(\tau+p)} \\ \hat{k}^{\Delta x_i(\tau+1)}_{\Delta x_i(\tau)} & \hat{k}^{\Delta x_i(\tau+1)}_{\Delta x_i(\tau+1)} & \cdots & \hat{k}^{\Delta x_i(\tau+1)}_{\Delta x_i(\tau+p)} \\ \vdots & \vdots & \ddots & \vdots \\ \hat{k}^{\Delta x_i(\tau+p)}_{\Delta x_i(\tau)} & \hat{k}^{\Delta x_i(\tau+p)}_{\Delta x_i(\tau+1)} & \cdots & \hat{k}^{\Delta x_i(\tau+p)}_{\Delta x_i(\tau+p)} \end{bmatrix} \tag{3.103} $$

$$ {}_{lr}\underline{\hat{\mathsf{K}}}_{\tilde{\underline{y}}\tilde{\underline{y}}}(j,j) = \begin{bmatrix} \hat{k}^{\Delta x_j(\tau+q+1)}_{\Delta x_j(\tau+q+1)} & \hat{k}^{\Delta x_j(\tau+q+1)}_{\Delta x_j(\tau+q+2)} & \cdots & \hat{k}^{\Delta x_j(\tau+q+1)}_{\Delta x_j(\tau+q+p+1)} \\ \hat{k}^{\Delta x_j(\tau+q+2)}_{\Delta x_j(\tau+q+1)} & \hat{k}^{\Delta x_j(\tau+q+2)}_{\Delta x_j(\tau+q+2)} & \cdots & \hat{k}^{\Delta x_j(\tau+q+2)}_{\Delta x_j(\tau+q+p+1)} \\ \vdots & \vdots & \ddots & \vdots \\ \hat{k}^{\Delta x_j(\tau+q+p+1)}_{\Delta x_j(\tau+q+1)} & \hat{k}^{\Delta x_j(\tau+q+p+1)}_{\Delta x_j(\tau+q+2)} & \cdots & \hat{k}^{\Delta x_j(\tau+q+p+1)}_{\Delta x_j(\tau+q+p+1)} \end{bmatrix} \tag{3.104} $$

$$ {}_{lr}\underline{\hat{\mathsf{K}}}_{\tilde{\underline{x}}\tilde{\underline{y}}}(i,j) = \begin{bmatrix} \hat{k}^{\Delta x_i(\tau)}_{\Delta x_j(\tau+q+1)} & \hat{k}^{\Delta x_i(\tau)}_{\Delta x_j(\tau+q+2)} & \cdots & \hat{k}^{\Delta x_i(\tau)}_{\Delta x_j(\tau+q+p+1)} \\ \hat{k}^{\Delta x_i(\tau+1)}_{\Delta x_j(\tau+q+1)} & \hat{k}^{\Delta x_i(\tau+1)}_{\Delta x_j(\tau+q+2)} & \cdots & \hat{k}^{\Delta x_i(\tau+1)}_{\Delta x_j(\tau+q+p+1)} \\ \vdots & \vdots & \ddots & \vdots \\ \hat{k}^{\Delta x_i(\tau+p)}_{\Delta x_j(\tau+q+1)} & \hat{k}^{\Delta x_i(\tau+p)}_{\Delta x_j(\tau+q+2)} & \cdots & \hat{k}^{\Delta x_i(\tau+p)}_{\Delta x_j(\tau+q+p+1)} \end{bmatrix} \tag{3.105} $$

The individual elements of the special empirical $l_\alpha r_\alpha$-covariance matrices ${}_{lr}\underline{\hat{\mathsf{K}}}_{\tilde{\underline{x}}\tilde{\underline{x}}}(i,i)$, ${}_{lr}\underline{\hat{\mathsf{K}}}_{\tilde{\underline{y}}\tilde{\underline{y}}}(j,j)$ and ${}_{lr}\underline{\hat{\mathsf{K}}}_{\tilde{\underline{x}}\tilde{\underline{y}}}(i,j)$ describe the empirical covariances $\hat{k}$ between the $l_\alpha r_\alpha$-increments $\Delta x_i(\tau)$, $\Delta x_i(\tau+1)$, ..., $\Delta x_i(\tau+p)$ and $\Delta x_j(\tau+q+1)$, $\Delta x_j(\tau+q+2)$, ..., $\Delta x_j(\tau+q+p+1)$ of the fuzzy variables $\tilde{x}_\tau$, $\tilde{x}_{\tau+1}$, ..., $\tilde{x}_{\tau+p}$ and $\tilde{x}_{\tau+q+1}$, $\tilde{x}_{\tau+q+2}$, ..., $\tilde{x}_{\tau+q+p+1}$, which are lumped together in the fuzzy vectors $\tilde{\underline{x}}$ and $\tilde{\underline{y}}$ according to Eqs. (3.99) and (3.100).

By arranging the empirical $l_\alpha r_\alpha$-vector correlation matrices ${}_{lr}\underline{\hat{\Lambda}}_{\tilde{x}_\tau}(p,q)$ in a matrix with column indexing q = 0, 1, 2, ... and row indexing p = 0, 1, 2, ... according to Eq. (3.106) the empirical $l_\alpha r_\alpha$-correlation table ${}_{lr}\underline{\hat{T}}_{\tilde{x}_\tau}$ for the given fuzzy time series is obtained.

$$ {}_{lr}\underline{\hat{T}}_{\tilde{x}_\tau} = \begin{bmatrix} {}_{lr}\underline{\hat{\Lambda}}_{\tilde{x}_\tau}(0,0) & {}_{lr}\underline{\hat{\Lambda}}_{\tilde{x}_\tau}(0,1) & {}_{lr}\underline{\hat{\Lambda}}_{\tilde{x}_\tau}(0,2) & \cdots \\ {}_{lr}\underline{\hat{\Lambda}}_{\tilde{x}_\tau}(1,0) & {}_{lr}\underline{\hat{\Lambda}}_{\tilde{x}_\tau}(1,1) & {}_{lr}\underline{\hat{\Lambda}}_{\tilde{x}_\tau}(1,2) & \cdots \\ {}_{lr}\underline{\hat{\Lambda}}_{\tilde{x}_\tau}(2,0) & {}_{lr}\underline{\hat{\Lambda}}_{\tilde{x}_\tau}(2,1) & {}_{lr}\underline{\hat{\Lambda}}_{\tilde{x}_\tau}(2,2) & \cdots \\ \vdots & \vdots & \vdots & \ddots \end{bmatrix} \tag{3.106} $$

With the aid of the empirical $l_\alpha r_\alpha$-correlation table ${}_{lr}\hat{\underline{T}}_{\tilde{x}_\tau}$ for a given fuzzy time series it is possible to specify the model order $[p, q]$ of an underlying fuzzy ARMA process. If the empirical $l_\alpha r_\alpha$-correlation table ${}_{lr}\hat{\underline{T}}_{\tilde{x}_\tau}$ is characterized by a zero block beginning at position $[p, q]$ and extending infinitely in the advancing row and column direction, the underlying fuzzy ARMA process is of order $[p, q]$.

Remark 3.37. Analogous to the BOX-JENKINS method, an exact zero block does not normally exist in the empirical $l_\alpha r_\alpha$-correlation table ${}_{lr}\hat{\underline{T}}_{\tilde{x}_\tau}$ of a given fuzzy time series. Similar to the procedure adopted in the BOX-JENKINS method, a tolerance interval is thus specified for the elements of the empirical $l_\alpha r_\alpha$-correlation table ${}_{lr}\hat{\underline{T}}_{\tilde{x}_\tau}$. The value zero is assigned to elements with a value less than the chosen tolerance interval. ◆

Specification of Model Order by the Maximum Likelihood Method

In the following a further method based on the maximum likelihood method is developed for determining the model order $[p, q]$.

For a fuzzy random process $(\tilde{X}_\tau)_{\tau \in \mathbf{T}}$ with known parameters $\underline{P}$ the probability distribution of realizations $\tilde{x}_1$, $\tilde{x}_2$, ..., $\tilde{x}_N$ of length N is given by the N-dimensional fuzzy probability density function form II.

$$f(\tilde{x}_1, \tilde{x}_2, ..., \tilde{x}_N \,|\, \underline{P}) \quad \underline{P} = \text{constant} \tag{3.107}$$

For each realization $\tilde{x}_1$, $\tilde{x}_2$, ..., $\tilde{x}_N$ of length N it is possible to state the value of the fuzzy probability density function form II by means of Eq. (3.107). The inverse problem must be solved in order to specify the model order. For a given fuzzy time series $\tilde{x}_1$, $\tilde{x}_2$, ..., $\tilde{x}_N$ of length N the parameters $\underline{P}$ of the underlying fuzzy random process model $(\tilde{X}_\tau)_{\tau \in \mathbf{T}}$ are sought. By means of the maximum likelihood method it seems reasonable to seek a parameter combination $\underline{P}$ whose fuzzy probability density function form II takes on a maximum value for a given realization $\tilde{x}_1$, $\tilde{x}_2$, ..., $\tilde{x}_N$. The fuzzy probability density function form II with a given realization $\tilde{x}_1$, $\tilde{x}_2$, ..., $\tilde{x}_N$ and unknown, variable parameters $\underline{P}$ is referred to as the $l_\alpha r_\alpha$-likelihood function ${}_{lr}l(\underline{P} \,|\, \tilde{x}_1, \tilde{x}_2, ..., \tilde{x}_N)$ (see Eq. (3.108)).

$${}_{lr}l(\underline{P} \,|\, \tilde{x}_1, \tilde{x}_2, ..., \tilde{x}_N) = f(\tilde{x}_1, \tilde{x}_2, ..., \tilde{x}_N \,|\, \underline{P}) \quad \tilde{x}_1, \tilde{x}_2, ..., \tilde{x}_N = \text{constant} \tag{3.108}$$

The sought parameter combination for the fuzzy random process model is thus obtained by maximizing the $l_\alpha r_\alpha$-likelihood function.

Several fuzzy ARMA processes of different order $[p, q]$ are taken as a basis for the given fuzzy time series. The maximum value of the $l_\alpha r_\alpha$-likelihood function is determined for each of these models. By using these values it is possible select the 'best' fuzzy random process model with the aid of the decision criterion formulated in the following.

The simple choice of model order according to the largest of the computed maximum values does not necessarily yield the best model variant, however. For $p' \geqslant p$ and $q' \geqslant q$ a fuzzy ARMA$[p,q]$ process is included in a fuzzy ARMA$[p',q']$ process as a special case if all elements of the additional parameter matrices $\underline{A}_j$ $(j = p+1, ..., p')$ and $\underline{B}_j$ $(j = q+1, ..., q')$ are equal to zero. The following inequality then holds for the maximum values of the $l_\alpha r_\alpha$-likelihood function:

$$\sup_{\underline{P}} \left[{}_{lr}l \left(\underline{P}(p', q') \,|\, \tilde{x}_1, \tilde{x}_2, ..., \tilde{x}_N \right) \right] \geqslant \sup_{\underline{P}} \left[{}_{lr}l \left(\underline{P}(p, q) \,|\, \tilde{x}_1, \tilde{x}_2, ..., \tilde{x}_N \right) \right] . \quad (3.109)$$

The model order $[p,q]$ should be chosen in such a way that the specified fuzzy ARMA process with as few parameter matrices $\underline{A}_j$ $(j = 1, ..., p)$ and $\underline{B}_j$ $(j = 1, ..., q)$ as possible represents the best fuzzy stochastic model of the given fuzzy time series. The choice of model order according to the largest of the computed maximum values does not fulfill this requirement because this always yields the fuzzy ARMA process with the greatest number of parameter matrices. A suitable selection criterion derived from classical time series analysis may be applied, however [2, 56, 61]. The best model order $[p,q]$ is given by the fuzzy ARMA process for which the so-called BIC criterion (BAYESian Information Criterion) according to Eq. (3.110) is a minimum.

$$\mathrm{BIC}(p,q) = \ln \left[\sup_{\underline{P}} \left[{}_{lr}l \left(\underline{P}(p, q) \,|\, \tilde{x}_1, \tilde{x}_2, ..., \tilde{x}_N \right) \right] \right] + \frac{(p+q)\ln N}{N} \quad (3.110)$$

By this means, higher preference is given to compact models with a small number of parameter matrices $\underline{A}_j$ $(j = 1, ..., p)$ and $\underline{B}_j$ $(j = 1, ..., q)$ whereas lower preference is given to fuzzy ARMA processes with a large number of parameters.

Remark 3.38. A precondition for the specification of model order by means of the maximum likelihood method is the a priori determination of fuzzy ARMA processes with different model orders $[p,q]$. The respective process parameters $\underline{P}$ are matched by maximizing the $l_\alpha r_\alpha$-likelihood function according to Eq. (3.108). Based on the determined values of the $l_\alpha r_\alpha$-likelihood function, the model order is finally obtained by means of the BIC criterion. Due to the fact the model order is first selected after the parameter estimation, this method is referred to as a posteriori specification. ◆

Numerical realization. A precondition frequently encountered in classical time series analysis is the assumption of a multivariate GAUSSian normal distribution for the possible realizations of the underlying random process (see e.g. [6, 8, 33, 60]). This assumption appears questionable for the majority of practical applications, however, and relies solely on the existence of a closed solution for the multivariate GAUSSian normal distribution. Due to the requirement of Eq. (2.47) the application of a GAUSSian normal distribution is not possible in the analysis of fuzzy time series. For this reason a generally valid approach for determining the $l_\alpha r_\alpha$-likelihood function is presented in the following.

For a fuzzy random process $(\tilde{X}_\tau)_{\tau \in \mathbf{T}}$ with given parameters $\underline{P}$ the N-dimensional fuzzy probability density function form II $f(\tilde{x}_1, \tilde{x}_2, ..., \tilde{x}_N \,|\, \underline{P})$ of the N fuzzy random variables $\tilde{X}_\tau$ $(\tau = 1, 2, ..., N)$ is estimated by a Monte Carlo simulation followed by a determination of the empirical fuzzy probability density function form II $\hat{f}(\tilde{x}_1, \tilde{x}_2, ..., \tilde{x}_N \,|\, \underline{P})$. For this purpose, s fuzzy time series of length N are computed with the aid of a Monte Carlo simulation and the underlying fuzzy random process $(\tilde{X}_\tau)_{\tau \in \mathbf{T}}$ (see Sect. 3.5.4). After subdividing the simulated fuzzy time series $\tilde{x}_1, \tilde{x}_2, ..., \tilde{x}_N$ into classes, the N-dimensional fuzzy probability density function form II may be estimated. This procedure is analogous to the determination of the empirical fuzzy probability density function form II described in Sect. 2.2.2, whereby we are here concerned with a sample of s fuzzy time series of length N rather than a sample of s fuzzy variables. The value of the estimated N-dimensional fuzzy probability density function form II $\hat{f}(\tilde{x}_1, \tilde{x}_2, ..., \tilde{x}_N \,|\, \underline{P})$ resulting from the given fuzzy time series $\tilde{x}_1$, $\tilde{x}_2$, ..., $\tilde{x}_N$ may be used as an estimator for the value of the $l_\alpha r_\alpha$-likelihood function ${}_{lr}l(\underline{P} \,|\, \tilde{x}_1, \tilde{x}_2, ..., \tilde{x}_N)$ resulting from the assumed $\underline{P}$ according to Eq. (3.108). Accordingly, one functional value of the $l_\alpha r_\alpha$-likelihood function is known.

In order to compute additional functional values of the $l_\alpha r_\alpha$-likelihood function this procedure is repeated for altered values of the parameter $\underline{P}$.

If it is intended to match the process parameters $\underline{P}$ of an underlying fuzzy ARMA process of given model order $[p, q]$ to the given fuzzy time series $\tilde{x}_1$, $\tilde{x}_2$, ..., $\tilde{x}_N$ by maximizing the $l_\alpha r_\alpha$-likelihood function, the optimization problem according to Eq. (3.111) must be solved.

$$ {}_{lr}l(\underline{P} \,|\, \tilde{x}_1, \tilde{x}_2, ..., \tilde{x}_N) \stackrel{!}{=} \max \tag{3.111} $$

This optimization problem may be solved by means of classical optimization methods, e.g. the modified evolution strategy after [36]. A determination of the value of the objective function at a position $\underline{P}$ hereby requires the computation of a functional value of the $l_\alpha r_\alpha$-likelihood function in each case.

3.5.6 Parameter Estimation

The fuzzy MA, fuzzy AR and fuzzy ARMA processes presented in Sects. 3.5.2, 3.5.3 and 3.5.4 are parametric fuzzy stochastic models based on fuzzy white-noise processes. An essential prerequisite for their numerical realization is the $l_\alpha r_\alpha$-discretization introduced in Sect. 2.2. If a fuzzy random process model is stipulated or assumed for a fuzzy time series, the model order (see Sect. 3.5.5) and the corresponding process parameters $\underline{A}_1$, ..., $\underline{A}_p$ and $\underline{B}_1$, ..., $\underline{B}_q$ must be determined as accurately as possible. It should hereby be noted that the estimation of the parameter matrices $\underline{A}_1$, ..., $\underline{A}_p$ and $\underline{B}_1$, ..., $\underline{B}_q$ is interactively coupled with the determination of the underlying fuzzy white-noise process. The term $\underline{P}$ again represents the parameter matrices $\underline{A}_1$, ..., $\underline{A}_p$ and $\underline{B}_1$, ..., $\underline{B}_q$ in an abridged form in the following.

Remark 3.39. A basic condition for the determination of the parameter matrices $\underline{A}_1$, ..., $\underline{A}_p$ and $\underline{B}_1$, ..., $\underline{B}_q$ is the permissibility of the fuzzy random process according to Remark 3.34. In other words, at each time point $\tau = 1, 2, \ldots$ the fuzzy random variables $\tilde{\mathrm{X}}_\tau$ of the underlying fuzzy AR, fuzzy MA or fuzzy ARMA process must satisfy the requirement given by Eq. (2.80). ◆

Different methods for estimating the parameters of fuzzy ARMA processes are presented in the following section. These may also be applied to fuzzy MA and fuzzy AR processes. In addition, special methods are presented for estimating the parameters of fuzzy MA and fuzzy AR processes. The methods outlined in the following for parameter estimation assume that Eq. (2.80) is fulfilled. A special verification of the latter may be required in certain cases.

Parameter Estimation in Fuzzy ARMA Processes

Characteristic value method. In order to apply the characteristic value method, stationary and ergodic fuzzy time series are assumed. If the fuzzy time series under investigation exhibits non-stationary properties, this may be converted into a stationary fuzzy time series with the aid of the fuzzy component model (see Sect. 3.2) or by filtration (see Sect. 3.4).

The characteristic value method is based on the requirement that the empirical characteristic values of the fuzzy time series concerned should match with the characteristic values of the hypothetically underlying stationary fuzzy ARMA process. In other words, the empirical characteristic values of the fuzzy time series serve as unbiased estimators for the characteristic values of the fuzzy random process. Of particular importance are the fuzzy expected value $\tilde{m}_{\tilde{\mathrm{X}}_\tau}(\underline{P})$ including its $l_\alpha r_\alpha$-increments $\Delta m_j(\underline{P})$) and $l_\alpha r_\alpha$-variance ${}_{lr}\underline{\sigma}^2_{\tilde{\mathrm{X}}_\tau}(\underline{P})$ as well as the $l_\alpha r_\alpha$-correlation function ${}_{lr}\underline{R}_{\tilde{\mathrm{X}}_\tau}(\Delta\tau, \underline{P})$ including its elements $r_{k,l}(\Delta\tau, \underline{P})$). For the optimum estimation of the parameters $\underline{P}$ of the underlying stationary fuzzy ARMA process and the included fuzzy white-noise process the optimization problem according to Eq. (3.112) must be solved.

$$\sum_{j=1}^{2n} \left(\Delta\overline{x}_j - \Delta m_j(\underline{P})\right)^2 + \sum_{\Delta\tau=-\infty}^{\infty} \sum_{k,l=1}^{2n} \left(\hat{r}_{k,l}(\Delta\tau) - r_{k,l}(\Delta\tau, \underline{P})\right)^2 \stackrel{!}{=} \min \tag{3.112}$$

The term $\Delta\overline{x}_j$ hereby represents the $l_\alpha r_\alpha$-increments of the empirical fuzzy mean value $\tilde{\overline{x}}$ (see Eq. (3.16)) while $\hat{r}_{k,l}(\Delta\tau)$ are the elements of the empirical $l_\alpha r_\alpha$-correlation function ${}_{lr}\underline{\hat{R}}_{\tilde{x}_\tau}(\Delta\tau)$ (see Eq. (3.22)).

A fundamental constraint of the optimization problem is compliance with Eq. (2.18) for the realizations $\tilde{\varepsilon}_\tau$ of the included fuzzy white-noise process.

This means that the non-negativity requirement must be fulfilled for the $l_\alpha r_\alpha$-increments $\Delta\varepsilon_j(\tau)$ of the realizations $\tilde{\varepsilon}_\tau$.

$$\Delta\varepsilon_j(\tau) \geqslant 0 \quad \text{for } j = 1, 2, ..., n-1, n+1, ...2n \tag{3.113}$$

An additional requirement is that the fuzzy variables $\tilde{\varepsilon}_\tau$ are realizations of a stationary fuzzy white-noise process.

The third basic constraint of the optimization problem is the stationarity of the fuzzy process model. In the case of fuzzy ARMA[0,q] processes this requirement is fulfilled per definition, whereas for fuzzy ARMA[p,0] and fuzzy ARMA[p,q] processes, the parameters to be estimated are subject to defined restrictions (see Sects. 3.5.3 and 3.5.4).

Numerical realization. Depending on the extent of the optimization problem, mesh search strategies, Monte Carlo methods or the modified evolution strategy after [36] are suggested as possible methods of solution. The elements of the parameter matrices $\underline{P}$ are the decision variables of the optimization problem. Depending on the applied optimization method, random or systematically selected starting points for the decision variables are specified. Each element of the parameter matrices $\underline{P}$ is initialized with a real value from the interval $[-1, 1]$.

The objective function given by Eq. (3.112) is set up as follows for given values of the parameter matrices $\underline{P}$: in order to check the permissibility of the selected $\underline{A}_1$, ..., $\underline{A}_p$ the given fuzzy time series $\tilde{x}_\tau$ is transformed into a realization ${}_{MA}\tilde{x}_\tau$ of a fuzzy MA[q] process (see Remark 3.35) by means of:

$$\begin{aligned} {}_{MA}\tilde{x}_\tau &= \tilde{x}_\tau \ominus \underline{A}_1 \odot \tilde{x}_{\tau-1} \ominus ... \ominus \underline{A}_p \odot \tilde{x}_{\tau-p} \\ &\text{with} \quad \tau = p+1, p+2, ..., N\,. \end{aligned} \tag{3.114}$$

The parameter matrices $\underline{A}_1$, ..., $\underline{A}_p$ are permissible if the realization ${}_{MA}\tilde{x}_\tau$ constitutes a stationary fuzzy time series (see Sect. 3.3). Compliance with the requirement given by Eq. (2.47) is not necessary for the ${}_{MA}\tilde{x}_\tau$, as these only represent an intermediate result.

The permissibility of the $\underline{B}_1$, ..., $\underline{B}_q$ is checked on the basis of the realizations $\tilde{\varepsilon}_\tau$ of the fuzzy white-noise process. These must fulfill the condition Eq. (3.113). A realization of this type may be determined from the given fuzzy time series by means of Eq. (3.115).

$$\tilde{\varepsilon}_\tau = \tilde{x}_\tau \ominus \underline{A}_1 \odot \tilde{x}_{\tau-1} \ominus ... \ominus \underline{A}_p \odot \tilde{x}_{\tau-p} \oplus \underline{B}_1 \odot \tilde{\varepsilon}_{\tau-1} \oplus ... \oplus \underline{B}_q \odot \tilde{\varepsilon}_{\tau-q} \tag{3.115}$$

$$\text{with} \quad \tilde{\varepsilon}_{\tau-i} = \begin{cases} \tilde{\varepsilon}_{\tau-i} & \text{for } \tau - i > p \\ \hat{E}[\tilde{\mathcal{E}}_\tau] & \text{for } \tau - i \leqslant p \end{cases} \quad \text{and} \quad i = 1, 2, ..., q$$

$\hat{E}[\tilde{\mathcal{E}}_\tau]$ is the estimated fuzzy expected value of the underlying fuzzy white-noise process $(\tilde{\mathcal{E}}_\tau)_{\tau\in\mathbf{T}}$, and is obtained by solving the system of linear equations

(3.116) for the unknown $l_\alpha r_\alpha$-increments of the fuzzy expected value $\hat{E}[\tilde{\mathcal{E}}_\tau]$. The fuzzy expected value $\hat{E}[\tilde{\mathcal{E}}_\tau]$ is used in Eq. (3.115) as the best possible estimator for the unknown realizations $\tilde{\varepsilon}_{\tau-i}$ at time points $\tau - i \leqslant p$.

$$\hat{E}[{}_{MA}\tilde{\mathrm{X}}_\tau] = \left[\sum_{j=0}^{q} (-\underline{B}_j)\right] \odot \hat{E}[\tilde{\mathcal{E}}_\tau] \tag{3.116}$$

The fuzzy expected value $E[{}_{MA}\tilde{\mathrm{X}}_\tau]$ of the fuzzy MA[q] process required by Eq. (3.116) may be estimated from the transformed fuzzy time series ${}_{MA}\tilde{x}_\tau$ according to Eq. (3.16). If the fuzzy variables $\tilde{\varepsilon}_\tau$ computed from Eq. (3.115) fulfill the requirements specified for the realizations of a fuzzy white-noise process (especially the condition given by Eq. (3.113)), the probability distribution function of the fuzzy white-noise variables $\tilde{\mathcal{E}}_\tau$ may be estimated on the basis of the $\tilde{\varepsilon}_\tau$. Realizations of the fuzzy white-noise variables $\tilde{\mathcal{E}}_\tau$ may be simulated with the aid of the estimated probability distribution function.

By means of Eq. (3.87) it is now possible to simulate the realizations of the assumed fuzzy ARMA[p, q] process, and also estimate the fuzzy expected value, the $l_\alpha r_\alpha$-variance and the $l_\alpha r_\alpha$-correlation function.

It is then possible to compute one value of the objective function by means of Eq. (3.112). Depending on the case in question, the chosen optimization strategy yields improved parameters $\underline{P}$.

Distance method. Estimation of the parameters of a fuzzy ARMA process by the characteristic value method assumes stationary and ergodic fuzzy time series. If the given fuzzy time series exhibits systematic changes such as fuzzy trends or fuzzy cycles, non-stationary fuzzy random process models must be assumed. A suitable approach for parameter estimation in such cases is to minimize the average distance $\overline{d}_F$ between the optimum single-step forecast $\mathring{\tilde{x}}_\tau(\underline{P})$ and the known values $\tilde{x}_\tau$ of the given fuzzy time series with $p < \tau \leqslant N$ according to Eq. (3.117). A decisive advantage of this method is that neither ergodic nor stationary fuzzy time series must be assumed. The method permits the modeling of non-stationary fuzzy time series with the aid of non-stationary process models without the need to specify (not meaningfully estimable for non-stationary fuzzy time series) empirical characteristic values.

$$\overline{d}_F(\underline{P}) = \frac{1}{N-p} \sum_{\tau=p+1}^{N} d_F\left(\tilde{x}_\tau; \mathring{\tilde{x}}_\tau(\underline{P})\right) \stackrel{!}{=} \min \tag{3.117}$$

Depending on the process parameters $\underline{P}$ to be determined, this method involves a determination of the optimum single-step forecasts $\mathring{\tilde{x}}_\tau(\underline{P})$ for each time point τ ($p < \tau \leqslant N$), whose distance d_F from the corresponding fuzzy variables $\tilde{x}_\tau$ are computed and averaged over time. The minimization of this average distance $\overline{d}_F$ yields an unbiased estimator for the process parameters $\underline{P}$. Regarding the determination of the optimum single-step forecasts $\mathring{\tilde{x}}_\tau(\underline{P})$, the reader is referred to Sect. 4. A definition of the distance d_F between two fuzzy variables is given in Sect. 2.1.4.

Numerical realization. Analogous to the characteristic value method, the minimization problem is solved by means of mesh search methods, Monte Carlo methods or the modified evolution strategy suggested by [36]. Depending on the optimization method used, randomly or systematically selected starting points from the interval $[-1, 1]$ are specified for the elements of the parameter matrices $\underline{P}$ and the decision variables. The objective function is set up and evaluated as in the characteristic value method. Firstly, the fuzzy time series ${}_{MA}\tilde{x}_\tau$ is constructed (see Remark 3.35). Provided this fuzzy time series is stationary (see Sect. 3.3), the selected $\underline{A}_1$, ..., $\underline{A}_p$ are permissible. The realizations $\tilde{\varepsilon}_\tau$ of the underlying fuzzy white-noise process $(\tilde{\mathcal{E}}_\tau)_{\tau\in\mathbf{T}}$ are computed from Eq. (3.115). Applying the estimators for the fuzzy expected value $E[{}_{MA}\tilde{\mathrm{X}}_\tau]$ of the fuzzy MA$[q]$ process, the fuzzy expected value $\hat{E}[\tilde{\mathcal{E}}_\tau]$ is obtained by solving the system of linear equations (3.116). The realizations $\tilde{\varepsilon}_\tau$ computed according to Eq. (3.115) must fulfill the condition given by Eq. (3.113). If the fuzzy variables $\tilde{\varepsilon}_\tau$ do not satisfy these requirements placed on the realizations of a fuzzy white-noise process, the specified parameter matrices $\underline{B}_1$, ..., $\underline{B}_q$ are impermissible and hence rejected. Applying the estimators for $E[\tilde{\mathcal{E}}_\tau]$, the optimum single-step forecasts $\mathring{\tilde{x}}_\tau(\underline{P})$ and the distance d_F between $\mathring{\tilde{x}}_\tau(\underline{P})$ and the corresponding fuzzy variables $\tilde{x}_\tau$ may now be computed according to Sect. 4.

Gradient method. Another effective method for estimating the parameter matrices $\underline{A}_1, ..., \underline{A}_p$ and $\underline{B}_1, ..., \underline{B}_q$ of fuzzy ARMA processes is the gradient method. This permits the modeling non-stationary fuzzy time series and hence does not rely on the assumption of ergodicity. The basic idea is again the matching of optimum single-step forecasts $\mathring{\tilde{x}}_\tau(\underline{P})$ to the known values $\tilde{x}_\tau$ of the given fuzzy time series with $p < \tau \leqslant N$. The definition of an error function analogous to Eq. (3.117) is not appropriate in this case, however, as the underlying integral for determining the distance d_F via the HAUSDORFF distance d_H (see Sect. 2.1.4) represents a function which is only differentiable over certain intervals. The square error between the forecasted increments $\Delta\mathring{x}_i(\tau, \underline{P})$ $(i = 1, 2, ..., 2n)$ and the given increments $\Delta x_i(\tau)$ of the fuzzy time series is defined according to Eq. (3.118) as the error function E to be optimized.

$$E = \frac{1}{2} \sum_{\tau=1+p}^{N} \sum_{i=1}^{2n} \left(\Delta x_i(\tau) - \Delta\mathring{x}_i(\tau, \underline{P})\right)^2 \stackrel{!}{=} \min \tag{3.118}$$

Numerical realization. The optimum single-step forecasts $\mathring{\tilde{x}}_\tau(\underline{P})$ are determined as in the distance method. Firstly, starting points are defined for the decision variables of the optimization problem, i.e. the parameter matrices $\underline{A}_1, ..., \underline{A}_p$ and $\underline{B}_1, ..., \underline{B}_q$ of the fuzzy ARMA$[p, q]$ process are initialized. This is carried out assigning random real values from the interval $[-1, 1]$ to the matrix elements. The objective function is set up and evaluated as in the characteristic value method. Firstly, a check is made to establish whether the selected $\underline{A}_1$, ..., $\underline{A}_p$ lead to a stationary realization of the fuzzy MA[q] process. For this purpose the realization ${}_{MA}\tilde{x}_\tau$ is computed (see Remark 3.35) and a check is made to establish whether this fuzzy time series is stationary (see Sect. 3.3). If the chosen $\underline{A}_1$, ..., $\underline{A}_p$ do not yield a stationary realization, the parameters are re-initialized. The realizations $\tilde{\varepsilon}_\tau$ of the underlying fuzzy white-noise process $(\tilde{\mathcal{E}}_\tau)_{\tau \in \mathbf{T}}$ are computed from Eq. (3.115). Using the estimators for the fuzzy expected value $E[{}_{MA}\tilde{\mathrm{X}}_\tau]$ of the fuzzy MA[q] process, the fuzzy expected value $\hat{E}[\tilde{\mathcal{E}}_\tau]$ is obtained by solving the system of linear equations (3.116). The realizations $\tilde{\varepsilon}_\tau$ computed according to Eq. (3.115) must fulfill the conditions given by Eq. (3.113). If the fuzzy variables $\tilde{\varepsilon}_\tau$ do not fulfill these conditions, the specified parameter matrices $\underline{B}_1$, ..., $\underline{B}_q$ are impermissible and hence rejected. With the aid of the estimators $\hat{E}[\tilde{\mathcal{E}}_\tau]$ the optimum single-step forecasts $\mathring{\tilde{x}}_\tau(\underline{P})$ are determined according to Sect. 4 and the square error is computed according to Eq. (3.118).

The parameter matrices $\underline{A}_1$, ..., $\underline{A}_p$ and $\underline{B}_1$, ..., $\underline{B}_q$ are improved by means of the correction matrices $\Delta\underline{A}_1$, ..., $\Delta\underline{A}_p$ and $\Delta\underline{B}_1$, ..., $\Delta\underline{B}_q$ according to Eqs. (3.119) and (3.120), respectively.

$$\underline{A}_r(\text{new}) = \underline{A}_r(\text{old}) + \Delta\underline{A}_r \quad \text{with} \quad r = 1, 2, ..., p \tag{3.119}$$

$$\underline{B}_s(\text{new}) = \underline{B}_s(\text{old}) + \Delta\underline{B}_s \quad \text{with} \quad s = 1, 2, ..., q \tag{3.120}$$

The correction matrices are determined according to Eqs. (3.121) and (3.122) by partial differentiation of the error function with respect to the parameters. The step length may be arbitrarily defined by means of the factor η with $\eta > 0$.

$$\Delta\underline{A}_r = -\eta \frac{\partial E}{\partial \underline{A}_r} = -\eta \begin{bmatrix} \frac{\partial E}{\partial a_{1,1}(r)} & \cdots & \frac{\partial E}{\partial a_{1,2n}(r)} \\ \vdots & \ddots & \vdots \\ \frac{\partial E}{\partial a_{2n,1}(r)} & \cdots & \frac{\partial E}{\partial a_{2n,2n}(r)} \end{bmatrix} \tag{3.121}$$

$$\Delta\underline{B}_s = -\eta \frac{\partial E}{\partial \underline{B}_s} = -\eta \begin{bmatrix} \frac{\partial E}{\partial b_{1,1}(s)} & \cdots & \frac{\partial E}{\partial b_{1,2n}(s)} \\ \vdots & \ddots & \vdots \\ \frac{\partial E}{\partial b_{2n,1}(s)} & \cdots & \frac{\partial E}{\partial b_{2n,2n}(s)} \end{bmatrix} \tag{3.122}$$

The partial derivatives $\frac{\partial E}{\partial a_{u,v}(r)}$ and $\frac{\partial E}{\partial b_{u,v}(s)}$ of the square error function E with respect to the elements of the parameter matrices are given by Eqs. (3.123) and (3.124), respectively, for $u, v = 1, 2, ..., 2n$.

$$\frac{\partial E}{\partial a_{u,v}(r)} = \sum_{\tau=1+p}^{N} \left(\Delta x_u(\tau) - \Delta \mathring{x}_u(\tau, \underline{P})\right) \Delta x_v(\tau - r) \tag{3.123}$$

$$\frac{\partial E}{\partial b_{u,v}(s)} = \sum_{\tau=1+p}^{N} \left(\Delta x_u(\tau) - \Delta \mathring{x}_u(\tau, \underline{P})\right) \Delta \varepsilon_v(\tau - s) \tag{3.124}$$

For time points $\tau - s \leqslant p$ the non-computable realizations $\Delta\varepsilon_v(\tau - s)$ in Eq. (3.124) of the fuzzy white-noise process $(\tilde{\mathcal{E}}_\tau)_{\tau\in\mathbf{T}}$ are again replaced by the $l_\alpha r_\alpha$-increments of the estimated fuzzy expected value $\hat{E}[\tilde{\mathcal{E}}_\tau]$.

$$\frac{\partial E}{\partial b_{u,v}(s)} = \sum_{\tau=1+p}^{N} \left(\Delta x_u(\tau) - \Delta \mathring{x}_u(\tau, \underline{P})\right) \hat{E}\left[\Delta \varepsilon_v\right] \tag{3.125}$$

An obligatory constraint for the optimization problem in this case is also the non-negativity of the $l_\alpha r_\alpha$-increments $\Delta\varepsilon_j(\tau)$ according to Eq. (3.113). Moreover, the fuzzy variables $\tilde{\varepsilon}_\tau$ must satisfy the requirements placed on the realizations of a fuzzy white-noise process.

Iteration method. A further, easily applicable method for the iterative estimation of the parameter matrices $\underline{A}_1$, ..., $\underline{A}_p$ and $\underline{B}_1$, ..., $\underline{B}_q$ of a fuzzy ARMA process may be derived from the regression approach of DURBIN [12]. Firstly, a fuzzy AR process of high order $p = k$ is presupposed for the fuzzy time series in question, i.e. the parameter matrices $\underline{B}_1$, ..., $\underline{B}_q$ are zero matrices. After estimating the parameter matrices $\underline{A}_1^*, \underline{A}_2^*, ..., \underline{A}_k^*$ (see *Parameter estimation in fuzzy AR processes* on p. 105) of the presupposed AR process, the realizations $\tilde{\varepsilon}_\tau$ of the fuzzy white-noise process $\tilde{\mathcal{E}}_\tau$ are estimated in iteration step zero at time point τ according to Eq. (3.126), whereby $k < \tau \leqslant N$ holds.

$$\tilde{\varepsilon}_\tau^{(0)} = \tilde{x}_\tau \ominus \hat{\underline{A}}^*{}_1 \odot \tilde{x}_{\tau-1} \ominus ... \ominus \hat{\underline{A}}^*{}_k \odot \tilde{x}_{\tau-k} \tag{3.126}$$

In the next step the parameter matrices $\underline{A}_1$, ..., $\underline{A}_p$ and $\underline{B}_1$, ..., $\underline{B}_q$ are improved by minimizing the average distance $\overline{d}_F$ according to Eq. (3.127). A definition of the distance d_F between two fuzzy variables is given in Sect. 2.1.4. In contrast to the distance method, the estimated realizations $\tilde{\varepsilon}_\tau$ of the fuzzy white-noise process $(\tilde{\mathcal{E}}_\tau)_{\tau\in\mathbf{T}}$ remain constant during the optimization. This means that the objective function may be evaluated directly, i.e. without additional computation of the fuzzy variables $\tilde{\varepsilon}_\tau$.

$$\overline{d}_F(\underline{P}) = \frac{1}{N-p} \sum_{\tau=p+1}^{N} d_F\left(\tilde{x}_\tau; \tilde{x}_\tau^*(\underline{P})\right) \stackrel{!}{=} \min \tag{3.127}$$

The term $\tilde{x}_\tau^*(\underline{P})$ of the objective function $\overline{d}_F(\underline{P})$ is determined according to Eq. (3.128).

$$\tilde{x}_\tau^*(\underline{P}) = \underline{A}_1 \odot \tilde{x}_{\tau-1} \oplus ... \oplus \underline{A}_p \odot \tilde{x}_{\tau-p} \ominus \underline{B}_1 \odot \tilde{\varepsilon}_{\tau-1}^{(0)} \ominus ... \ominus \underline{B}_q \odot \tilde{\varepsilon}_{\tau-q}^{(0)} \tag{3.128}$$

Using the parameter matrices $\underline{A}_1, ..., \underline{A}_p$ and $\underline{B}_1, ..., \underline{B}_q$ determined from Eq. (3.127), the realizations $\tilde{\varepsilon}_\tau$ of the included fuzzy white-noise process $\tilde{\mathcal{E}}_\tau$ are recomputed in the next iteration step by means of Eq. (3.129).

$$\tilde{\varepsilon}_\tau^{(1)} = \tilde{x}_\tau \ominus \hat{\underline{A}}_1 \odot \tilde{x}_{\tau-1} \ominus ... \ominus \hat{\underline{A}}_p \odot \tilde{x}_{\tau-p} \oplus \hat{\underline{B}}_1 \odot \tilde{\varepsilon}_{\tau-1}^{(0)} \oplus ... \oplus \hat{\underline{B}}_q \odot \tilde{\varepsilon}_{\tau-q}^{(0)} \quad (3.129)$$

Inserting the obtained fuzzy variables $\tilde{\varepsilon}_\tau$ into Eq. (3.128) and resolving the minimization problem by means of Eq. (3.127) yields improved values of $\underline{A}_1, ..., \underline{A}_p$ and $\underline{B}_1, ..., \underline{B}_q$. The iteration is continued until the parameter matrices to be estimated converge.

A basic precondition for the applicability of the method is again that the determined fuzzy variables $\tilde{\varepsilon}_\tau$ satisfy the requirements placed on the realizations of a fuzzy white-noise process in each iteration step, i.e. they especially fulfill the condition given by Eq. (3.113).

Parameter Estimation in Fuzzy MA Processes

The algorithm suggested by WILSON [68] for estimating the parameters of a classical moving average process with crisp realizations is modified and extended in the following to deal with fuzzy moving average processes. The parameter matrices $\underline{B}_1$, ..., $\underline{B}_q$ of a fuzzy MA process $(\tilde{\mathrm{X}}_\tau)_{\tau \in \mathbf{T}}$ are hereby estimated from the empirical $l_\alpha r_\alpha$-covariance function ${}_{lr}\hat{\underline{K}}_{\tilde{x}_\tau}(\Delta\tau_k)$ of a given fuzzy time series as described in the following, whereby $\Delta\tau_0 = 0$, $\Delta\tau_1 = 1$, ..., $\Delta\tau_q = q$ holds.

The $l_\alpha r_\alpha$-covariance function ${}_{lr}\underline{K}_{\tilde{\mathrm{X}}}(\Delta\tau_k)$ of a fuzzy moving average process $(\tilde{\mathrm{X}}_\tau)_{\tau \in \mathbf{T}}$ according to Eq. (3.74) is given by Eq. (3.130) for $k = 0, 1, ..., q$. The term ${}_{lr}\underline{K}_{\tilde{\mathcal{E}}_\tau}(\Delta\tau = 0)$ thereby represents the $l_\alpha r_\alpha$-covariance function of the corresponding fuzzy white-noise process $(\tilde{\mathcal{E}}_\tau)_{\tau \in \mathbf{T}}$ for $\Delta\tau = 0$.

$${}_{lr}\underline{K}_{\tilde{\mathrm{X}}_\tau}(\Delta\tau_k) = \sum_{c=0}^{q-k} \underline{B}_{c+k} \; {}_{lr}\underline{K}_{\tilde{\mathcal{E}}_\tau}(\Delta\tau = 0) \; \underline{B}_c^T \quad (3.130)$$

If the $l_\alpha r_\alpha$-covariance function ${}_{lr}\underline{K}_{\tilde{\mathrm{X}}_\tau}(\Delta\tau_k)$ is replaced by the empirical $l_\alpha r_\alpha$-covariance function ${}_{lr}\hat{\underline{K}}_{\tilde{x}_\tau}(\Delta\tau_k)$, implicit conditional equations are obtained for the unknown parameter matrices $\underline{B}_k$. Approximation values for the $\underline{B}_k$ are determined iteratively with the aid of correction matrices. If $\underline{B}_k^{(i)}$ is used to denote the estimation of $\underline{B}_k$ in the i-th iteration step, correction matrices $\underline{\Delta}_k^{(i)}$ of size $[2n, 2n]$ are sought which yield the best possible approximation of the empirical $l_\alpha r_\alpha$-covariance function ${}_{lr}\hat{\underline{K}}_{\tilde{x}_\tau}(\Delta\tau_k)$ according to Eq. (3.131). In other words, correction matrices $\underline{\Delta}_k^{(i)}$ are determined in each iteration step i such that $\underline{B}_k^{(i+1)} = \underline{B}_k^{(i)} + \underline{\Delta}_k^{(i)}$ yields an improved approximation of the empirical $l_\alpha r_\alpha$-covariance function according to Eq. (3.131). In the first iteration step, random starting values within the interval $[-0,1; 0,1]$ are assigned to the elements of the parameter matrices $\underline{B}_k^{(1)}$ for $k > 0$. The parameter matrix $\underline{B}_0$ is the negative unit matrix according to the definition of a fuzzy MA process. In each iteration step i the following equations must be evaluated for $k = 1, 2, ..., q$.

$$_{lr}\hat{\underline{K}}_{\tilde{x}_\tau}(\Delta\tau_k) = \sum_{c=0}^{q-k} \left(\underline{B}_{c+k}^{(i)} + \underline{\Delta}_{c+k}^{(i)}\right) {}_{lr}\underline{K}_{\tilde{\mathcal{E}}_\tau}(\Delta\tau = 0) \left(\underline{B}_c^{T^{(i)}} + \underline{\Delta}_c^{T^{(i)}}\right) \quad (3.131)$$

By applying the approximations of the parameter matrices $\underline{B}_k^{(i)}$ in each iteration step i obtained from the $(q+1)$-th solution of Eq. (3.132) and computing the arithmetic mean according to Eq. (3.133), it is possible to estimate the unknown $l_\alpha r_\alpha$-covariance function ${}_{lr}\underline{K}_{\tilde{\mathcal{E}}_\tau}(\Delta\tau = 0)$ of the fuzzy white-noise process $(\tilde{\mathcal{E}}_\tau)_{\tau\in\mathbf{T}}$.

$$_{lr}\hat{\underline{K}}_{\tilde{x}_\tau}(\Delta\tau_k) = \sum_{c=0}^{q-k} \underline{B}_{c+k}^{(i)} \; {}_{lr}\underline{K}_{\tilde{\mathcal{E}}_\tau}^{(i),k}(\Delta\tau = 0) \, \underline{B}_c^{T^{(1)}} \quad (3.132)$$

$$_{lr}\underline{K}_{\tilde{\mathcal{E}}_\tau}^{(i)}(\Delta\tau = 0) = \frac{1}{q+1} \sum_{c=0}^{q-k} {}_{lr}\underline{K}_{\tilde{\mathcal{E}}_\tau}^{(i),c}(\Delta\tau = 0) \quad (3.133)$$

$$\text{with} \quad k = 0, 1, ..., q$$

The $(q+1)$-th solution of Eq. (3.132) and the use of the arithmetic mean as an estimator for the $l_\alpha r_\alpha$-covariance function ${}_{lr}\underline{K}_{\tilde{\mathcal{E}}}(\Delta\tau = 0)$ promote the convergence behavior. The use of only a specific solution of Eq. (3.132) as an estimator for ${}_{lr}\underline{K}_{\tilde{\mathcal{E}}}(\Delta\tau = 0)$ leads to a trivial solution of Eq. (3.131); this results in numerical instabilities.

By means of Eq. (3.134) the element-by-element formulation of Eq. (3.131) is given for $i, j = 1, 2, ..., 2n$. The various terms are defined as follows:

$\hat{k}_{\tilde{x}_\tau}[i,j](\Delta\tau_k)$: the elements of the empirical $l_\alpha r_\alpha$-covariance function ${}_{lr}\hat{\underline{K}}_{\tilde{x}_\tau}(\Delta\tau_k)$,
$b_c^{(i)}[j,b]$ and $b_{c+k}^{(i)}[i,a]$: the elements of the parameter matrices $\underline{B}_c^{(i)}$ and $\underline{B}_{c+k}^{(i)}$,
$\delta_c^{(i)}[j,b]$ and $\delta_{c+k}^{(i)}[i,a]$: the elements of the correction matrices $\underline{\Delta}_c^{(i)}$ and $\underline{\Delta}_{c+k}^{(i)}$, and
$k_{\tilde{\mathcal{E}}_\tau}^{(i)}[a,b](\Delta\tau_k = 0)$: the elements of the estimated $l_\alpha r_\alpha$-covariance function ${}_{lr}\underline{K}_{\tilde{\mathcal{E}}_\tau}^{(i)}(\Delta\tau = 0)$ in the i-th iteration step.

$$\hat{k}_{\tilde{x}_\tau}[i,j](\Delta\tau_k) = \sum_{c=0}^{q-k} \sum_{b=1}^{2n} \sum_{a=1}^{2n} \left(b_c^{(i)}[j,b] + \delta_c^{(i)}[j,b]\right) \ldots \quad (3.134)$$

$$\ldots k_{\tilde{\mathcal{E}}_\tau}^{(i)}[a,b](\Delta\tau_k = 0) \left(b_{c+k}^{(i)}[i,a] + \delta_{c+k}^{(i)}[i,a]\right)$$

A linearization of Eq. (3.134) by neglecting the quadratic terms of the elements $\delta_c^{(i)}[j,b]$ and $\delta_{c+k}^{(i)}[i,a]$ leads to the iteration equation (3.135).

$$\begin{aligned}\hat{k}_{\tilde{x}_\tau}[i,j](\Delta\tau_k) - \sum_{c=0}^{q-k}\sum_{b=1}^{2n}\sum_{a=1}^{2n} b_c^{(i)}[j,b]\, k_{\tilde{\mathcal{E}}_\tau}^{(i)}[a,b](\Delta\tau_k=0)\, b_{c+k}^{(i)}[i,a] = \\ \sum_{c=0}^{q-k}\sum_{b=1}^{2n}\sum_{a=1}^{2n} b_c^{(i)}[j,b]\, k_{\tilde{\mathcal{E}}_\tau}^{(i)}[a,b](\Delta\tau_k=0)\, \delta_{c+k}^{(i)}[i,a] + \\ \sum_{c=0}^{q-k}\sum_{b=1}^{2n}\sum_{a=1}^{2n} \delta_c^{(i)}[j,b]\, k_{\tilde{\mathcal{E}}_\tau}^{(i)}[a,b](\Delta\tau_k=0)\, b_{c+k}^{(i)}[i,a]\end{aligned} \tag{3.135}$$

As this equation is linear with respect to the elements $\delta_c^{(i)}[j,b]$ and $\delta_{c+k}^{(i)}[i,a]$, a determination of the correction matrices is unproblematic. The elements of the improved approximation solutions $\underline{B}_k^{(i+1)} = \underline{B}_k^{(i)} + \underline{\Delta}_k^{(i)}$ are inserted into Eq. (3.135) for the subsequent iteration step $i+1$.

After determining the parameter matrices $\underline{B}_1, ..., \underline{B}_q$, the underlying fuzzy white-noise process is estimated according to Eqs. (3.71) to (3.74).

Parameter Estimation in Fuzzy AR Processes

If a given, stationary fuzzy time series is modeled as a fuzzy autoregressive process of order p, the corresponding parameter matrices $\underline{A}_1, ..., \underline{A}_p$ may be estimated using the empirical $l_\alpha r_\alpha$-correlation function ${}_{lr}\hat{\underline{R}}_{\tilde{x}_\tau}(\Delta\tau)$. According to the classical YULE-WALKER equations after [32], the elements of $\underline{A}_1, ..., \underline{A}_p$ may be determined with the aid of Eqs. (3.136) and (3.137).

$$[\underline{A}_1, ..., \underline{A}_p]\begin{bmatrix} {}_{lr}\hat{\underline{R}}_{\tilde{x}_\tau}(\Delta\tau_{1-t}) \\ \vdots \\ {}_{lr}\hat{\underline{R}}_{\tilde{x}_\tau}(\Delta\tau_{p-t}) \end{bmatrix} = {}_{lr}\hat{\underline{R}}_{\tilde{x}_\tau}(\Delta\tau_{-t}) \quad \text{with} \quad t = 1, ..., p \tag{3.136}$$

$$[\underline{A}_1, ..., \underline{A}_p]\begin{bmatrix} {}_{lr}\hat{\underline{R}}_{\tilde{x}_\tau}(\Delta\tau_0) & \cdots & {}_{lr}\hat{\underline{R}}_{\tilde{x}_\tau}(\Delta\tau_{1-p}) \\ \vdots & \ddots & \vdots \\ {}_{lr}\hat{\underline{R}}_{\tilde{x}_\tau}(\Delta\tau_{p-1}) & \cdots & {}_{lr}\hat{\underline{R}}_{\tilde{x}_\tau}(\Delta\tau_0) \end{bmatrix} = \begin{bmatrix} {}_{lr}\hat{\underline{R}}_{\tilde{x}_\tau}(\Delta\tau_{-1}) \\ \vdots \\ {}_{lr}\hat{\underline{R}}_{\tilde{x}_\tau}(\Delta\tau_{-p}) \end{bmatrix} \tag{3.137}$$

With the aid of the parameter matrices $\underline{A}_1, ..., \underline{A}_p$ estimated by Eqs. (3.136) and (3.137) it is possible to determine the realizations $\tilde{\varepsilon}_\tau$ of the fuzzy white-noise variables $\tilde{\mathcal{E}}_\tau$ for each time point τ according to Eq. (3.138).

$$\tilde{\varepsilon}_\tau = \tilde{x}_\tau \ominus \hat{\underline{A}}_1 \odot \tilde{x}_{\tau-1} \ominus ... \ominus \hat{\underline{A}}_p \odot \tilde{x}_{\tau-p} \tag{3.138}$$

Owing to the use of the empirical $l_\alpha r_\alpha$-correlation function ${}_{lr}\hat{R}_{\tilde{x}_\tau}(\Delta\tau)$ for computing the $\underline{A}_1, ..., \underline{A}_p$, the non-negativity of the $l_\alpha r_\alpha$-increments $\Delta\varepsilon_j(\tau)$

according to Eq. (3.113) is not necessarily complied with. In this case it is necessary to revert to the methods of parameter estimation for fuzzy ARMA processes, e.g. the distance method. In all events, the stationarity of the fuzzy AR process must be verified (see Sect. 3.3).

3.6 Modeling on the Basis of Artificial Neural Networks

The methods presented in Sect. 3.5 interpret a time series comprised of fuzzy data as a realization of a fuzzy random process. After analyzing the fuzzy time series, the process parameters corresponding to the chosen model are estimated according to Sect. 3.5.6. With the aid of the underlying fuzzy random process it is subsequently possible to make forecasts in accordance with Sect. 4.

Alternatively, it is possible to forecast time series comprised of fuzzy data by means of artificial neural networks for fuzzy variables. In contrast to the analytical regression methods mentioned in Sect. 3.5, the use of artificial neural networks for fuzzy variables for analyzing and forecasting fuzzy time series does not require the specification of a functional type. Artificial neural networks are not only capable of learning the characteristics of a given fuzzy time series but are also able to simulate nonlinear fuzzy random processes and derive forecasts. A basic precondition for the latter is an appropriate network architecture. Artificial neural networks for analyzing and forecasting time series comprised of fuzzy data are developed in the following section.

3.6.1 The Basics of Artificial Neural Networks

The creation of artificial neural networks is an attempt to mathematically model the performance capability of the human brain. Human beings possess the ability to learn and apply what they have learnt. Moreover, they are able to find solutions to new problems intuitively. Intuition is interpreted as the ability to make decisions based on the 'inner' logic of prevailing circumstances without the need to explicitly understand the underlying relationships. Intuition is thus obviously closely linked to previous learning experiences.

The functionality of the human brain is essentially based on the interaction between the brain's highly cross-linked nerve cells also called natural neurons. Communication within a natural neural network of this type takes place via signals. A neuron serves for receiving, processing and passing on incoming signals. The signals passed on from neighboring nerve cells are summated in the neuron. If the stimulation exceeds a particular threshold value, a further signal is activated and transmitted to the adjoining neurons. The information to be transferred is coded as the sum of the signals and their repetition rate.

This basic structure is imitated by artificial neural networks. These are used to realize complex mappings of input variables on output variables. Artificial neural networks consist of cross-linked computational nodes or artificial neurons. Communication hereby takes place via numerical values. These are real numbers in the classical sense. An extension of the latter to fuzzy variables is presented in this section. An artificial neuron receives numerical values from neighboring neurons (input signals), which are combined to form a weighted sum. The determined sum is compared with a threshold value (*bias*) and used as the argument of a so-called activation function. The activation function yields a value (output signal) which is an input signal for connected artificial neurons.

A large number of different types of artificial neural networks exist for widely varying fields of application. Besides their use for investigating natural neural networks [29, 70], artificial neural networks are mainly applied for cognitive purposes, e.g. in medical diagnostics [1, 30, 49], for optimizing production processes [31] or for analyzing economic time series [9, 34], particularly those of the stock market [54, 55]. A coarse subdivision of artificial neural networks into methods for approximating functions, for classification purposes and as associative memory units has been undertaken by [55]. A special type of artificial neural network is the multilayer perceptron. Multilayer perceptrons are universally applicable in all of the areas mentioned. These are adapted in the following for forecasting time series comprised of fuzzy data.

3.6.2 Multilayer Perceptron for Fuzzy Variables

A multilayer perceptron is a special type of artificial neural network. The artificial neurons are hereby arranged in layers. Starting from an input layer, numerical values are transferred to an output layer via one or more hidden layers. The output layer provides the input data with corresponding result data. Both the input and output data are usually real numbers. The mapping of fuzzy variables with the aid of a multilayer perceptron or an artificial neural network is only described to a limited extent in the literature [13, 19]. By applying the extension principle and restricting the analysis to fuzzy triangular numbers it is not possible to obtain generally applicable solutions. In the following, the multilayer perceptron is modified in such a way as to permit the mapping of arbitrary fuzzy variables. A precondition for the latter is a suitable representation of the fuzzy variables to be mapped by means of the $l_\alpha r_\alpha$-discretization technique introduced in Sect. 2.1.1.

The following explanations are given for the example of a multilayer perceptron with one hidden layer. An extension to several hidden layers is also possible. A so-called two-layered multilayer perceptron for fuzzy variables is shown in Fig. 3.14. Only the output layer and the hidden layers are counted. For a more detailed description of artificial neural networks for real numbers, which forms the basis of the extension developed here for the mapping of fuzzy variables, the reader is referred to [70] and [22].

A fuzzy variable $\tilde{x}_r$ is assigned to each artificial neuron r of the input layer I. The counters $r = 1, 2, ..., n_I$ and n_I hereby denote the number of neurons in the input layer. The task of the input layer is to receive the input data $\tilde{x}_r$

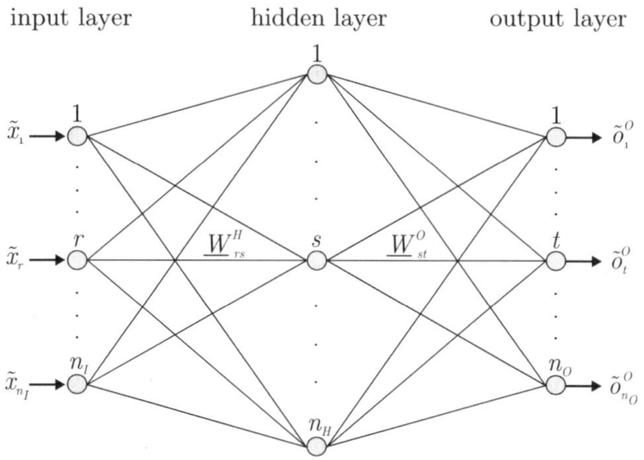

Fig. 3.14. A typical two-layer multilayer perceptron for fuzzy variables

and to pass these data on to the hidden layer as fuzzy output variables $\tilde{o}^I_r$. Each neuron s of the hidden layer H lumps together the fuzzy input variables $\tilde{x}_r$ weighted by the matrix $\underline{W}^H_{rs}$ according to Eq. (3.139). The counters $s = 1, 2, ..., n_H$ and n_H thereby denote the number of neurons in the hidden layer. The dimension $[2n, 2n]$ of the weighting matrices $\underline{W}^H_{rs}$ is dependent on the chosen $l_\alpha r_\alpha$-discretization. The intermediate result obtained is referred to as the fuzzy net input $\tilde{net}^H_s$ of the neuron s, and is not subject to the non-negativity requirement given by Eq. (2.47).

$$\tilde{net}^H_s = \bigoplus_{r=1}^{n_I} \underline{W}^H_{rs} \odot \tilde{x}_r = \bigoplus_{r=1}^{n_I} \underline{W}^H_{rs} \odot \tilde{o}^I_r \tag{3.139}$$

The task of the artificial neurons s in the hidden layer is to map the fuzzy net inputs $\tilde{net}^H_s$ onto fuzzy output variables $\tilde{o}^H_s$ and transfer these to the next layer. This is performed with the aid of the fuzzy activation function $\tilde{f}_A(\cdot)$. The fuzzy activation function $\tilde{f}_A(\cdot)$ after [36] is introduced to denote the mapping of the fuzzy variables $\tilde{net}^H_s$ onto the fuzzy variables $\tilde{o}^H_s$ according to Eq. (3.140).

$$\tilde{o}^H_s = \tilde{f}_A(\tilde{net}^H_s) \tag{3.140}$$

The following then holds for the $l_\alpha r_\alpha$-increments (see Eq. (2.32)) of the two fuzzy variables $\tilde{net}^H_s$ and $\tilde{o}^H_s$:

$$\Delta o^H_j(s) = f_A\left(\Delta net^H_j(s)\right) \quad \text{for} \quad j = 1, 2, ..., 2n\,. \tag{3.141}$$

The real function $f_A(\cdot)$ is a trajectory of the fuzzy function $\tilde{f}_A(\cdot)$ and is equivalent to a deterministic activation function according to classical artificial neural networks. The sigmoidal activation function according to Eq. (3.142) is used in the following.

$$f_A(x) = \frac{1}{1 + e^{-x}} \tag{3.142}$$

With the aid of the weighting matrices $\underline{W}_{st}^O$ the fuzzy output variables $\tilde{o}_s^H$ of the hidden layer H are lumped together in each case by the artificial neurons t of the output layer O to yield fuzzy net input variables $\tilde{net}_t^O$ according to Eq. (3.143). The counters $t = 1, 2, ..., n_O$ and n_O hereby denote the number of neurons in the output layer.

$$\tilde{net}_t^O = \bigoplus_{s=1}^{n_H} \underline{W}_{st}^O \odot \tilde{o}_s^H \tag{3.143}$$

Subsequent mapping of the fuzzy net input variables $\tilde{net}_t^O$ onto the fuzzy output variables $\tilde{o}_t^O$ according to Eq. (3.144) yields the result data of the multilayer perceptron. In contrast to Eq. (3.140), the fuzzy output function $\tilde{f}_O(\cdot)$ is hereby necessary in order to fulfill the non-negativity requirement according to Eq. (2.47) on the one hand, and on the other hand, not to restrict the values of the $l_\alpha r_\alpha$-increments of the fuzzy output variables $\tilde{o}_t^O$ to the interval [0,1].

$$\tilde{o}_t^O = \tilde{f}_O\left(\tilde{f}_A(\tilde{net}_t^O)\right) \tag{3.144}$$

Analogous to Eq. (3.141), the following holds for the $l_\alpha r_\alpha$-increments:

$$\Delta o_j^O(t) = f_O\left(f_A\left(\Delta net_j^O(t)\right)\right) \quad \text{for} \quad j = 1, 2, ..., 2n\,. \tag{3.145}$$

A necessary condition for the deterministic output function $f_O(\cdot)$ is that the functional values according to Eq. (3.146) are restricted. If transformed fuzzy variables are used as input data for the multilayer perceptron (see Sect. 3.6.5), the lower bound must be matched accordingly.

$$f_O(x) \geqslant 0 \quad \forall\ x \in [0, 1] \tag{3.146}$$

The inverse logarithmic normal distribution or the inverse exponential distribution may be chosen to represent the output function $f_O(\cdot)$. In particular cases, non-zero upper constraints and lower bounds may also exist for the $l_\alpha r_\alpha$-increments of the fuzzy result variables. In such cases an inverse sine function according to Eq. (3.147) is recommended to represent the output function $f_O(\cdot)$.

$$f_O(x) = \left(\arcsin(x) + \frac{\pi}{2}\right) \frac{x_2 - x_1}{\pi} \tag{3.147}$$

with x_2 ... upper bound

and x_1 ... lower bound

Further considerations regarding the fuzzy output function $\tilde{f}_O(\cdot)$ are dealt with in Sect. 3.6.5 in connection with the conditioning of fuzzy data.

In the case of classical artificial neural networks threshold values are usually specified for the individual neurons. These define the threshold above which the particular neuron becomes (highly) active. Applying monotonically increasing activation functions, this is the point of maximum ascent. In the case of artificial neural networks for fuzzy variables, fuzzy threshold values (fuzzy bias) are prespecified. The fuzzy threshold values $\tilde{\theta}_s^H$ and $\tilde{\theta}_t^O$ for the neurons $s = 1, 2, ..., n_H$ and $t = 1, 2, ..., n_O$, respectively, may be accounted for directly either in the fuzzy activation functions according to Eq. (3.148).

$$\tilde{o}_s^H = \tilde{f}_A(\tilde{net}_s^H \ominus \tilde{\theta}_s^H) \quad \text{and} \quad \tilde{o}_t^O = \tilde{f}_O\left(\tilde{f}_A(\tilde{net}_t^O \ominus \tilde{\theta}_t^O)\right) \tag{3.148}$$

or by an additional neuron in the hidden or input layer. The latter variant is especially suitable if the backpropagation algorithm is used (see Sect. 3.6.3). Eqs. (3.139) and (3.143) then reduce to Eqs. (3.149) and (3.150).

$$\tilde{net}_s^H = \bigoplus_{r=1}^{n_I+1} \underline{W}_{rs}^H \odot \tilde{o}_r^I \tag{3.149}$$

$$\tilde{net}_t^O = \bigoplus_{s=1}^{n_H+1} \underline{W}_{st}^O \odot \tilde{o}_s^H \tag{3.150}$$

The additional weighting matrices $\underline{W}_{rs}^H$ ($r = n_I + 1$ and $s = 1, 2, ..., n_H$) and $\underline{W}_{st}^O$ ($s = n_H + 1$ and $t = 1, 2, ..., n_O$) are diagonal matrices. The fuzzy output variables $\tilde{o}_r^I$ and $\tilde{o}_s^H$ of the additional neurons $r = n_I + 1$ and $s = n_H + 1$, respectively, are thereby constant in accordance with Eq. (3.151).

$$\tilde{o}_r^I = \text{constant} \quad \text{and} \quad \tilde{o}_s^H = \text{constant} \tag{3.151}$$

The following holds for the $l_\alpha r_\alpha$-increments:

$$\Delta o_j^I(r) = 1 \quad \text{and} \quad \Delta o_j^H(s) = 1 \quad \text{for} \quad j = 1, 2, ..., 2n\,. \tag{3.152}$$

The fuzzy threshold values are then given by Eqs. (3.153) and (3.154), and are accounted for in the fuzzy net input variables $\tilde{net}_s^H$ and $\tilde{net}_t^O$ of the hidden and output layers according to Eqs. (3.149) and (3.150). This means that adaptation of the fuzzy activation functions given by Eq. (3.148) is not necessary.

$$\tilde{\theta}_s^H = -\underline{W}_{rs}^H \odot \tilde{o}_r^I \quad \text{with} \quad r = n_I + 1 \quad \text{and} \quad s = 1, 2, ..., n_H \tag{3.153}$$

$$\tilde{\theta}_t^O = -\underline{W}_{st}^O \odot \tilde{o}_s^H \quad \text{with} \quad s = n_H + 1 \quad \text{and} \quad t = 1, 2, ..., n_O \tag{3.154}$$

The algorithm formulated for a two-layered multilayer perceptron may be generalized if several hidden layers are present. The fuzzy output variables

$\tilde{o}_s^H$ obtained from Eq. (3.140) are then transferred to the next hidden layer rather than the output layer, and are processed according to Eqs. (3.139) and (3.140).

Processing of the fuzzy variables in an artificial neuron of the hidden layer(s) is shown schematically in Fig. 3.15.

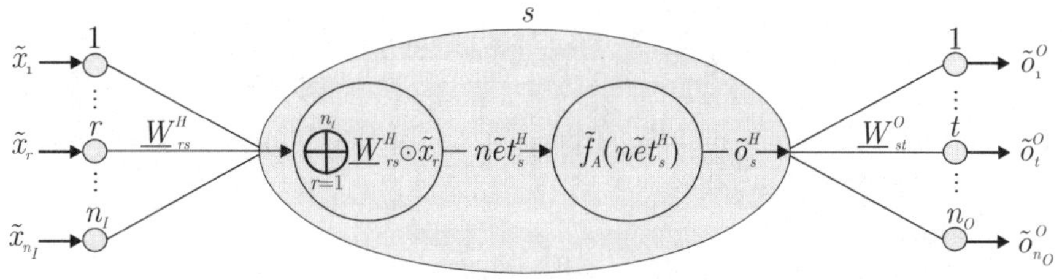

Fig. 3.15. Processing of fuzzy variables in an artificial neuron

3.6.3 Backpropagation Algorithm

Using predefined weighting matrices $\underline{W}_{rs}^H$ and $\underline{W}_{st}^O$ with $r = 1, 2, ..., n_I$ and $s = 1, 2, ..., n_H$, the multilayer perceptron for fuzzy variables presented in Sect. 3.6.2 maps a given sequence of fuzzy input variables $\tilde{x}_1$, $\tilde{x}_2$, ..., $\tilde{x}_{n_I}$ on sequence of fuzzy output variables $\tilde{o}_1^O$, $\tilde{o}_2^O$, ..., $\tilde{o}_{n_O}^O$ to which the same fuzzy values are always assigned.

Several sequences of fuzzy input variables $\tilde{x}_1$, $\tilde{x}_2$, ..., $\tilde{x}_{n_I}$ and several sequences of fuzzy control variables $\tilde{y}_1$, $\tilde{y}_2$, ..., $\tilde{y}_{n_O}$ may be extracted from a given fuzzy time series. With the aid of the latter it is possible to train the weighting matrices. Fuzzy training patterns and fuzzy training sets are constructed for this purpose.

Definition 3.40. *A fuzzy training pattern consists of a fuzzy input vector* $\underline{\tilde{x}} = (\tilde{x}_1, ..., \tilde{x}_r, ..., \tilde{x}_{n_I})^T$ *and a fuzzy control vector* $\underline{\tilde{y}} = (\tilde{y}_1, ..., \tilde{y}_t, ..., \tilde{y}_{n_O})^T$. ◆

If several fuzzy training patterns are available, these form a fuzzy training set.

Definition 3.41. *A fuzzy training set consists of m fuzzy training patterns. This set is defined by the vectors* $\underline{\tilde{x}}_k = (\tilde{x}_1(k), ..., \tilde{x}_r(k), ..., \tilde{x}_{n_I}(k))^T$ *and* $\underline{\tilde{y}}_k = (\tilde{y}_1(k), ..., \tilde{y}_t(k), ..., \tilde{y}_{n_O}(k))^T$ *with* $k = 1, 2, ..., m$. ◆

For each fuzzy input vector $\underline{\tilde{x}}_k = (\tilde{x}_1(k), ..., \tilde{x}_r(k), ..., \tilde{x}_{n_I}(k))^T$ of the fuzzy training set the multilayer perceptron yields a fuzzy output vector $\underline{\tilde{o}}_k^O = (\tilde{o}_1^O(k), ..., \tilde{o}_t^O(k), ..., \tilde{o}_{n_O}^O(k))^T$. The output error of the multilayer perceptron is determined by comparing the fuzzy output vector $\underline{\tilde{o}}_k^O$ with the known fuzzy control vector $\underline{\tilde{y}}_k$ with the aid of of the square error E_k given by Eq. (3.155).

$$E_k = f_E(\underline{\tilde{y}}_k, \underline{\tilde{o}}_k^O) \tag{3.155}$$

Using the $l_\alpha r_\alpha$-increments yields the following:

$$E_k = \frac{1}{2} \sum_{t=1}^{n_O} \sum_{j=1}^{2n} \left(\Delta y_j(t,k) - \Delta o_j^O(t,k) \right)^2 . \tag{3.156}$$

The output error of the multilayer perceptron for the complete fuzzy training set is defined as the mean square error MSE according to Eq. (3.157).

$$MSE = \frac{1}{m} \sum_{k=1}^{m} f_E(\underline{\tilde{y}}_k, \underline{\tilde{o}}_k^O) \tag{3.157}$$

With the aid of the backpropagation algorithm the elements of the weighting matrices are determined in such a way as to minimize their mean square error MSE. Optimization of the weighting matrices is also referred to as training or learning of the multilayer perceptron. The backpropagation algorithm for fuzzy variables is described in the following.

Firstly, the weighting matrices of the multilayer perceptron are initialized. This is accomplished by randomly assigning real values from the interval $[-1, 1]$ to the matrix elements. In the second step a given fuzzy input vector $\underline{\tilde{x}}_k$ is transferred to the input layer of the multilayer perceptron, and the corresponding fuzzy output vector $\underline{\tilde{o}}_k^O$ is computed. The computed fuzzy output vector $\underline{\tilde{o}}_k^O$ is compared with the corresponding given fuzzy control vector $\underline{\tilde{y}}_k$ and the square error is determined according to Eq. (3.155). In the next step the correction matrices ${}^{\Delta}\underline{W}_{st}^O(k)$ and ${}^{\Delta}\underline{W}_{rs}^H(k)$ are determined according to Eqs. (3.158) and (3.159). The correction matrices are defined in each case as being proportional to the partial derivatives of the errors with respect to the weighting matrices. As in the case of $\underline{W}_{st}^O$ and $\underline{W}_{rs}^H$, these matrices are also of dimension $[2n, 2n]$. Accordingly, the backpropagation algorithm for fuzzy variables is equivalent to a gradient descent method. The factor η, with $\eta > 0$, is referred to as the learning rate.

$${}^{\Delta}\underline{W}_{st}^O(k) = -\eta \frac{\partial E_k}{\partial \underline{W}_{st}^O} = -\eta \begin{bmatrix} \frac{\partial E_k}{\partial w_{1,1}^O(st)} & \cdots & \frac{\partial E_k}{\partial w_{1,2n}^O(st)} \\ \vdots & \ddots & \vdots \\ \frac{\partial E_k}{\partial w_{2n,1}^O(st)} & \cdots & \frac{\partial E_k}{\partial w_{2n,2n}^O(st)} \end{bmatrix} \tag{3.158}$$

$${}^{\Delta}\underline{W}_{rs}^H(k) = -\eta \frac{\partial E_k}{\partial \underline{W}_{rs}^H} = -\eta \begin{bmatrix} \frac{\partial E_k}{\partial w_{1,1}^H(rs)} & \cdots & \frac{\partial E_k}{\partial w_{1,2n}^H(rs)} \\ \vdots & \ddots & \vdots \\ \frac{\partial E_k}{\partial w_{2n,1}^H(rs)} & \cdots & \frac{\partial E_k}{\partial w_{2n,2n}^H(rs)} \end{bmatrix} \tag{3.159}$$

The weighting matrices may be corrected at different points in the backpropagation algorithm. In *online* training the weighting matrices $\underline{W}^O_{st}$ and $\underline{W}^H_{rs}$ are modified according to Eq. (3.160) immediately after the processing of a fuzzy training model. Thus corresponds to a descent in the gradient direction of the error function given by Eq. (3.155).

$$\underline{W}^O_{st}(\text{new}) = \underline{W}^O_{st}(\text{old}) + {}^{\Delta}\underline{W}^O_{st}(k) \tag{3.160}$$

$$\text{and} \quad \underline{W}^H_{rs}(\text{new}) = \underline{W}^H_{rs}(\text{old}) + {}^{\Delta}\underline{W}^H_{rs}(k)$$

In *offline* training the weighting matrices $\underline{W}^O_{st}$ and $\underline{W}^H_{rs}$ are modified according to Eq. (3.161) after first taking account of all given m fuzzy training patterns.

$$\underline{W}^O_{st}(\text{new}) = \underline{W}^O_{st}(\text{old}) + \frac{1}{m}\sum_{k=1}^{m} {}^{\Delta}\underline{W}^O_{st}(k) \tag{3.161}$$

$$\text{and} \quad \underline{W}^H_{rs}(\text{new}) = \underline{W}^H_{rs}(\text{old}) + \frac{1}{m}\sum_{k=1}^{m} {}^{\Delta}\underline{W}^H_{rs}(k)$$

The partial derivatives in Eqs. (3.158) and (3.159) are determined on the level of the $l_\alpha r_\alpha$-increments in order that the classical backpropagation algorithm (see e.g. [22]) may be extended. A distinction must thereby be made between the output layer and the hidden layers.

In order to compute the partial derivatives $\frac{\partial E_k}{\partial w^O_{j,i}(st)}$ of the errors E_k with respect to the elements of the output layer weighting matrices the chain rule according to Eq. (3.162) must be applied, whereby $i, j = 1, 2, ..., 2n$ holds. In order to improve transparency the index k is dispensed with in the following.

$$\frac{\partial E}{\partial w^O_{j,i}(st)} = \frac{\partial E}{\partial \Delta net^O_j(t)} \, \frac{\partial \Delta net^O_j(t)}{\partial w^O_{j,i}(st)} \tag{3.162}$$

Using Eq. (3.143), the term $\frac{\partial \Delta net^O_j(t)}{\partial w^O_{j,i}(st)}$ according to Eq. (3.163) may be simplified. In order to improve transparency the symbols s' and i' are used as the incrementation variables of the summation operators.

$$\begin{aligned}\frac{\partial \Delta net^O_j(t)}{\partial w^O_{j,i}(st)} &= \frac{\partial}{\partial w^O_{j,i}(st)} \sum_{s'=1}^{n_H}\sum_{i'=1}^{2n} w^O_{j,i'}(s't)\, \Delta o^H_{i'}(s') \\ &= \frac{\partial}{\partial w^O_{j,i}(st)} w^O_{j,i}(st)\, \Delta o^H_i(s) \\ &= \Delta o^H_i(s)\end{aligned} \tag{3.163}$$

The term $\frac{\partial E}{\partial \Delta net^O_j(t)}$ is again expanded with the aid of the chain rule according to Eq. (3.164).

$$\frac{\partial E}{\partial \Delta net_j^O(t)} = \frac{\partial E}{\partial \Delta o_j^O(t)} \frac{\partial \Delta o_j^O(t)}{\partial \Delta net_j^O(t)} \tag{3.164}$$

By inserting Eqs. (3.155) and (3.144) the two partial derivatives in Eq. (3.164) are evaluated.

$$\begin{aligned} \frac{\partial E}{\partial \Delta o_j^O(t)} &= \frac{\partial}{\partial \Delta o_j^O(t)} \frac{1}{2} \sum_{t'=1}^{n_O} \sum_{j'=1}^{2n} \left(\Delta y_{j'}(t') - \Delta o_{j'}^O(t')\right)^2 \\ &= \frac{\partial}{\partial \Delta o_i^O(t)} \frac{1}{2} \left(\Delta y_j(t) - \Delta o_j^O(t)\right)^2 \\ &= -\left(\Delta y_j(t) - \Delta o_j^O(t)\right) \end{aligned} \tag{3.165}$$

$$\begin{aligned} \frac{\partial \Delta o_j^O(t)}{\partial \Delta net_j^O(t)} &= \frac{\partial}{\partial \Delta net_j^O(t)} f_O\left(f_A\left(\Delta net_j^O(t)\right)\right) \\ &= f_O'\left(f_A\left(\Delta net_j^O(t)\right)\right) f_A'\left(\Delta net_j^O(t)\right) \end{aligned} \tag{3.166}$$

If the sigmoidal function given by Eq. (3.142) is chosen for the activation function of the output layer, Eq. (3.166) reduces to Eq. (3.167).

$$\frac{\partial \Delta o_j^O(t)}{\partial \Delta net_j^O(t)} = f_O'\left(f_A(\Delta net_j^O(t))\right) f_A(\Delta net_j^O(t)) \left(1 - f_A(\Delta net_j^O(t))\right) \tag{3.167}$$

In summarizing, the partial derivatives of Eq. (3.158) may be reproduced by Eq. (3.168).

$$\begin{aligned} \frac{\partial E}{\partial w_{j,i}^O(st)} = &-\Delta o_i^H(s) \left(\Delta y_j(t) - \Delta o_j^O(t)\right) f_O'\left(f_A\left(\Delta net_j^O(t)\right)\right) ... \\ &... f_A\left(\Delta net_j^O(t)\right) \left(1 - f_A\left(\Delta net_j^O(t)\right)\right) \end{aligned} \tag{3.168}$$

According to the usual notation adopted in the literature, the abbreviation $\delta_j^O(t)$ is used for the term $\frac{\partial E}{\partial \Delta net_j^O(t)}$ in Eq. (3.162).

$$\delta_j^O(t) = -\frac{\partial E}{\partial \Delta net_j^O(t)} \tag{3.169}$$

The correction matrices ${}^{\Delta}\underline{W}_{st}^O$ may then be computed by means of Eq. (3.170).

$${}^{\Delta}\underline{W}_{st}^O = \eta \begin{bmatrix} \delta_1^O(t)\Delta o_1^H(s) & \cdots & \delta_1^O(t)\Delta o_{2n}^H(s) \\ \vdots & \ddots & \vdots \\ \delta_{2n}^O(t)\Delta o_1^H(s) & \cdots & \delta_{2n}^O(t)\Delta o_{2n}^H(s) \end{bmatrix} \tag{3.170}$$

In order to compute the partial derivatives $\frac{\partial E_k}{\partial w_{j,i}^H(rs)}$ of the errors E_k with respect to the elements of the weighting matrices of the hidden layer the chain rule according to Eq. (3.171) must also be applied. The index k is again dispensed with, whereby $i, j = 1, 2, ..., 2n$ holds.

$$\frac{\partial E}{\partial w_{j,i}^H(rs)} = \frac{\partial E}{\partial \Delta net_j^H(s)} \frac{\partial \Delta net_j^H(s)}{\partial w_{j,i}^H(rs)} \tag{3.171}$$

Analogous to the procedure adopted for the output layer, but using Eq. (3.139), Eq. (3.171) simplifies as follows:

$$\frac{\partial E}{\partial w_{j,i}^H(rs)} = \frac{\partial E}{\partial \Delta net_j^H(s)} \Delta o_i^I(r)\,. \tag{3.172}$$

With the aid of the chain rule, the term $\frac{\partial E}{\partial \Delta net_j^H(s)}$ is described by Eq. (3.173).

$$\frac{\partial E}{\partial \Delta net_j^H(s)} = \frac{\partial E}{\partial \Delta o_j^H(s)} \frac{\partial \Delta o_j^H(s)}{\partial \Delta net_j^H(s)} \tag{3.173}$$

The two partial derivatives of Eq. (3.173) are obtained by repeated application of the chain rule, and by substitution of Eqs. (3.140), (3.143) and (3.169).

$$\begin{aligned} \frac{\partial E}{\partial \Delta o_j^H(s)} &= \sum_{t'=1}^{n_O} \sum_{i'=1}^{2n} \frac{\partial E}{\partial \Delta net_{i'}^O(t')} \frac{\partial \Delta net_{i'}^O(t')}{\partial \Delta o_j^H(s)} \\ &= \sum_{t'=1}^{n_O} \sum_{i'=1}^{2n} \underbrace{\frac{\partial E}{\partial \Delta net_{i'}^O(t')}}_{-\delta_{i'}^O(t')} \frac{\partial \Delta}{\partial \Delta o_j^H(s)} \sum_{s'=1}^{n_H} \sum_{j'=1}^{2n} w_{i',j'}^O(s't') \Delta o_{j'}^H(s') \\ &= -\sum_{t'=1}^{n_O} \sum_{i'=1}^{2n} \delta_{i'}^O(t')\, w_{i',j}^O(st') \end{aligned} \tag{3.174}$$

$$\frac{\partial \Delta o_j^H(s)}{\partial \Delta net_j^H(s)} = f_A'\left(\Delta net_j^H(s)\right) \tag{3.175}$$

If the sigmoidal function according to Eq. (3.142) is chosen to represent the activation function of the hidden layer neurons, Eq. (3.175) may be replaced by Eq. (3.176).

$$\frac{\partial \Delta o_j^H(s)}{\partial \Delta net_j^H(s)} = f_A'\left(\Delta net_j^H(s)\right) = \Delta o_j^H(s)\left(1 - \Delta o_j^H(s)\right) \tag{3.176}$$

By means of these simplifications the partial derivatives of Eq. (3.172) may be lumped together in Eq. (3.177).

$$\frac{\partial E}{\partial w_{j,i}^H(rs)} = -\Delta o_i^I(r)\, \Delta o_j^H(s)\left(1 - \Delta o_j^H(s)\right) \sum_{t'=1}^{n_O} \sum_{i'=1}^{2n} \delta_{i'}^O(t')\, w_{i',j}^O(st') \tag{3.177}$$

Analogous to the output layer, the abbreviated notation $\delta_j^H(s)$ is used for the term $\frac{\partial E}{\partial \Delta net_j^H(s)}$ in Eq. (3.172), thereby resulting in Eq. (3.178).

$$\delta_j^H(s) = -\frac{\partial E}{\partial \Delta net_j^H(s)} = f'_A\left(\Delta net_j^H(s)\right) \sum_{t'=1}^{n_O} \sum_{i'=1}^{2n} \delta_{i'}^O(t')\, w_{i',j}^O(st') \tag{3.178}$$

The correction matrices ${}^{\Delta}\underline{W}_{rs}^H$ for the hidden layer are thus given by Eq. (3.179).

$${}^{\Delta}\underline{W}_{rs}^H = \eta \begin{bmatrix} \delta_1^H(s)\Delta o_1^I(r) & \cdots & \delta_1^H(s)\Delta o_{2n}^I(r) \\ \vdots & \ddots & \vdots \\ \delta_{2n}^H(s)\Delta o_1^I(r) & \cdots & \delta_{2n}^H(s)\Delta o_{2n}^I(r) \end{bmatrix} \tag{3.179}$$

The $\delta_{i'}^O(t')$ terms of the output layer are necessary in order to determine the $\delta_j^H(s)$ terms according to Eq. (3.178). For this reason the determination of the correction matrices (and modification of the weighting matrices) always begins with the output layer and proceeds in the direction of the input layer (backpropagation). In the case of a multilayer perceptron for fuzzy variables with several hidden layers the corresponding correction matrices are determined analogous to Eqs. (3.170) and (3.179). The terms $\delta_{i'}^O(t')$ and $w_{i',j}^O(st')$ are then replaced by the variables of the corresponding hidden layer. The same holds for the fuzzy output variables $\tilde{o}_r^I$.

Because the backpropagation algorithm for fuzzy variables is de facto equivalent to a gradient descent method, the problems which arise in gradient descent methods must be avoided by adopting suitable strategies. An overview of the most frequent problems encountered when applying the classical backpropagation algorithm for real-valued data is given in [70]. The greatest danger in gradient descent methods is that it is not possible to depart from a local minimum. On the other hand it possible to depart from a detected (global) optimum in favor of a suboptimum minimum. Moreover, flat plateaus or steep ravines in the error function given by Eq. (3.155) may lead to stagnation or oscillation of the learning process. When applying the backpropagation algorithm for fuzzy variables the damping or elimination of these problems may be achieved by a simple modification of the method. An advantageous possibility is the use of a momentum term γ after [58]. The correction matrices ${}^{\Delta}\underline{W}_i$ given by Eqs. (3.158) and (3.159) in learning step i are thereby supplemented by the corresponding correction matrices ${}^{\Delta}\underline{W}_{(i-1)}$ of the $(i-1)$-th learning step according to Eq. (3.180).

$${}^{\Delta}\underline{W}_i = -\eta \frac{\partial E}{\partial \underline{W}} + \gamma\, {}^{\Delta}\underline{W}_{(i-1)} \tag{3.180}$$

Introduction of the momentum γ counteracts stagnations on flat plateaus as well as oscillations between steeply descending regions of the error function. Values between 0 and 1 are recommended for the momentum γ. A random assignment of values to the learning rate η and the momentum γ is advisable in each learning step i. By this means, however, it is not possible to prevent the departure from a detected (global) optimum. In order to avoid oscillations in steep ravines the learning rate η may be linearly coupled to the gradients. Large values of the gradients then result in a lower learning learning rate. The disadvantage of such a coupling is the possible departure from an optimum due to a high learning rate η combined with small gradients. According to [70], it is not possible offer practical tips regarding the choice of the learning rate η. In the case of large values of η no narrow valleys are detected, and departure from the global optimum as well as oscillations may occur. On the other hand, small values of η may lead to stagnations and the risk of not being able to depart from a local minimum. The choice of η is thus highly dependent on the given training data and the architecture of the artificial neural network. In order to overcome this seeming dilemma the following approach is suggested.

A randomly varying assignment of values to the learning rate η as well as the momentum γ combined with an evolutive adaptation of the learning process depending on the value of the error function is chosen. At the beginning in learning step $i = 1$ random values are assigned to η and γ. If a reduction in the error value according to Eq. (3.155) is achieved in the subsequent learning step $i+1$ using the correction matrices ${}^{\Delta}\underline{W}_i$, random values are again assigned to the learning rate η as well as the momentum γ, and the correction matrices ${}^{\Delta}\underline{W}_{(i+1)}$ are determined. If a reduction in the error value is not achieved, the correction matrices ${}^{\Delta}\underline{W}_i$ are recomputed with new random values of η and γ in conjunction with a new calculation of the error function. This approach firstly ensures that departure from a detected (global) optimum in favor of a suboptimum local minimum is avoided. Secondly, the random assignment of values to the learning rate η and the momentum γ ensures the departure from local minima in favor of a better (local/global) solution. This approach is far superior to the classical gradient descent method in which constant values of η and γ are used.

When applying sigmoidal activation functions $f_A(\cdot)$ according to Eq. (3.142), fuzzy input data $\tilde{x}_r$ with $l_\alpha r_\alpha$-increments $\Delta x_i(r) \gg 1$ may lead to slower convergence of the backpropagation algorithm. As, according to Eq. (3.139), the $l_\alpha r_\alpha$-increments $\Delta net_j^H(s)$ of the fuzzy net input variables $\tilde{net}_s^H$ are computed by summation of the $\Delta x_i(r)$, the condition $\Delta net_j^H(s) \gg 1$ may also hold for these $l_\alpha r_\alpha$-increments. The expression $f_A\left(\Delta net_j^H(s)\right)$ in Eq. (3.141) then yields values lying in a region where the ascent of the sigmoidal activation function $f_A(\cdot)$ is very slight, thereby resulting in slow convergence. In order to avoid this effect, conditioning of the fuzzy input variables is recommended. Conditioning of the fuzzy input variables is described in detail in Sect. 3.6.5.

3.6.4 Neural Network Architecture for Fuzzy Time Series

The architecture of artificial neural networks for fuzzy time series is essentially characterized by the number of layers, the number of neurons and the fuzzy activation function. The network architecture depends decisively on the chosen forecasting objective and may be specified with the aid of a training strategy. Artificial neural networks may be applied to stationary and non-stationary fuzzy time series without a priori distinction.

In the following a distinction is made between optimum forecasting network architecture and optimum simulation network architecture.

Optimum forecasting network architecture permits the determination of an optimum forecast. The optimum forecast for time points $\tau = N + h$ represents the fuzzy average of all potential future realizations of the fuzzy random variable at the same time point $\tau = N + h$ (see Def. 4.2). A network architecture is thus sought which contains sufficient neurons to account for all systematic effects of the fuzzy time series and permits the computation of an estimator for the fuzzy expected value function.

The optimum simulation network architecture serves for simulating realizations of the fuzzy random variables to be expected at future time points $\tau = N + h$ (see Def. 4.6). By means of this network architecture it is possible to take account of random fluctuations of the fuzzy time series.

Every artificial neural network consists of an input layer, one or more hidden layers and an output layer. The fuzzy training pattern (see Def. 3.40) selected for the particular fuzzy time series concerned also defines the number of neurons in the input and output layers. For practical reasons a fuzzy training pattern for fuzzy time series consists of a sequence of the fuzzy time series of n_I fuzzy variables (lumped together in the fuzzy input vector $\underline{\tilde{x}}$ of length n_I) and the fuzzy variable following the sequence, which serves as a fuzzy control vector $\underline{\tilde{y}}$ of unit length, i.e. $n_O = 1$. By this means one neuron is assigned to the output layer. A fuzzy input vector $\underline{\tilde{x}}$ containing n_I fuzzy elements requires n_I neurons in the input layer. The number of required hidden layers and neurons per layer is determined by means of a training strategy.

Training Strategy for an Optimum Forecasting Network Architecture

Structuring of the fuzzy time series. The given fuzzy time series containing N fuzzy variables is subdivided into a fuzzy training series and a fuzzy validation series (Fig. 3.16). In order that all systematic effects of the given fuzzy time series are accounted for, the fuzzy training series chosen for an optimum forecasting network architecture should not be too short. If an optimum h-step forecast (see Sect. 4.1) is planned, the validation series should contain at least h fuzzy variables.

After specifying the fuzzy training series, m fuzzy training patterns with n_I fuzzy variables $\tilde{x}_1(k)$, $\tilde{x}_2(k)$, ..., $\tilde{x}_{n_I}(k)$ in the fuzzy input vector $\underline{\tilde{x}}_k$ and

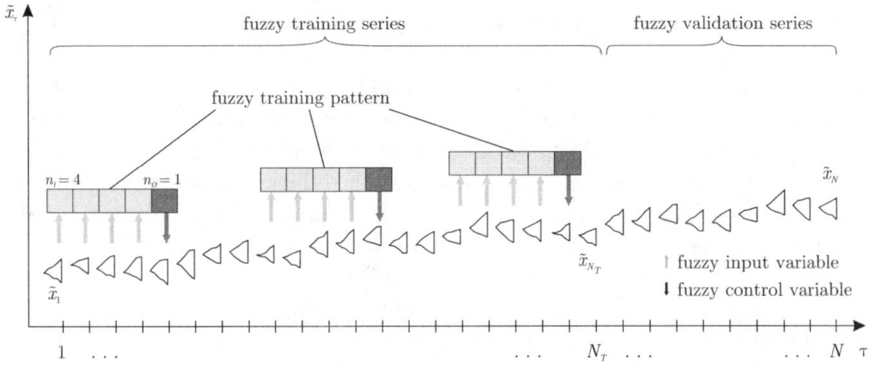

Fig. 3.16. Structuring of a fuzzy time series for determining the optimum forecasting network architecture

(advantageously) one fuzzy variable $\tilde{y}(k)$ in the fuzzy control vector $\underline{\tilde{y}}_k$ may be chosen, whereby $k = 1, 2, ..., m$. The number of fuzzy training patterns with $n_O = 1$ fuzzy control variables that may be extracted from a fuzzy training series containing N_T fuzzy variables is

$$m = N_T - n_I \,. \tag{3.181}$$

These constitute a fuzzy training set according to Def. 3.41. Selected fuzzy training patterns with $n_I = 4$ and $n_O = 1$ are shown in Fig. 3.16.

Choice of the network architecture. It is now possible to specify a starting variant of the network architecture. The number of neurons n_I in the input layer and n_O in the output layer are already defined. It is now necessary to select the number of hidden layers and the number of neurons per layer. As a tendency, more hidden layer neurons should be chosen with increasing complexity and nonlinearity of the fuzzy time series to be analyzed because more neurons imply more weighting matrices. The problem of so-called *overfitting*, which is often encountered in classical fields of application of artificial neural networks, may also arise in the analysis and forecasting of fuzzy time series.

Whereas overfitting must be avoided in optimum forecasting network architectures, it is purposely exploited in optimum simulation network architectures. In an optimum forecasting network architecture the number of chosen layers and neurons should guarantee a 'smooth' approximation, i.e. an approximation in which random fluctuations of the fuzzy time series are absent.

Training of the artificial neural network. After selecting a particular network architecture, the artificial neural network must be trained. The mean square error MSE according to Eq. (3.157) for the chosen number of hidden layers and neurons is minimized by means of the backpropagation algorithm.

Modification of the weighting matrices may be achieved either by online training or by offline training (see Sect. 3.6.3).

In *online* training the m fuzzy training patterns are advantageously selected in a random order. This approach prevents the occurrence of repetitive learning processes which may result in 'going round in circles' in the search for the optimum. Moreover, a random selection reduces the probability of getting trapped in a local minimum. In online training, an improvement is only obtained on average along the gradients of the mean square error MSE to be minimized according to Eq. (3.157). The formulation of a convergence criterion is hence more complicated than in offline training. As an alternative, the training may be terminated after a prescribed number of iterations. An advantage of online training compared with offline training is the lower storage requirement.

In *offline* training the entire fuzzy training set must first be processed before it is possible to correct the weighting matrices. The search for a minimum thus takes place along the gradients of the mean square error MSE according to Eq. (3.157). Compared with online training, this increases the probability that a local minimum is accepted as the solution. The convergence criterion may be formulated as the non-exceedance of a maximum mean square error $MSE_{\max}$. Whether non-exceedance of the prescribed error $MSE_{\max}$ is due to the detection of the global minimum or the attainment of an acceptable local minimum is of secondary importance, especially considering the fact that the global minimum is non-verifiable. Alternatively, the training process may be terminated after a prescribed number of iterations.

The application of online training as well as offline training may result in oscillations between regions marked by a sharp decrease in the error function. Possible strategies for damping or eliminating such problems usually encountered in gradient descent methods have already been outlined in Sect. 3.6.3.

Modification of the network architecture. The trained artificial neural network is now checked and assessed with the aid of the fuzzy validation series. With the aid of this network and the fuzzy variables of the fuzzy training series, an optimum h-step forecast is firstly made for time points $N_T + 1$ to N of the fuzzy validation series (see Sect. 4.3.1). The forecasted fuzzy variables are then compared with the given fuzzy variables of the validation series. On the basis of the differences, an error is computed which permits an assessment of the quality of the forecast. In order to formulate this error the fuzzy variables of the fuzzy training series and those of the fuzzy validation series are denoted by $\tilde{x}_i^T$ $(i = 1, 2, ..., N_T)$ and $\tilde{x}_j^V$ $(j = 1, 2, ..., N_V$ and $N_V = N - N_T)$, respectively. The mean forecast error MPF_{NN} of the artificial neural network is introduced as a measure of the quality of the forecast, and hence also a measure of the quality of the network architecture. The mean forecast error MPF_{NN} is defined as the average value of the distances d_F between the observed fuzzy data $\tilde{x}_j^V$ and the forecasted fuzzy variables $\mathring{\tilde{x}}_j^V$ of the fuzzy validation series according to Eq. (3.182).

$$MPF_{NN} = \frac{\sum_{j=1}^{N_V} d_F(\tilde{x}_j^V ; \mathring{\tilde{x}}_j^V)}{N_V} \tag{3.182}$$

The network architecture with a minimum forecast error MPF_{NN} is referred to as the optimum forecasting network architecture. The mean square error MSE according to Eq. (3.157) must thereby be sufficiently small.

The detection of a network architecture with a minimum forecast error MPF_{NN} is an optimization problem with the objective function given by Eq. (3.182) and the layers and neurons as decision variables.

The network architecture may be optimized, for example, using Monte Carlo methods, network search techniques or the modified evolution strategy after [36]. The genetic methods described in [4, 16, 25] for solving the optimization problem posed by artificial neural networks for real numbers may also be applied to artificial neural networks for fuzzy variables. A simple means of approximately detecting the minimum of MPF_{NN} is by heuristic modification of the network architecture. After modifying the network architecture it is necessary to re-train the artificial neural network. The network is modified repeatedly until such time as the mean forecast error MPF_{NN} takes on a minimum value. The efficiency of the optimization depends on the method used. An acceptable local minimum is attained depending on the case in question.

The forecast computed by the optimized network architecture is chosen as an estimator for the fuzzy expected value function. The optimum forecasting network architecture found in this way is suitable for simulating non-stationary as well as stationary fuzzy time series.

Training Strategy for an Optimum Simulation Network Architecture

Identification of the random components of the given fuzzy time series. The detection of an optimum simulation network architecture presupposes an optimum forecasting network architecture. If the latter has been determined (see *Training strategy for an optimum forecasting network architecture* on p. 118), the optimum forecasting network architecture is used to compute optimum single-step forecasts $\mathring{\tilde{x}}_\tau$ (see Sect. 4.3.1) for the time period $\tau = n_I + 1, ..., N$. Each single-step forecast is thereby based on the measured values of the fuzzy time series. The optimum single-step forecasts represent estimators for the conditional fuzzy expected values at time points $\tau = n_I+1, ..., N$. If the fuzzy variables of the optimum single-step forecasts are subtracted from the fuzzy variables of the given fuzzy time series at each time point, a new stationary fuzzy time series $(\tilde{e}_\tau)_{\tau \in \mathbf{T}}$ is obtained which reproduces the random components of the original time series.

$$\tilde{e}_\tau = \tilde{x}_\tau \ominus \mathring{\tilde{x}}_\tau \tag{3.183}$$

The determined fuzzy variables $\tilde{e}_\tau$ are fuzzy variables in the improper sense because they do not fulfill the requirement according to Eq. (2.47). They are considered to be a realization of the fuzzy random process $(\tilde{\mathsf{E}}_\tau)_{\tau \in \mathbf{T}}$ with the following characteristics:

$$E[\tilde{\mathsf{E}}_\tau] = \tilde{m}_{\tilde{\mathsf{E}}_\tau} = 0 \quad \forall\ \tau \in \mathbf{T} \tag{3.184}$$

$$_{lr}Var[\tilde{\mathsf{E}}_\tau] = {}_{lr}\underline{\sigma}^2_{\tilde{\mathsf{E}}_\tau} \tag{3.185}$$

$$_{lr}K_{\tilde{\mathsf{E}}_\tau}(\Delta\tau) = \begin{cases} _{lr}K_{\tilde{\mathsf{E}}_\tau}(\tau_a, \tau_b) & \text{for} \quad \Delta\tau = 0 \\ \underline{0} & \text{for} \quad \Delta\tau \neq 0 \end{cases} . \tag{3.186}$$

The empirical moments $_{lr}\underline{s}^2_{\tilde{e}_\tau}$ and $_{lr}\underline{\hat{K}}_{\tilde{e}_\tau}(0)$ of the fuzzy time series $(\tilde{e}_\tau)_{\tau \in \mathbf{T}}$ are used as estimators for $_{lr}\underline{\sigma}^2_{\tilde{\mathsf{E}}_\tau}$ and $_{lr}K_{\tilde{\mathsf{E}}_\tau}(0)$. The empirical characteristic values provide a means of determining the optimum simulation network architecture.

Selection of the optimum simulation network architecture for the fuzzy time series $(\tilde{e}_\tau)_{\tau \in \mathbf{T}}$. A suitable starting variant is chosen for the optimum simulation network architecture, which is subsequently improved in an optimization process. The optimum simulation network architecture to be developed for the fuzzy time series $(\tilde{e}_\tau)_{\tau \in \mathbf{T}}$ is independent of the optimum forecasting network architecture. As the simulation of random effects is achieved with the aid of overfitting of artificial neural networks, the number of chosen layers and neurons, and hence the number of weighting matrices, must be large enough to permit a mapping of the fluctuations of the fuzzy time series $(\tilde{e}_\tau)_{\tau \in \mathbf{T}}$.

Example 3.42. This is illustrated by the example of a realization of a fuzzy white-noise process shown in Fig. 3.17. The optimum forecasting network architecture yields the optimum fuzzy forecast at time points $\tau = N+1,\ N+2,\ \ldots$ This is the fuzzy expected value of the fuzzy white-noise process. A continuation of the fuzzy time series by an artificial neural network with *overfitting* yields realizations which exhibit the properties of the fuzzy white-noise process. ◆

Training of the artificial neural network. After selecting a starting variant for the optimum simulation network architecture the artificial neural network is trained with the aid of the backpropagation algorithm. For training purposes the fuzzy training patterns are extracted from the fuzzy time series $(\tilde{e}_\tau)_{\tau \in \mathbf{T}}$. The weighting matrices may be modified either by online training or offline training. The training is terminated when the mean square error MSE according to Eq. (3.157) falls below a predefined maximum error $MSE_{\max}$. Whether non-exceedance of the prescribed error $MSE_{\max}$ is due to the detection of the global minimum or the attainment of an acceptable local minimum is of secondary importance, especially considering the fact that the global minimum is non-verifiable. Alternatively, the training may be terminated after a specified number of iterations.

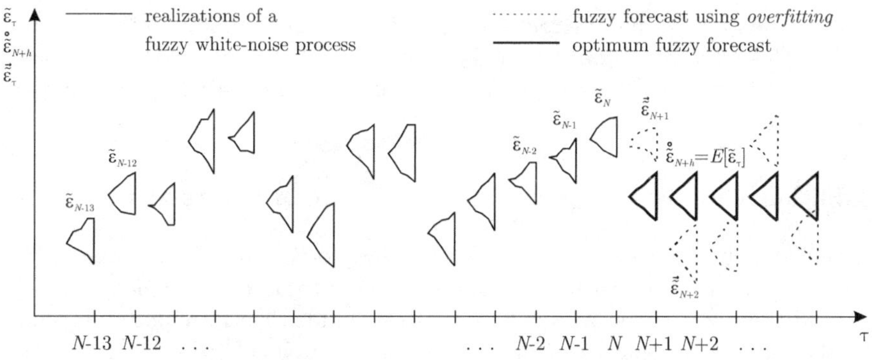

Fig. 3.17. Fuzzy forecast using *overfitting* and optimum fuzzy forecastings

Modification of the network architecture. The trained artificial neural network is now assessed using an error function and improved with the aid of an optimization strategy. Firstly, the trained optimum simulation artificial neural network is used to simulate a fuzzy time series $({}_s\tilde{e}_\tau)_{\tau\in\mathbf{T}}$ of preferably long length (see Sect. 4.3). The simulation may be started with an arbitrary fuzzy training pattern extracted from the fuzzy time series $(\tilde{e}_\tau)_{\tau\in\mathbf{T}}$. The simulated fuzzy realizations ${}_s\tilde{e}_\tau$ are evaluated statistically. The temporal fuzzy average according to Eq. (3.16) and the empirical $l_\alpha r_\alpha$-covariance function according to Eq. (3.20) are computed. Because the fuzzy time series $(\tilde{e}_\tau)_{\tau\in\mathbf{T}}$ is stationary and the fuzzy expected value $E[\tilde{\mathsf{E}}_\tau]$ is zero, the temporal fuzzy average ${}_s\tilde{\overline{e}}$ of the fuzzy realizations ${}_s\tilde{e}_\tau$ must also be zero. In addition, the empirical $l_\alpha r_\alpha$-covariance function ${}_{lr}\underline{\hat{K}}_{{}_s\tilde{e}_\tau}(\Delta\tau)$ of the fuzzy time series $({}_s\tilde{e}_\tau)_{\tau\in\mathbf{T}}$ is compared with the theoretical $l_\alpha r_\alpha$-covariance function ${}_{lr}K_{\tilde{\mathsf{E}}_\tau}(\Delta\tau)$. The forecast error is computed from the difference as follows:

$$PF_{so} = \sum_{j=1}^{2n} (0 - \Delta_s\overline{e}_j)^2 + \sum_{\Delta\tau=-\infty}^{\infty} \sum_{k,l=1}^{2n} \left(\hat{k}_{\tilde{\mathsf{E}}_\tau}[k,l](\Delta\tau) - \hat{k}_{{}_s\tilde{e}_\tau}[k,l](\Delta\tau)\right)^2 . \quad (3.187)$$

The term $\Delta_s\overline{e}_j$ hereby denotes the $l_\alpha r_\alpha$-increments of the empirical fuzzy average ${}_s\tilde{\overline{e}}$, and $\hat{k}_{\tilde{\mathsf{E}}_\tau}[k,l](\Delta\tau)$ and $\hat{k}_{{}_s\tilde{e}_\tau}[k,l](\Delta\tau)$ represent the elements of the $l_\alpha r_\alpha$-covariance functions ${}_{lr}K_{\tilde{\mathsf{E}}_\tau}(\Delta\tau)$ and ${}_{lr}\underline{\hat{K}}_{{}_s\tilde{e}_\tau}(\Delta\tau)$, respectively.

In an optimization process in which the objective function is given by Eq. (3.187) the optimum simulation network architecture is improved. This may be achieved by applying optimization methods such as Monte Carlo methods or network search methods, or by means of a heuristic modification.

If, for example, the optimum simulation network architecture has been modified heuristically, it must be re-trained using the backpropagation algorithm. The network modification is repeated until such time as forecast error PF_{so} takes on a minimum values. The efficiency of the optimization depends on the method used. An acceptable local minimum is attained depending on the case in question.

Additional Remarks Concerning the Choice of Layers and Neurons

Artificial neural networks for analyzing and forecasting fuzzy time series should have at least two hidden layers in order to account for nonlinear effects of the fuzzy time series. The number of hidden layers required depends on the number of neurons in the hidden layers. As a rule, the fewer neurons present per hidden layer then the more hidden layers are required in order to guarantee functional equivalence.

An artificial neural network with too few neurons in the hidden layers is unsuitable for forecasting purposes. If it is not possible to adequately model the complexity of a fuzzy time series due to too few neurons and hence an insufficient number of weighting matrices, it is not possible to obtain an optimum forecast; i.e. *underfitting* occurs.

The number of hidden layers and neurons depends only slightly on the fuzzy activation function $\tilde{f}_A(\cdot)$. The sigmoidal fuzzy activation function $\tilde{f}_A(\cdot)$ according to Eq. (3.188) is especially suitable for artificial neural networks intended for the analysis and forecasting of fuzzy time series.

$$\tilde{o} = \tilde{f}_A(\tilde{net}) \tag{3.188}$$

$$\text{with} \quad \Delta o_i = \frac{1}{1 + e^{-\Delta net_i}} \quad \text{for} \quad i = 1, 2, ..., 2n$$

The advantage of using this fuzzy activation function is that the requirement according to Eq. (2.18) is complied with. Moreover, the trajectories of the fuzzy activation function, i.e. the deterministic sigmoidal functions $f_A(\cdot)$, are differentiable.

The use of a linear fuzzy activation function for the analysis and forecasting of fuzzy time series is not appropriate. In the case of a linear fuzzy activation function each series connection of artificial neurons may be replaced by a single neuron, i.e. each multilayer perceptron for fuzzy variables may be reduced to a functionally equivalent single-layer perceptron. Furthermore, compliance with the requirement given by Eq. (2.18) is not possible using linear fuzzy activation functions.

3.6.5 Conditioning of the Fuzzy Data

By means of conditioning, the $l_\alpha r_\alpha$-increments of the fuzzy input and fuzzy output variables of an artificial neural network are transformed into the number range of the interval [0,1]. After this transformation it is possible to use multilayer perceptrons without the need to classify a fuzzy time series as being stationary or non-stationary a priori. Conditioning is also a precondition for good convergence of the backpropagation algorithm.

Conditioning for an Optimum Forecasting Network Architecture

Fuzzy input variables are lumped together in fuzzy training patterns (see Def. 3.40) and transferred to the multilayer perceptron. Because the transformation depends on the position of each individual fuzzy input variable within the fuzzy time series, the following indexing is introduced for the fuzzy input vectors and their fuzzy input variables. Each fuzzy element $\tilde{x}_{k-r}$ of the k-th fuzzy input vector $\underline{\tilde{x}}_k = (\tilde{x}_{k-n_I}, ..., \tilde{x}_{k-r}, ..., \tilde{x}_{k-1})^T$ (with $r = n_I, n_I - 1, ..., 1$ and $k = n_I + 1, n_I + 2, ..., N$) which is represented by means of $l_\alpha r_\alpha$-discretization is transformed with the aid of the diagonal matrix $\underline{T}_k$ and the moving fuzzy average $\tilde{\overline{x}}_k$ according to Eq. (3.189).

$$\tilde{x}^*_{k-r} = \underline{T}_k \odot \left(\tilde{x}_{k-r} \ominus \tilde{\overline{x}}_k\right) \tag{3.189}$$

The $\tilde{x}^*_{k-r}$ constitute the transformed fuzzy input vector $\underline{\tilde{x}}^*_k = (\tilde{x}^*_{k-n_I}, ..., \tilde{x}^*_{k-r}, ..., \tilde{x}^*_{k-1})^T$. The moving fuzzy average $\tilde{\overline{x}}_k$ is defined as the elementary special case of the linear filter according to Sect. 3.4.1, and is determined by means of Eq. (3.190). The moving fuzzy average $\tilde{\overline{x}}_k$ provides an approximation of the fuzzy trend of a fuzzy time series.

$$\tilde{\overline{x}}_k = \frac{1}{n_I} \bigoplus_{r=1}^{n_I} \tilde{x}_{k-r} \tag{3.190}$$

The elements $t_{i,j}(k)$ of the real-valued diagonal matrix $\underline{T}_k$ are determined for $i, j = 1, 2, ..., 2n$ using Eq. (3.191). The terms $\Delta\overline{x}_i(k)$ are hereby the $l_\alpha r_\alpha$-increments of the moving fuzzy average $\tilde{\overline{x}}_k$. The transformation matrix $\underline{T}_k$ may be interpreted as an indicator of the scatter of the fuzzy time series on a segmental basis.

$$t_{i,j}(k) = \begin{cases} \left(\max_{r=0,\,1,\,...,\,n_I-1} |\Delta x_i(k-r) - \Delta\overline{x}_i(k)|\right)^{-1} & \text{for } i = j \\ 0 & \text{for } i \neq j \end{cases} \tag{3.191}$$

The elements $\tilde{x}^*_{k-r}$ of the transformed fuzzy input vectors $\underline{\tilde{x}}^*_k$, which represent intermediate values according to Sect. 2.1.2, are hereby not subject to the requirement given by Eq. (2.47). For the fuzzy output variable $\tilde{o}^O_k$ of the multilayer perceptron, on the other hand, the requirement of non-negativity of the $l_\alpha r_\alpha$-increments holds. For this reason the fuzzy output function $\tilde{f}_{O,k}(\cdot)$ in combination with the sigmoidal fuzzy activation function $\tilde{f}_A(\cdot)$ is introduced for transforming the fuzzy net input variable $\tilde{net}$ of the output layer. The fuzzy output function $\tilde{f}_{O,k}(\cdot)$ replaces $\tilde{f}_O(\cdot)$ in Eq. (3.144). The transformation of the fuzzy net input variable $\tilde{net}$ of the output neuron is defined by Eq. (3.192).

$$\tilde{o}_k^{O*} = \tilde{f}_{O,k}\left(\tilde{f}_A(\tilde{net})\right) \tag{3.192}$$

The term $\tilde{f}_A(\cdot)$ is hereby the sigmoidal fuzzy activation function according to Eq. (3.188). The fuzzy output function $\tilde{f}_{O,k}(\cdot)$ is advantageously defined by Eq. (3.193), whereby $i = 1, 2, ..., 2n$.

$$\tilde{o}_k^{O*} = \tilde{f}_{O,k}\left(\tilde{f}_A(\tilde{net})\right) \tag{3.193}$$

$$\text{with} \quad \Delta o_i^{O*}(k) = f_{O,i,k}(f_A(\Delta net_i))$$

$$= -\frac{\Delta\overline{x}_i(k)}{t_{i,i}(k)} + e^{\left(p_1(i,k)+p_2(i,k)\cdot\Phi^{-1}(f_A(\Delta net_i))\right)}$$

The term $\Phi^{-1}(\cdot)$ is hereby the inverse deterministic distribution function of the normalized normal distribution. The inverse three-parameter logarithmic normal distribution is chosen here to represent the deterministic function $f_{O,i,k}(\cdot)$. A major advantage of this function is that it is possible a define a lower bound for the functional values for the case of best possible congruence with the inverse deterministic sigmoidal function $f_A(\cdot)$ according to Eq. (3.142). In other words, the fuzzy output function $\tilde{f}_{O,k}(\cdot)$ according to Eq. (3.192) yields fuzzy output variables $\tilde{o}_k^{O*}$ whose $l_\alpha r_\alpha$-increments are restricted by a lower bound and lie in the effective region of the deterministic functions $f_{O,i,k}(\cdot)$. The effective region of the functions is understood here to be the region of steepest ascent of the functional curve. The subsequent transformation according to Eq. (3.194) guarantees compliance with the requirement given by Eq. (2.47) for the fuzzy output variable $\tilde{o}_k^O$ of the multilayer perceptron.

$$\tilde{o}_k^{O*} = \underline{T}_k \odot \left(\tilde{o}_k^O \ominus \tilde{\overline{x}}_k\right) \quad \text{and} \quad \tilde{o}_k^O = \underline{T}_k^{-1} \odot \tilde{o}_k^{O*} \oplus \tilde{\overline{x}}_k \tag{3.194}$$

The parameters $p_1(i,k)$ and $p_2(i,k)$ of the deterministic output function $f_{O,i,k}(\cdot)$ are given by Eqs. (3.195) and (3.196), respectively. These result from the requirement of the best possible congruence between $f_{O,i,k}(\cdot)$ and $f_A(\cdot)$. This requirement is fulfilled when the mean and the variance of both functions are in agreement.

$$p_2(k) = \sqrt{\ln\left(1 + 3.29\left(\frac{t_{i,i}(k)}{\Delta m_i(k)}\right)^2\right)} \tag{3.195}$$

$$p_1(k) = \ln\left(\frac{\Delta\overline{x}_i(k)}{t_{i,i}(k)}\right) - \frac{p_2(k)}{2} \tag{3.196}$$

The factor 3.29 in Eq. (3.195) corresponds to the variance of the differentiated sigmoidal activation function $f_A(\cdot)$ according to Eq. (3.142). This factor guarantees the best possible congruence between the inverse three-parameter logarithmic normal distribution $f_{O,i,k}(\cdot)$ and the function $f_A(\cdot)$. An approximation formula after [64] is given for the inverse function of the normalized normal distribution $\Phi^{-1}(\cdot)$ by Eq. (3.197).

$$\Phi^{-1}\left(f_A(\Delta net_i)\right) = \begin{cases} -u + \frac{c_0+c_1u+c_2u^2}{1+d_1u+d_2u^2+d_3u^3} & \text{for } 0 < f_A(\Delta net_i) \leqslant 0.5 \\ -\Phi^{-1}(1 - f_A(\Delta net_i)) & \text{for } 0.5 < f_A(\Delta net_i) \leqslant 1 \end{cases} \tag{3.197}$$

$$\text{with} \qquad u = \sqrt{\ln \frac{1}{\left(f_A(\Delta net_i)\right)^2}}$$

$$\text{and} \qquad \begin{matrix} c_0 = 2.515517 & c_1 = 0.802853 & c_2 = 0.010328 \\ d_1 = 1.432788 & d_2 = 0.189269 & d_3 = 0.001308 \end{matrix}$$

The linear transformations of the fuzzy input variables and the fuzzy result variables according to Eqs. (3.189) and (3.194), respectively, guarantee that the fuzzy input variables and the fuzzy result variables of the multilayer perceptron lie in the effective region of the fuzzy activation functions. Applying the two-step fuzzy activation function according to Eq. (3.192), compliance with the requirement given by Eq. (2.47) is also ensured for the fuzzy output variable $\tilde{o}_k^O$ of the multilayer perceptron. This means that it is possible to obtain robust fuzzy forecasts regardless of the value range of the $l_\alpha r_\alpha$-increments.

Conditioning for an Optimum Simulation Network Architecture

In order to condition the fuzzy input data of the optimum simulation network architecture the fuzzy variables $\tilde{e}_\tau$ are transformed. The elements of the k-th fuzzy input vector $\underline{\tilde{e}}_k^* = (\tilde{e}^*_{k-n_I^{(s)}}, \ldots, \tilde{e}^*_{k-r}, \ldots, \tilde{e}^*_{k-1})^T$, with $r = n_I^{(s)}, n_I^{(s)} - 1, \ldots, 1$ and $k = n_I^{(p)} + 1, n_I^{(p)} + 2, \ldots$, result from the linear transformation according to Eq. (3.198).

$$\tilde{e}^*_{k-r} = \underline{T}^*_k \odot (\tilde{e}_{k-r}) \tag{3.198}$$

The elements $t^*_{i,j}(k)$ of the diagonal matrix $\underline{T}^*_k$ are given by Eq. (3.199) for each fuzzy input vector $\underline{\tilde{e}}^*_k$.

$$t^*_{i,j}(k) = \begin{cases} \left(\max_{r=0,1,\ldots,n_I^{(s)}-1} |\Delta e_i(k-r)|\right)^{-1} & \text{for } i = j \\ 0 & \text{for } i \neq j \end{cases} \tag{3.199}$$

The fuzzy net input variable $\tilde{net}$ of the output layer is mapped onto the fuzzy output variable $\tilde{o}_k^{O*}$ with the aid of the fuzzy activation function and the fuzzy output function according to Eq. (3.200). As the fuzzy output variables $\tilde{o}_k^{O*}$ of the optimum simulation multilayer perceptron represent intermediate results, non-negativity of the $l_\alpha r_\alpha$-increments is not required. The sigmoidal fuzzy activation function according to Eq. (3.188) is chosen as the fuzzy activation function $\tilde{f}_A(\cdot)$. The fuzzy output function $\tilde{f}_{O,k}(\cdot)$ of the output neuron is given by Eq. (3.193), whereby $i = 1, 2, ..., 2n$ holds.

$$\tilde{o}_k^{O*} = \tilde{f}_{O,k}(\tilde{f}_A(\tilde{net})) \tag{3.200}$$

$$\text{with} \quad \Delta o_i^{O*}(k) = f_{O,i,k}(f_A(\Delta_i))$$

$$= -\frac{\Delta \mathring{x}_i(k)}{t_{i,i}^*(k)} + e^{\left(p_1^*(i,k)+p_2^*(i,k)\cdot\Phi^{-1}(\Delta net_i)\right)}$$

An approximation formula for the inverse distribution function of the normalized normal distribution $\Phi^{-1}(\cdot)$ is given by Eq. (3.197) (see also [64]). The parameters $p_1^*(i,k)$ and $p_2^*(i,k)$ are given by Eqs. (3.201) and (3.202), respectively.

$$p_2^*(i,\, k) = \sqrt{\ln\left(1 + 3,29\left(\frac{t_{i,i}^*(k)}{\Delta \mathring{x}_i(k)}\right)^2\right)} \tag{3.201}$$

$$p_1^*(i,\, k) = \ln\left(\frac{\Delta \mathring{x}_i(k)}{t_{i,i}^*(k)}\right) - \frac{p_2^*(k)}{2} \tag{3.202}$$

The terms $\Delta \mathring{x}_i(k)$ are hereby the $l_\alpha r_\alpha$-increments of the optimum single-step forecast $\mathring{\tilde{x}}_k$. The optimum single-step forecast $\mathring{\tilde{x}}_k$ is computed from the measured fuzzy variables $\tilde{x}_\tau$ ($\tau \leqslant N$) or the simulated realizations $\vec{\tilde{x}}_\tau$ ($\tau > N$) of the previous time points (see Sect. 4.3.1). The fuzzy output variable $\tilde{o}_k^{O*}$ of the artificial neural network is subsequently mapped onto the fuzzy variable $\tilde{o}_k^{O}$ with the aid of the transformation given by Eq. (3.203).

$$\tilde{o}_k^{O*} = \underline{T}_k^* \odot \left(\tilde{o}_k^{O} \ominus \mathring{\tilde{x}}_k\right) \quad \text{and} \quad \tilde{o}_k^{O} = \underline{T}_k^{*^{-1}} \odot \tilde{o}_k^{O*} \oplus \mathring{\tilde{x}}_k \tag{3.203}$$

The use of the fuzzy output function according to Eq. (3.200) always guarantees compliance of the fuzzy variable $\tilde{o}_k^{O}$ with the requirement given by Eq. (2.47).

Initialization of the Weighting Matrices

Before starting the backpropagation algorithm the weighting matrices must be initialized according to the particular architecture chosen for the multilayer perceptron. The initialization may be carried out in accordance with the number of artificial neurons and the chosen $l_\alpha r_\alpha$-discretization. The larger the number of artificial neurons and the chosen α-level sets then the more $l_\alpha r_\alpha$-increments must be lumped together in the neurons to form the fuzzy net input variables $\tilde{net}$ (see Eq. (3.139)). In order to ensure that the fuzzy net input variables $\tilde{net}$ lie in the effective region of the fuzzy activation functions, small values of the weighting matrices should be chosen initially. The initialization values of the weighting matrices are thus chosen to be inversely proportional to the number of artificial neurons in the layers as well as the number n of the selected α-level sets. When the backpropagation algorithm is started, the matrix elements of the first hidden layer are assigned random real values from the interval

$$\left[-\frac{1}{2 \cdot n \cdot n_I} ; \frac{1}{2 \cdot n \cdot n_I}\right] . \tag{3.204}$$

The term n_I hereby denotes the number of artificial neurons in the input layer. For subsequent hidden layers and the output layer, initialization is carried out using real values from the interval

$$\left[-\frac{1}{2 \cdot n \cdot n_H} ; \frac{1}{2 \cdot n \cdot n_H}\right] . \tag{3.205}$$

The term n_H hereby denotes the number of artificial neurons in the preceding layer. By means of this procedure the occurrence of too large or too small fuzzy net input variables $\tilde{net}$ is avoided. The latter would otherwise lie in the non-effective flat region of the fuzzy activation functions. A possible consequence of this would be insufficient convergence of the backpropagation algorithm, as this tends to stagnate in flat regions of the error function.

4

Forecasting of Time Series Comprised of Uncertain Data

4.1 Underlying Concept

The aim of forecasting a time series containing fuzzy data is to obtain an assessed prognosis of the follow-up or future values $\tilde{x}_{N+h}$ $(h = 1, 2, ...)$ of an observed fuzzy time series $\tilde{x}_1, \tilde{x}_2, ..., \tilde{x}_N$. A precondition for this is the assumption and matching of an underlying fuzzy random process $(\tilde{\mathrm{X}}_\tau)_{\tau \in \mathbf{T}}$ according to Sect. 3.5 or modeling of the fuzzy time series with aid of an artificial neural network according to Sect. 3.6.

The basic idea behind the modeling of fuzzy time series is to treat a given fuzzy time series as a realization of a fuzzy random process $(\tilde{\mathrm{X}}_\tau)_{\tau \in \mathbf{T}}$. In order to forecast a fuzzy time series $\tilde{x}_1, \tilde{x}_2, ..., \tilde{x}_N$ future realizations $\tilde{x}_{N+h}$ of the fuzzy random process $(\tilde{\mathrm{X}}_\tau)_{\tau \in \mathbf{T}}$ must be determined. As a rule, the future realizations $\tilde{x}_{N+h}$ are dependent on the given realizations, i.e. the observed fuzzy time series $\tilde{x}_1, \tilde{x}_2, ..., \tilde{x}_N$. An exception to this are fuzzy white-noise processes. The fuzzy random variables within the forecasting period are therefore conditional fuzzy random variables $(\tilde{\mathrm{X}}_{N+h} \,|\, \tilde{x}_1, \tilde{x}_2, ..., \tilde{x}_N)$.

Analogous to classical time series analysis according to [60], the fuzzy random forecasting process $(\vec{\tilde{\mathrm{X}}}_\tau)_{\tau \in \mathbf{T}}$ is introduced for forecasting a fuzzy time series $\tilde{x}_1, \tilde{x}_2, ..., \tilde{x}_N$.

Definition 4.1. *The fuzzy random forecasting process* $(\vec{\tilde{\mathrm{X}}}_\tau)_{\tau \in \mathbf{T}}$ *is the continuation of the underlying fuzzy random process* $(\tilde{\mathrm{X}}_\tau)_{\tau \in \mathbf{T}}$ *within the forecasting period* $\tau = N + 1,\ N + 2,\ ...,$ *taking into consideration the known realizations* $\tilde{x}_1, \tilde{x}_2, ..., \tilde{x}_N$. *The fuzzy random variables* $\vec{\tilde{\mathrm{X}}}_{N+h}$ *of the fuzzy random forecasting process* $(\vec{\tilde{\mathrm{X}}}_\tau)_{\tau \in \mathbf{T}}$ *are hence conditional fuzzy random variables* $(\tilde{\mathrm{X}}_{N+h} \,|\, \tilde{x}_1, \tilde{x}_2, ..., \tilde{x}_N)$ *of the underlying fuzzy random process* $(\tilde{\mathrm{X}}_\tau)_{\tau \in \mathbf{T}}$. *At time points* $\tau = N + 1,\ N + 2,\ ...$*the fuzzy random forecasting process* $(\vec{\tilde{\mathrm{X}}}_\tau)_{\tau \in \mathbf{T}}$ *yields the discrete fuzzy random variables* $\vec{\tilde{\mathrm{X}}}_{N+h}(\tilde{x}_1, \tilde{x}_2, ..., \tilde{x}_N)$, *which are dependent on the realizations* $\tilde{x}_1, \tilde{x}_2, ..., \tilde{x}_N$ *of the fuzzy random variables* $\tilde{\mathrm{X}}_1, \tilde{\mathrm{X}}_2, ..., \tilde{\mathrm{X}}_N$. ◆

The forecasting of fuzzy time series may take the form of an optimum forecast, a fuzzy interval forecast or a fuzzy random forecast. In order to distinguish between these different forecast forms, the following definitions are introduced.

Definition 4.2. *The fuzzy mean value of all potential future realizations of the fuzzy random variable* $\vec{\tilde{X}}_{N+h}$ *is referred to as the optimum forecast* $\mathring{\tilde{x}}_{N+h}$ *at time point* $\tau = N + h$. *The optimum forecast* $\mathring{\tilde{x}}_{N+h}$ *corresponds to the conditional fuzzy expected value according to Eq. (4.1).*

$$\mathring{\tilde{x}}_{N+h}(\tilde{x}_1, \tilde{x}_2, ..., \tilde{x}_N) = E[\tilde{X}_{N+h} \,|\, \tilde{x}_1, \tilde{x}_2, ..., \tilde{x}_N] = E[\vec{\tilde{X}}_{N+h}] \tag{4.1}$$

◆

The optimum forecasts $\mathring{\tilde{x}}_{N+1}$, $\mathring{\tilde{x}}_{N+2}$, ..., $\mathring{\tilde{x}}_{N+h}$ are specific sequential values of the observed fuzzy time series with a minimum forecast error.

It is also possible to specify fuzzy forecast intervals within which the expected fuzzy variables with the given probability κ (confidence level) will lie.

Definition 4.3. *A fuzzy interval* $\tilde{x}_I$ *is referred to as a fuzzy forecast interval* $\tilde{x}^{\kappa}_{N+h}$ *if the realization* $\vec{\tilde{x}}_{N+h}$ *of* $\vec{\tilde{X}}_{N+h}$ *of a fuzzy random forecasting process* $(\vec{\tilde{X}}_\tau)_{\tau\in\mathbf{T}}$ *will be completely included in the interval with the probability* κ. *A fuzzy variable* $\vec{\tilde{x}}_{N+h}$ *is completely included in the fuzzy interval* $\tilde{x}_I$ ($\vec{\tilde{x}}_{N+h} \subseteq \tilde{x}_I$) *if the inequality (4.2) holds for the membership functions of the fuzzy variables* $\vec{\tilde{x}}_{N+h}$ *and* $\tilde{x}_I$.

$$\mu_{\vec{\tilde{x}}_{N+h}}(x) \leqslant \mu_{\tilde{x}_I}(x) \quad \forall\, x \in \mathbb{R} \tag{4.2}$$

◆

Example 4.4. The fuzzy forecast interval $\tilde{x}^{0,95}_{N+1}$ and the optimum forecast $\mathring{\tilde{x}}_{N+1}$ for a fuzzy AR[2] process are shown by way of example in Fig. 4.1. Whereas the forecast $\mathring{\tilde{x}}_{N+1}$ satisfies Eq. (4.2), this does not hold for the realization $\vec{\tilde{x}}_{N+1}$ shown in Fig. 4.2. ◆

A fuzzy forecast interval $\tilde{x}^{\kappa}_{N+h}$ at time point $\tau = N + h$ may be estimated by a Monte Carlo simulation of potential future progressions of the given time series containing fuzzy data $(\tilde{x}_\tau)_{\tau\in\mathbf{T}}$. If the underlying fuzzy random process $(\tilde{X}_\tau)_{\tau\in\mathbf{T}}$ (see Sect. 3.5.5) of the fuzzy time series $(\tilde{x}_\tau)_{\tau\in\mathbf{T}}$ is known or may justifiably be assumed, a Monte Carlo simulation of realizations $\tilde{x}_{N+h}$ at each time point $\tau = N + h$ is possible.

Remark 4.5. With the aid of a Monte Carlo simulation it is possible to estimate forecast intervals for arbitrary fuzzy random processes. In classical time series analysis (e.g. [7, 32]), on the other hand, GAUSS processes are usually stipulated for determining forecast intervals. ◆

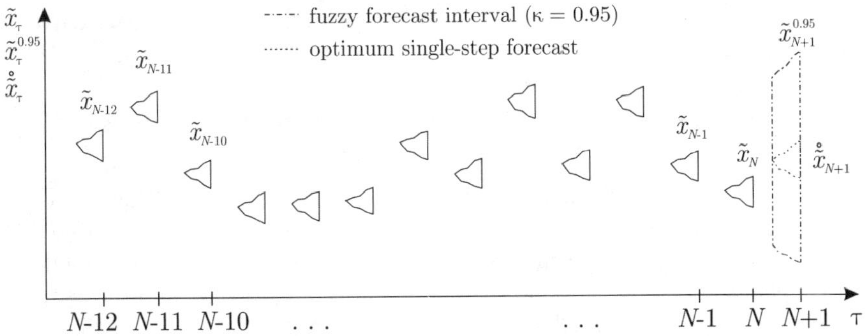

Fig. 4.1. Realization of a fuzzy AR[2] process with a 95 % fuzzy forecast interval

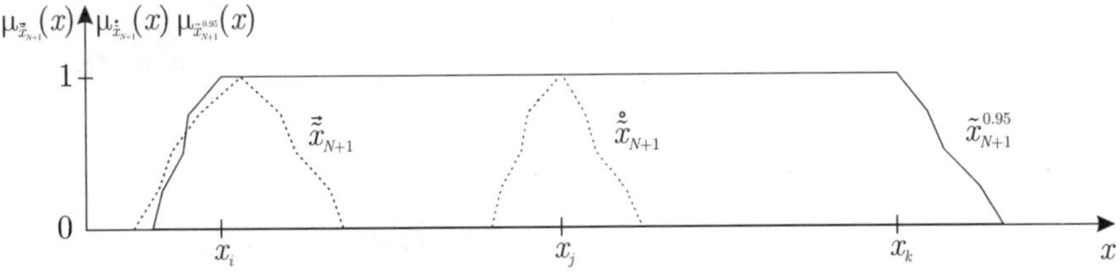

Fig. 4.2. 95 % fuzzy forecast interval $\tilde{x}^{0.95}_{N+1}$ and optimum forecast $\mathring{\tilde{x}}_{N+1}$

A fuzzy random forecast permits a determination of the fuzzy random variables $\vec{\tilde{\mathrm{X}}}_\tau$ within the forecasting period $\tau = N+1,\ N+2,\ ...,\ N+h$. If the $\vec{\tilde{\mathrm{X}}}_\tau$ are known, it is possible to state the probability of occurrence of the expected fuzzy realizations within the forecasting period.

Definition 4.6. *A fuzzy random forecast yields the expected fuzzy random variables* $\vec{\tilde{\mathrm{X}}}_\tau$ *of the fuzzy random forecasting process* $(\vec{\tilde{\mathrm{X}}}_\tau)_{\tau \in \mathbf{T}}$ *at future time points* $\tau = N + h$. ◆

If the underlying fuzzy random process $(\tilde{\mathrm{X}}_\tau)_{\tau \in \mathbf{T}}$ of a fuzzy time series is known, the fuzzy random variables $\vec{\tilde{\mathrm{X}}}_{N+h}$ are estimated by a Monte Carlo simulation of potential future conditional realizations $(\tilde{x}_{N+h} \,|\, \tilde{x}_1, \tilde{x}_2, ..., \tilde{x}_N)$ of $(\tilde{\mathrm{X}}_\tau)_{\tau \in \mathbf{T}}$. The fuzzy probability distribution functions form II $F_{\vec{\tilde{\mathrm{X}}}_{N+h}}(\tilde{x})$ and ${}_{lr}F_{\vec{\tilde{\mathrm{X}}}_{N+h}}(\tilde{x})$ (see Sect. 2.2.2) and the characteristic moments (see Sect. 2.2.3) may be used to characterize the fuzzy random variables $\vec{\tilde{\mathrm{X}}}_{N+h}$ at future time points $\tau = N + h$.

4.2 Forecasting on the Basis of Specific Fuzzy Random Processes

The determination of optimum forecasts $\mathring{\tilde{x}}_{N+h}$, fuzzy forecast intervals $\tilde{x}^{\kappa}_{N+h}$, and forecasted fuzzy random variables $\vec{\tilde{X}}_{N+h}$ for fuzzy MA processes, fuzzy AR processes and fuzzy ARMA processes is presented in the following. Each forecast relies on the specification and parameter estimation of the underlying fuzzy random process in each case. In order to perform the forecast the theoretical characteristic values and parameters of the fuzzy random process must be replaced by the empirical or estimated values, and the observed realizations and computed realizations according to Eq. (3.115) must be substituted for $\tilde{X}_\tau$ and $\tilde{\mathcal{E}}_\tau$, respectively.

In order to forecast a given fuzzy time series $\tilde{x}_1, \tilde{x}_2, ..., \tilde{x}_N$ the fuzzy random forecasting process $(\vec{\tilde{X}}_\tau)_{\tau\in\mathbf{T}}$ according to Def. 4.1 is applied. If the underlying process is a fuzzy ARMA[p, q] process $(\tilde{X}_\tau)_{\tau\in\mathbf{T}}$, the fuzzy random forecasting process $(\vec{\tilde{X}}_\tau)_{\tau\in\mathbf{T}}$ is given by Eq. (4.3) with $h = 1, 2, ...$

$$\begin{aligned}\vec{\tilde{X}}_{N+h} = {} & \underline{A}_1 \odot \vec{\tilde{X}}_{N+h-1} \oplus ... \oplus \underline{A}_p \odot \vec{\tilde{X}}_{N+h-p} \oplus \tilde{\mathcal{E}}_{N+h} \ominus \\ & \underline{B}_1 \odot \tilde{\mathcal{E}}_{N+h-1} \ominus ... \ominus \underline{B}_q \odot \tilde{\mathcal{E}}_{N+h-q}\end{aligned} \tag{4.3}$$

$$\text{with } \vec{\tilde{X}}_{N+h-u} = \begin{cases} \tilde{x}_{N+h-u} & \text{for } N+h-u \leqslant N \\ \vec{\tilde{X}}_{N+h-u} & \text{for } N+h-u > N \end{cases}, \quad u = 1, 2, ..., p \tag{4.4}$$

$$\text{and } \tilde{\mathcal{E}}_{N+h-v} = \begin{cases} \tilde{\varepsilon}_{N+h-v} & \text{for } N+h-v \leqslant N \\ \tilde{\mathcal{E}}_{N+h-v} & \text{for } N+h-v > N \end{cases}, \quad v = 1, 2, ..., q \tag{4.5}$$

For each time point $\tau = N + h - u \leqslant N$ the fuzzy random variable $\vec{\tilde{X}}_{N+h-u}$ is hereby replaced by the observed fuzzy variable $\tilde{x}_{N+h-u}$ of the fuzzy time series, whereas for time points $\tau = N + h - u > N$, the fuzzy random variable $\vec{\tilde{X}}_{N+h-u}$ is retained. For time points $\tau = N+h-v \leqslant N$ the realization $\tilde{\varepsilon}_{N+h-v}$ of the fuzzy white-noise process $(\tilde{\mathcal{E}}_\tau)_{\tau\in\mathbf{T}}$ computed according to Eq. (3.115) is inserted for the fuzzy white-noise variable $\tilde{\mathcal{E}}_{N+h-v}$, whereas for time points $\tau = N + h - v > N$, the fuzzy white-noise variable $\tilde{\mathcal{E}}_{N+h-v}$ is retained.

If the underlying process is a fuzzy MA process or a fuzzy AR process $(\tilde{X}_\tau)_{\tau\in\mathbf{T}}$, the fuzzy random forecasting process $(\vec{\tilde{X}}_\tau)_{\tau\in\mathbf{T}}$ follows from Eq. (4.3) as a special case with $p = 0$ or $q = 0$.

4.2.1 Optimum Forecast

Optimum h-step forecast. In the case of an underlying fuzzy ARMA[p, q] process $(\tilde{X}_\tau)_{\tau\in\mathbf{T}}$ the optimum forecasts $\mathring{\tilde{x}}_{N+1}$, $\mathring{\tilde{x}}_{N+2}$, ..., $\mathring{\tilde{x}}_{N+h}$ for a total of h steps are computed by means of the optimum h-step forecast according to

Eq. (4.6). Using the optimum forecasts obtained for earlier time points, the optimum forecasts are determined recursively by means of Eq. (4.6).

$$\mathring{\tilde{x}}_{N+h} = \underline{A}_1 \odot \tilde{x}_{N+h-1} \oplus ... \oplus \underline{A}_p \odot \tilde{x}_{N+h-p} \oplus E[\tilde{\mathcal{E}}_\tau] \ominus \underline{B}_1 \odot \tilde{\varepsilon}_{N+h-1} \ominus ... \ominus \underline{B}_q \odot \tilde{\varepsilon}_{N+h-q} \tag{4.6}$$

with $$\tilde{x}_{N+h-u} = \begin{cases} \tilde{x}_{N+h-u} & \text{for } N+h-u \leqslant N \\ \mathring{\tilde{x}}_{N+h-u} & \text{for } N+h-u > N \end{cases}, \quad u = 1, 2, ..., p \tag{4.7}$$

and $$\tilde{\varepsilon}_{N+h-v} = \begin{cases} \tilde{\varepsilon}_{N+h-v} & \text{for } N+h-v \leqslant N \\ E[\tilde{\mathcal{E}}_\tau] & \text{for } N+h-v > N \end{cases}, \quad v = 1, 2, ..., q \tag{4.8}$$

For each time point $\tau = N + h - v > N$ the optimum forecast of the fuzzy white-noise variable $\tilde{\mathcal{E}}_{N+h-v}$, as expressed by the fuzzy expected value $E[\tilde{\mathcal{E}}_\tau]$ of the fuzzy white-noise process $(\tilde{\mathcal{E}}_\tau)_{\tau \in \mathbf{T}}$ according to Definition 4.2, is inserted for $\tilde{\varepsilon}_{N+h-v}$ in Eq. (4.6). For time points $\tau = N + h - v \leqslant N$ the computed realization $\tilde{\varepsilon}_{N+h-v}$ of the fuzzy white-noise process $(\tilde{\mathcal{E}}_\tau)_{\tau \in \mathbf{T}}$ is inserted for $\tilde{\varepsilon}_{N+h-v}$. For each time point $\tau = N + h - u \leqslant N$ the observed fuzzy variable $\tilde{x}_{N+h-u}$ of the fuzzy time series is inserted for $\tilde{x}_{N+h-u}$. For time points $\tau = N + h - u > N$ the optimum forecast $\mathring{\tilde{x}}_{N+h-u}$ is inserted for $\tilde{x}_{N+h-u}$. The optimum h-step forecast according to Eq. (4.6) is thus equivalent to a recursive procedure. Because the fuzzy expected value $E[\tilde{\mathcal{E}}_\tau]$ of the fuzzy white-noise process is always inserted in Eq. (4.6) for $\tau = N + h - v > N$, the forecasted fuzzy variables converge on the fuzzy expected value function $E[\tilde{\mathrm{X}}_\tau]$ as the number of forecasting steps h increases.

The optimum h-step forecasts for a fuzzy MA[q] process or a fuzzy AR[p] process are included as a special case in Eq. (4.6). The optimum h-step forecast for a fuzzy MA[q] process $(\tilde{\mathrm{X}}_\tau)_{\tau \in \mathbf{T}}$ is computed according to Eq. (4.9). The terms $\tilde{\varepsilon}_{N+h-v}$ are hereby chosen according to Eq. (4.8).

$$\mathring{\tilde{x}}_{N+h} = E[\tilde{\mathcal{E}}_\tau] \ominus \underline{B}_1 \odot \tilde{\varepsilon}_{N+h-1} \ominus ... \ominus \underline{B}_q \odot \tilde{\varepsilon}_{N+h-q} \tag{4.9}$$

The optimum h-step forecast for a fuzzy AR[p] process $(\tilde{\mathrm{X}}_\tau)_{\tau \in \mathbf{T}}$ may be computed using Eq. (4.10). The realizations $\tilde{x}_{N+h-u}$ are chosen as in a similar way by means of Eq. (4.7).

$$\mathring{\tilde{x}}_{N+h} = \underline{A}_1 \odot \tilde{x}_{N+h-1} \oplus ... \oplus \underline{A}_p \odot \tilde{x}_{N+h-p} \oplus E[\tilde{\mathcal{E}}_\tau] \tag{4.10}$$

Optimum single-step forecast. The optimum single-step forecast is a special case of the optimum h-step forecast. For a fuzzy ARMA[p, q] process $(\tilde{\mathrm{X}}_\tau)_{\tau \in \mathbf{T}}$ the optimum single-step forecast according to Eq. (4.11) is retained. The optimum single-step forecast computes the optimum forecast $\mathring{\tilde{x}}_{N+1}$ at time point $\tau = N + 1$ using only the observed values $\tilde{x}_\tau$ ($\tau \leqslant N$) of the fuzzy time series and the computed realizations $\tilde{\varepsilon}_\tau$ ($\tau \leqslant N$) of the fuzzy white-noise process.

$$\mathring{\tilde{x}}_{N+1} = \underline{A}_1 \odot \tilde{x}_N \oplus ... \oplus \underline{A}_p \odot \tilde{x}_{N+1-p} \oplus E[\tilde{\mathcal{E}}_\tau] \ominus \underline{B}_1 \odot \tilde{\varepsilon}_N \ominus ... \ominus \underline{B}_q \odot \tilde{\varepsilon}_{N+1-q} \tag{4.11}$$

For a fuzzy AR[p] process $(\tilde{\text{X}}_\tau)_{\tau \in \mathbf{T}}$ the optimum single-step forecast is given by Eq. (4.12).

$$\mathring{\tilde{x}}_{N+1} = \underline{A}_1 \odot \tilde{x}_N \oplus ... \oplus \underline{A}_p \odot \tilde{x}_{N+1-p} \oplus E[\tilde{\mathcal{E}}_\tau] \tag{4.12}$$

The optimum single-step forecast for a fuzzy MA[q] process $(\tilde{\text{X}}_\tau)_{\tau \in \mathbf{T}}$ is retained according to Eq. (4.13).

$$\mathring{\tilde{x}}_{N+1} = E[\tilde{\mathcal{E}}_\tau] \ominus \underline{B}_1 \odot \tilde{\varepsilon}_N \ominus ... \ominus \underline{B}_q \odot \tilde{\varepsilon}_{N+1-q} \tag{4.13}$$

Remark 4.7. Optimum single-step forecasts are applied in the distance method and the gradient method for estimating the parameters of fuzzy ARMA processes (see Sect. 3.5.6). The optimum single-step forecasts $\mathring{\tilde{x}}_\tau$ are computed for the observation period $p < \tau \leqslant N$ and compared with the observed values. The term $\mathring{\tilde{x}}_{N+1}$ is thereby replaced by $\mathring{\tilde{x}}_\tau$, $\tilde{x}_N$ by $\tilde{x}_{\tau-1}$ etc. for $\tau = p+1, p+2, ..., N$ in Eq. (4.11). In contrast to the optimum h-step forecast, only the observed values $\tilde{x}_\tau$ of the fuzzy time series and the computed realizations $\tilde{\varepsilon}_\tau$ of the fuzzy white-noise process are used for computing the optimum single-step forecasts. ◆

Example 4.8. For a fuzzy time series with $N = 100$ realizations it is necessary to compute the optimum forecasts $\mathring{\tilde{x}}_{101}$, $\mathring{\tilde{x}}_{102}$, ..., $\mathring{\tilde{x}}_{110}$ as an h-step forecast with $h = 10$ for ten subsequent time points. An analysis of the given fuzzy time series indicated that the underlying process is a fuzzy ARMA[3,2] process. The parameter matrices $\underline{A}_1$, $\underline{A}_2$, $\underline{A}_3$ and $\underline{B}_1$, $\underline{B}_2$ have already been determined.

The optimum forecasted values $\mathring{\tilde{x}}_{N+h}$ are computed using Eq. (4.6). These are dependent on the realizations $\tilde{x}_\tau$ and the optimum forecasts $\mathring{\tilde{x}}_\tau$ as well as on the realizations $\tilde{\varepsilon}_\tau$ of the fuzzy white-noise process and the fuzzy expected value $E[\tilde{\mathcal{E}}_\tau]$. These dependencies are listed in Table 4.1. The realizations $\tilde{\varepsilon}_\tau$ of the fuzzy white-noise process are computed according to Eq. (3.115), beginning with $\tilde{\varepsilon}_1$. Based on the realizations $\tilde{\varepsilon}_\tau$ the fuzzy mean value may be computed as an estimator for the fuzzy expected value $E[\tilde{\mathcal{E}}_\tau]$ using Eq. (3.16).

For example, the optimum forecast $\mathring{\tilde{x}}_{102}$ is given by the following equation.

$$\mathring{\tilde{x}}_{102} = \underline{A}_1 \odot \mathring{\tilde{x}}_{101} \oplus \underline{A}_2 \odot \tilde{x}_{100} \oplus \underline{A}_3 \odot \tilde{x}_{99} \oplus E[\tilde{\mathcal{E}}_\tau] \ominus \underline{B}_1 \odot E[\tilde{\mathcal{E}}_\tau] \ominus \underline{B}_2 \odot \tilde{\varepsilon}_{100} \tag{4.14}$$

The last six values of the given fuzzy time series as well as the optimum forecasted values $\mathring{\tilde{x}}_{101}$, $\mathring{\tilde{x}}_{102}$ and $\mathring{\tilde{x}}_{103}$ are shown in Fig. 4.3. ◆

Table 4.1. Progression of the optimum h-step forecast

	optimum fuzzy forecast	realizations to be taken into consideration $\tilde{x}_\tau$, $\mathring{\tilde{x}}_\tau$	realizations to be taken into consideration $\tilde{\varepsilon}_\tau$, $E[\tilde{\mathcal{E}}_\tau]$
$h = 1$	$\mathring{\tilde{x}}_{101}$	$\tilde{x}_{100}$, $\tilde{x}_{99}$, $\tilde{x}_{98}$	$\tilde{\varepsilon}_{100}$, $\tilde{\varepsilon}_{99}$
$h = 2$	$\mathring{\tilde{x}}_{102}$	$\mathring{\tilde{x}}_{101}$, $\tilde{x}_{100}$, $\tilde{x}_{99}$	$E[\tilde{\mathcal{E}}_\tau]$, $\tilde{\varepsilon}_{100}$
$h = 3$	$\mathring{\tilde{x}}_{103}$	$\mathring{\tilde{x}}_{102}$, $\mathring{\tilde{x}}_{101}$, $\tilde{x}_{100}$	$E[\tilde{\mathcal{E}}_\tau]$, $E[\tilde{\mathcal{E}}_\tau]$
$\vdots$	$\vdots$	$\vdots$	$\vdots$
$h = 10$	$\mathring{\tilde{x}}_{110}$	$\mathring{\tilde{x}}_{109}$, $\mathring{\tilde{x}}_{108}$, $\mathring{\tilde{x}}_{107}$	$E[\tilde{\mathcal{E}}_\tau]$, $E[\tilde{\mathcal{E}}_\tau]$

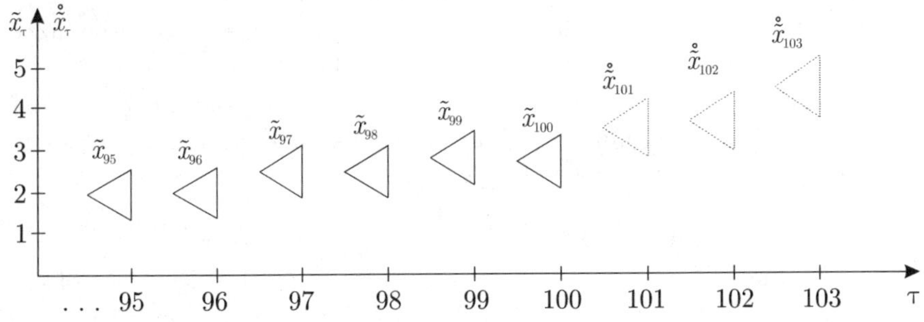

Fig. 4.3. Fuzzy time series and optimum forecast

4.2.2 Fuzzy Forecast Intervals

In order to compute fuzzy forecast intervals $\tilde{x}^{\kappa}_{N+h}$ (see Def. 4.3) at time points $\tau = N + h$ several sequences of future realizations of the underlying fuzzy random process are simulated by means of a Monte Carlo simulation.

If model matching for the given fuzzy time series yields a fuzzy ARMA[p, q] process, a sequence of future realizations $\vec{\tilde{x}}_{N+1}$, $\vec{\tilde{x}}_{N+2}$, ... of the fuzzy ARMA process for $h = 1$, 2, ... may be computed by means of Eq. (4.3). The computation begins with $h = 1$, as given by Eq. (4.15).

$$\vec{\tilde{\mathrm{X}}}_{N+1} = \underline{A}_1 \odot \tilde{x}_N \oplus ... \oplus \underline{A}_p \odot \tilde{x}_{N+1-p} \oplus \tilde{\mathcal{E}}_{N+1} \ominus \\ \underline{B}_1 \odot \tilde{\varepsilon}_N \ominus ... \ominus \underline{B}_q \odot \tilde{\varepsilon}_{N+1-q} \tag{4.15}$$

Because the fuzzy random variable $\vec{\tilde{\mathrm{X}}}_{N+1}$ is dependent on the fuzzy white-noise variable $\tilde{\mathcal{E}}_{N+1}$ in Eq. (4.15), a realization $\vec{\tilde{x}}_{N+1}$ can only be computed if a realization $\tilde{\varepsilon}_{N+1}$is known beforehand. From model matching, the fuzzy probability distribution function form II of the fuzzy white-noise variables $\tilde{\mathcal{E}}_\tau$ are known (see Sect. 3.5.6). Using this fuzzy probability distribution function for $\tilde{\mathcal{E}}_\tau$ in conjunction with a Monte Carlo simulation (see Sect. 2.2.4), it is possible to simulate a realization $\tilde{\varepsilon}_{N+1}$. The realization $\vec{\tilde{x}}_{N+1}$ is subsequently computed according to Eq. (4.15). Once $\vec{\tilde{x}}_{N+1}$ has been simulated, this may be used to simulate the realization $\vec{\tilde{x}}_{N+2}$ according to Eq. (4.16).

$$\vec{\tilde{X}}_{N+2} = \underline{A}_1 \odot \vec{\tilde{x}}_{N+1} \oplus \underline{A}_2 \odot \tilde{x}_N ... \oplus \underline{A}_p \odot \tilde{x}_{N+2-p} \oplus \tilde{\mathcal{E}}_{N+2} \ominus \\ \underline{B}_1 \odot \tilde{\varepsilon}_{N+1} \ominus ... \ominus \underline{B}_q \odot \tilde{\varepsilon}_{N+2-q} \tag{4.16}$$

Because the fuzzy random variable $\vec{\tilde{X}}_{N+2}$ is dependent on the fuzzy white-noise variable $\tilde{\mathcal{E}}_{N+2}$ in Eq. (4.16) , it is also necessary to simulate the realization $\tilde{\varepsilon}_{N+2}$. Once this has been done, it is then possible to compute the realization $\vec{\tilde{x}}_{N+2}$. The realizations $\vec{\tilde{x}}_{N+3}$, $\vec{\tilde{x}}_{N+4}$, ... are simulated in a similar way. A total of s sequences of realizations $\vec{\tilde{x}}_{N+1}$, $\vec{\tilde{x}}_{N+2}$, ... are computed.

The simulation of the sequences $\vec{\tilde{x}}_{N+1}$, $\vec{\tilde{x}}_{N+2}$, ... presented for fuzzy ARMA processes may also be applied to fuzzy MA and fuzzy AR processes.

With the aid of the s simulated, potential future progressions of the fuzzy time series the fuzzy forecast intervals $\tilde{x}^{\kappa}_{N+h}$ may be estimated as follows. The interval limits $\vec{x}_{\alpha_i l}(N+h)$ and $\vec{x}_{\alpha_i r}(N+h)]$ of the α-level sets $\vec{X}_{\alpha_i}(N+h)$ of all s simulated fuzzy variables $\vec{\tilde{x}}_{N+h}$ are sorted and indexed from the smallest to the largest value according to Eq. (4.17).

$$\begin{gathered} \vec{x}^1_{\alpha_i l}(N+h) \leqslant \vec{x}^2_{\alpha_i l}(N+h) \leqslant ... \leqslant \vec{x}^s_{\alpha_i l}(N+h) \\ \text{and} \\ \vec{x}^1_{\alpha_i r}(N+h) \leqslant \vec{x}^2_{\alpha_i r}(N+h) \leqslant ... \leqslant \vec{x}^s_{\alpha_i r}(N+h) \end{gathered} \tag{4.17}$$

For the confidence level κ the interval limits $x^{\kappa}_{\alpha_i l}(N+h)$ and $x^{\kappa}_{\alpha_i r}(N+h)$ of the α-level sets $X^{\kappa}_{\alpha_i}(N+h)$ of a fuzzy forecast interval $\tilde{x}^{\kappa}_{N+h}$ at time point $\tau = N+h$ may be estimated by means of Eq. (4.18). Eq. (4.18) only holds provided s is an even number.

$$\begin{gathered} x^{\kappa}_{\alpha_i l}(N+h) = \begin{cases} \leqslant \vec{x}^1_{\alpha_i l}(N+h) \text{ for } & a = 0 \\ \vec{x}^a_{\alpha_i l}(N+h) \text{ for } & 0 < a \leqslant \frac{s}{2} \end{cases} \quad \text{with}\, a = \operatorname{int}\left[s \cdot \left(\frac{1}{2} - \frac{\kappa}{2}\right)\right] \\ \text{and} \\ x^{\kappa}_{\alpha_i r}(N+h) = \begin{cases} \vec{x}^{b+1}_{\alpha_i r}(N+h) \text{ for } & \frac{s}{2} \leqslant b < s \\ \geqslant \vec{x}^s_{\alpha_i r}(N+h) \text{ for } & b = s \end{cases} \quad \text{with}\, b = \frac{s}{2} + \operatorname{int}\left[s \cdot \left(\frac{\kappa}{2}\right)\right] \end{gathered} \tag{4.18}$$

The interval limits of the α-level sets of the fuzzy forecast intervals $\tilde{x}^{\kappa}_{N+h}$ according to Eq. (4.18) thus correspond to the upper and lower quantile values of the empirical distributions of the interval limits for the simulated realizations $\vec{\tilde{x}}_{N+h}$. Future realizations $\vec{\tilde{x}}_{N+h}$ of a fuzzy time series $(\tilde{x}_\tau)_{\tau \in \mathbf{T}}$ are thus completely included in the estimated fuzzy forecast interval $\tilde{x}^{\kappa}_{N+h}$ with a probability κ.

The fuzzy forecast interval $\tilde{x}^{\kappa}_{N+h}$ only holds for the realization of the fuzzy random variable $\vec{\tilde{X}}_{N+h}$ at time point $\tau = N+h$. The fact that conditional probabilities are not considered in Eqs. (4.17) and (4.18) means that the $\tilde{x}^{\kappa}_{N+h}$ for different time points $\tau = N+h$ cannot be united to form a fuzzy forecast hose containing all realizations with a probability κ.

For a fuzzy forecast hose it is necessary to determine the conditional probabilities $P^*_{\mathrm{X}_{N+h}}$ with $h = 1, 2, ...$ according to Eq. (4.19).

$$P^*_{\mathrm{X}_{N+h}} = P[\vec{\tilde{\mathrm{X}}}_{N+h} \subseteq \tilde{x}^*_{N+h} \,|\, \vec{\tilde{\mathrm{X}}}_j \subseteq \tilde{x}^*_j,\ j = N+1, N+2, ..., N+h-1] \quad (4.19)$$

The term $P^*_{\mathrm{X}_{N+h}}$ expresses the probability that the realization $\vec{\tilde{x}}_{N+h}$ of $\vec{\tilde{\mathrm{X}}}_{N+h}$ at time point $\tau = N + h$ lies within a prescribed (conditional) fuzzy forecast interval $\tilde{x}^*_{N+h}$. A precondition for this is that the realizations $\vec{\tilde{x}}_{N+j}$ of $\vec{\tilde{\mathrm{X}}}_{N+j}$ at the preceding time points $\tau = N+j$ with $j = N+1, N+2, ..., N+h-1$ also lie within prescribed (conditional) fuzzy forecast intervals $\tilde{x}^*_{N+j}$. Alternatively, the conditional probabilities $P^*_{\mathrm{X}_{N+h}}$ at time points $\tau = N+h$ with $h = 1, 2, ...$ may be prescribed, and the corresponding conditional fuzzy forecast intervals $\tilde{x}^*_{N+h}$ (within which the realizations with a probability$P^*_{\mathrm{X}_{N+h}}$ lie) determined. Merging of the conditional fuzzy forecast intervals $\tilde{x}^*_{N+h}$ to form a forecast hose is permissible. The assertion that the realizations of the fuzzy random process lie within this hose with a probability κ^* is possible by multiplicative combination of the conditional probabilities $P^*_{\mathrm{X}_{N+h}}$according to Eq. (4.20).

$$\kappa^* = \prod_{j=N+1}^{N+h} P^*_{\mathrm{X}_j} \quad (4.20)$$

Example 4.9. Given is the realization $\tilde{x}_1, \tilde{x}_2, ..., \tilde{x}_N$ of a fuzzy AR[2] process $(\tilde{\mathrm{X}}_\tau)_{\tau\in\mathbf{T}}$. In order to determine the fuzzy forecast intervals $\tilde{x}^{0.95}_{N+h}$ for time points $\tau = N + h$ (h = 1, 2, ..., 10), s = 1000 sequences of future realizations are simulated and evaluated according to Eq. (4.17). The computed fuzzy forecast intervals $\tilde{x}^{0.95}_{N+h}$ are shown in Fig. 4.4. In the next step the fuzzy forecast intervals $\tilde{x}^{0.95}_{N+h}$ are prescribed as conditional fuzzy forecast intervals $\tilde{x}^*_{N+h}$. The aim is to compute the probability κ^* with which the realizations $\vec{\tilde{x}}_{N+1}, \vec{\tilde{x}}_{N+21}, ..., \vec{\tilde{x}}_{N+10}$ of $(\vec{\tilde{\mathrm{X}}}_\tau)_{\tau\in\mathbf{T}}$ lie within the prescribed conditional fuzzy forecast intervals $\tilde{x}^*_{N+1}, \tilde{x}^*_{N+2}, ..., \tilde{x}^*_{N+10}$. The probability κ^* may be advantageously estimated with the aid of the s = 1000 simulated sequences. For each simulated sequence a check is made to ascertain whether their realizations lie within the corresponding fuzzy forecast intervals $\tilde{x}^*_{N+h}$ at each time point $\tau = N+h$ (h = 1, 2, ..., 10) (see Eq. (4.2)). The fact that this condition is fulfilled by 530 of the 1000 simulated sequences leads to $\kappa^* = \frac{530}{1000} = 0.53$. This result implies that future realizations $\tilde{x}_{N+h}$ of the fuzzy AR[2] process lie within the corresponding fuzzy forecast interval at time point $\tau = N + h$ with a probability of $\kappa = 0.95$, and within the fuzzy forecast hose extending over the period $\tau = N+1,\ N+2,\ ...,\ N+10$ with a probability of $\kappa^* = 0.53$. Fig. 4.4 shows the forecast hose for the α-level $\alpha = 0$. ♦

Example 4.10. Based on the fuzzy time series of Example 4.8 it is intended to compute fuzzy forecast intervals for time points $\tau = N+1,\ N+2,\ N+3$ with a

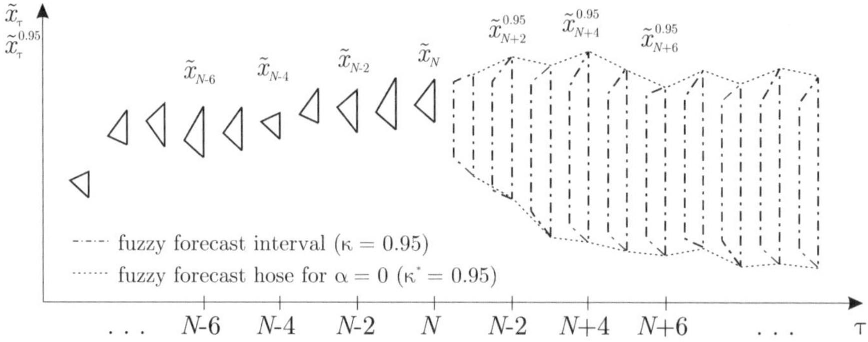

Fig. 4.4. Realization of a fuzzy AR[2] process and 95 % fuzzy forecast intervals $\tilde{x}_{N+h}^{0.95}$

confidence level of $\kappa = 0.95$. For this purpose s sequences of future realizations $\vec{\tilde{x}}_{N+1}$, $\vec{\tilde{x}}_{N+2}$, $\vec{\tilde{x}}_{N+3}$ are simulated. The realizations $\vec{\tilde{x}}_\tau(k)$ ($k = 1, 2, ..., s$) are computed recursively, beginning in each case with $\vec{\tilde{x}}_1(k)$ given by Eq. (4.15). These are dependent on the measured fuzzy variables $\tilde{x}_\tau$ or the realizations $\vec{\tilde{x}}_\tau(k)$ as well as on the realizations $\tilde{\varepsilon}_\tau$ or the simulated fuzzy variables $\tilde{\varepsilon}_\tau(k)$ of the fuzzy white-noise process corresponding to the preceding time points. These dependencies are listed in Table 4.2. The realizations $\tilde{\varepsilon}_\tau$ of the fuzzy white-noise process are computed according to Eq. (3.115), beginning with $\tilde{\varepsilon}_1$. The fuzzy variables $\tilde{\varepsilon}_\tau(k)$ of the fuzzy white-noise process are then simulated using the fuzzy probability distribution function form II of the fuzzy white-noise variables $\tilde{\mathcal{E}}_\tau$.

Table 4.2. Progression of the simulation of s sequences of future realizations

simulated realization		realizations to be taken into consideration $\tilde{x}_\tau$, $\vec{\tilde{x}}_\tau$	realizations to be taken into consideration $\tilde{\varepsilon}_\tau$
$k = 1$:			
$h = 1$	$\vec{\tilde{x}}_{101}(k{=}1)$	$\tilde{x}_{100}$, $\tilde{x}_{99}$, $\tilde{x}_{98}$	$\tilde{\varepsilon}_{100}$, $\tilde{\varepsilon}_{99}$
$h = 2$	$\vec{\tilde{x}}_{102}(k{=}1)$	$\vec{\tilde{x}}_{101}(k{=}1)$, $\tilde{x}_{100}$, $\tilde{x}_{99}$	$\tilde{\varepsilon}_{101}(k{=}1)$, $\tilde{\varepsilon}_{100}$
$h = 3$	$\vec{\tilde{x}}_{103}(k{=}1)$	$\vec{\tilde{x}}_{102}(k{=}1)$, $\vec{\tilde{x}}_{101}(k{=}1)$, $\tilde{x}_{100}$	$\tilde{\varepsilon}_{102}(k{=}1)$, $\tilde{\varepsilon}_{101}(k{=}1)$
$k = 2$:			
$h = 1$	$\vec{\tilde{x}}_{101}(k{=}2)$	$\tilde{x}_{100}$, $\tilde{x}_{99}$, $\tilde{x}_{98}$	$\tilde{\varepsilon}_{100}$, $\tilde{\varepsilon}_{99}$
$h = 2$	$\vec{\tilde{x}}_{102}(k{=}2)$	$\vec{\tilde{x}}_{101}(k{=}2)$, $\tilde{x}_{100}$, $\tilde{x}_{99}$	$\tilde{\varepsilon}_{101}(k{=}2)$, $\tilde{\varepsilon}_{100}$
$h = 3$	$\vec{\tilde{x}}_{103}(k{=}2)$	$\vec{\tilde{x}}_{102}(k{=}2)$, $\vec{\tilde{x}}_{101}(k{=}2)$, $\tilde{x}_{100}$	$\tilde{\varepsilon}_{102}(k{=}2)$, $\tilde{\varepsilon}_{101}(k{=}2)$
⋮	⋮	⋮	⋮

The interval limits of the α-level sets of all s simulated fuzzy variables $\vec{\tilde{x}}_{N+h}$ are sorted and indexed according to Eq. (4.17) from the smallest to the largest value. The interval limits of the α-level sets of the fuzzy forecast intervals $\tilde{x}^{\kappa}_{N+h}$ at time points $\tau = N+1$, $N+2$, $N+3$ may subsequently be estimated according to Eq. (4.18).

The last six values of the given fuzzy time series as well as the fuzzy forecast intervals $\tilde{x}^{\kappa}_{101}$, $\tilde{x}^{\kappa}_{102}$ and $\tilde{x}^{\kappa}_{103}$ are shown in Fig. 4.5.

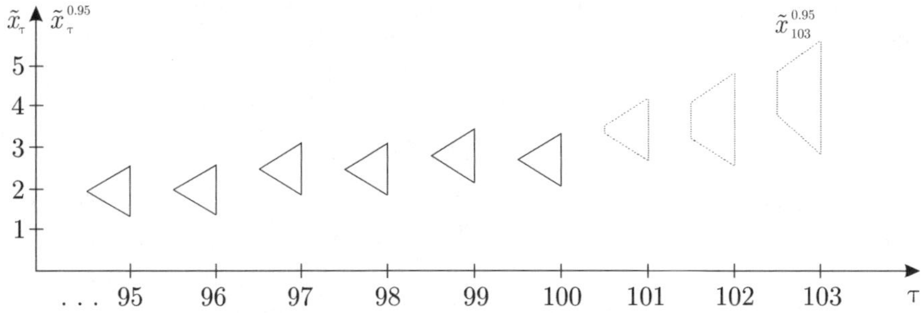

Fig. 4.5. Fuzzy time series and fuzzy forecast intervals

◆

4.2.3 Fuzzy Random Forecast

The fuzzy random variables $\vec{\tilde{X}}_{N+h}$ for the future time points $\tau = N+h$ are determined by means of a fuzzy random forecast (see Def. 4.6). For this purpose, sequences of future conditional realizations $(\tilde{x}_{N+h} \,|\, \tilde{x}_1, \tilde{x}_2, ..., \tilde{x}_N)$ of $(\tilde{X}_\tau)_{\tau\in\mathbf{T}}$ are simulated by means of repeated Monte Carlo simulations. This results in several realizations $\vec{\tilde{x}}_{N+h}$ for each time point $\tau = N+h$. These realizations are evaluated statistically.

If the underlying process of a given fuzzy time series $(\tilde{x}_\tau)_{\tau\in\mathbf{T}}$ is taken to be a fuzzy MA, AR or ARMA process $(\tilde{X}_\tau)_{\tau\in\mathbf{T}}$, the Monte Carlo simulation of future progressions of the fuzzy time series is carried out according to the recursive procedure described in Sect. 4.2.2. The fuzzy random variable $\vec{\tilde{X}}_{N+1}$ of the fuzzy random forecasting process $(\vec{\tilde{X}}_\tau)_{\tau\in\mathbf{T}}$ is given by Eq. (4.21).

$$\begin{aligned}\vec{\tilde{X}}_{N+1} = {} & \underline{A}_1 \odot \tilde{x}_N \oplus ... \oplus \underline{A}_p \odot \tilde{x}_{N+1-p} \oplus \tilde{\mathcal{E}}_{N+1} \ominus \\ & \underline{B}_1 \odot \tilde{\varepsilon}_N \ominus ... \ominus \underline{B}_q \odot \tilde{\varepsilon}_{N+1-q}\end{aligned} \tag{4.21}$$

The fuzzy variables $\tilde{x}_\tau$ at time points $\tau \leqslant N$ are given by the fuzzy time series, the fuzzy variables $\tilde{\varepsilon}_\tau$ are given by model matching. Following successive Monte Carlo simulations of the realizations $\tilde{\varepsilon}_{N+1}$, $\tilde{\varepsilon}_{N+2}$, ... of the fuzzy white-noise process $(\tilde{\mathcal{E}}_\tau)_{\tau\in\mathbf{T}}$ it is possible to compute potential future realizations $\vec{\tilde{x}}_{N+1}$, $\vec{\tilde{x}}_{N+2}$, ... of the fuzzy random variables $\vec{\tilde{X}}_{N+1}$, $\vec{\tilde{X}}_{N+2}$, ...

On the basis of s simulated realizations $\vec{\tilde{x}}^{c}_{\tau}$ for $\tau = N+1$, $N+2$, ... with $c = 1, 2, ..., s$ the empirical fuzzy probability distribution function form II

${}_{lr}\hat{F}_{\vec{\tilde{X}}_{N+h}}(\tilde{x})$ may be determined for each time point $\tau = N + h$ according to Sect. 2.2.2. For this purpose the simulated fuzzy variables $\vec{\tilde{x}}^c_{N+h}$ are statistically evaluated according to Eq. (2.115). The resulting empirical fuzzy probability distribution functions form II ${}_{lr}\hat{F}_{\vec{\tilde{X}}_{N+h}}(\tilde{x})$ are in each case unbiased estimators for the distributions of the fuzzy random variables $\vec{\tilde{X}}_{N+h}$ at time points $\tau = N + h$. Theoretical fuzzy probability distribution functions form II may be derived from the latter if required.

With the aid of the simulated fuzzy variables $\vec{\tilde{x}}^c_{N+h}$ with $c = 1, 2, ..., s$, estimators for the characteristic moments of the fuzzy random variables $\vec{\tilde{X}}_{N+h}$ at each time point $\tau = N+h$ are obtained as follows. The estimator $\hat{E}[\vec{\tilde{X}}_{N+h}]$ for the fuzzy expected value $E[\vec{\tilde{X}}_{N+h}]$ is defined according to Eq. (4.22) as the fuzzy mean value $\tilde{\overline{x}}_{N+h}$ of the fuzzy variables $\vec{\tilde{x}}^c_{N+h}$ simulated at time point $\tau = N + h$.

$$\hat{E}[\vec{\tilde{X}}_{N+h}] = \tilde{\overline{x}}_{N+h} = \frac{1}{s} \bigoplus_{c=1}^{s} \vec{\tilde{x}}^c_{N+h} \tag{4.22}$$

According to Eq. (4.1) the fuzzy expected value $E[\vec{\tilde{X}}_{N+h}]$ is equal to the fuzzy variable $\mathring{\tilde{x}}_{N+h}$ of the optimum forecast. This condition may be used to assess the quality of the Monte Carlo simulation. A criterion for the quality of the Monte Carlo simulation may be derived from the $l_\alpha r_\alpha$-subtraction $\hat{E}[\vec{\tilde{X}}_{N+h}] \ominus E[\vec{\tilde{X}}_{N+h}]$ or $\tilde{\overline{x}}_{N+h} \ominus \mathring{\tilde{x}}_{N+h}$. According to Eq. (4.23) the absolute value of the empirical $l_\alpha r_\alpha$-variance ${}_{lr}Var$ for the $l_\alpha r_\alpha$-subtraction $\tilde{\overline{x}}_{N+h}(c) \ominus \mathring{\tilde{x}}_{N+h}$ with $c = 1, 2, ..., s$ approaches zero as the number of realizations s increases. The Monte Carlo simulation hence yields an improved mapping of the characteristics of the fuzzy random forecasting process $(\vec{\tilde{X}}_\tau)_{\tau\in\mathbf{T}}$ as the number of realizations s increases.

$$\lim_{s\to\infty} \left| {}_{lr}Var \left[\tilde{\overline{x}}_{N+h}(c) \ominus \mathring{\tilde{x}}_{N+h} \mid c = 1, 2, ..., s \right] \right| = 0 \tag{4.23}$$

By specifying a maximum value η for the absolute value of the empirical $l_\alpha r_\alpha$-variance ${}_{lr}Var$ it is possible to determine a minimum number s_m of realizations according to Eq. (4.24). This is achieved by checking whether Eq. (4.24) is satisfied after s_m simulations.

$$\left| {}_{lr}Var \left[\tilde{\overline{x}}_{N+h}(c) \ominus \mathring{\tilde{x}}_{N+h} \mid c = 1, 2, ..., s_m \right] \right| \leqslant \eta \tag{4.24}$$

In order to evaluate Eq. (4.24) the elements of the $l_\alpha r_\alpha$-covariance function ${}_{lr}\underline{K}_{\vec{\tilde{X}}_\tau}(\tau_a, \tau_b)$ of the fuzzy random forecasting process $(\vec{\tilde{X}}_\tau)_{\tau\in\mathbf{T}}$ are estimated for $\tau_a, \tau_b = N + 1, N + 2, ...$ by means of Eq. (4.25) with $i, j = 1, 2, ..., n$, $l^* = l, r$, and $r^* = l, r$.

$$\hat{k}^{\alpha_i l*}_{\alpha_j r*}(\tau_a,\tau_b) = \frac{1}{s-1}\sum_{c=1}^{s}\Big[(\Delta\vec{x}^{c}_{\alpha_i l*}(\tau_a) - \Delta\mathring{x}_{\alpha_i l*}(\tau_a)) \ldots \\ \ldots (\Delta\vec{x}^{c}_{\alpha_j r*}(\tau_b) - \Delta\mathring{x}_{\alpha_j r*}(\tau_b))\Big] \quad (4.25)$$

The terms $\Delta\vec{x}^{c}_{\alpha_i l*}(\tau)$ denote the $l_\alpha r_\alpha$-increments of the simulated fuzzy variables $\tilde{\vec{x}}^{c}_{\tau}$ at time point $\tau > N$, whereas the terms $\Delta\mathring{x}_{\alpha_i l*}(\tau)$ denote the $l_\alpha r_\alpha$-increments of the optimum fuzzy forecast $\tilde{\mathring{x}}_\tau$ or the fuzzy expected value $E[\tilde{\vec{X}}_\tau]$. The elements of the estimator for the $l_\alpha r_\alpha$-variance ${}_{lr}\underline{\sigma}^2_{\tilde{\vec{X}}_\tau}$ correspond to the leading diagonal elements of the estimated $l_\alpha r_\alpha$-covariance function ${}_{lr}\underline{\hat{K}}_{\tilde{\vec{X}}_\tau}(\tau_a,\tau_b)$ for $\tau_a = \tau_b = \tau$.

The fuzzy random forecast yields the fuzzy random variables $\tilde{\vec{X}}_\tau$ of the fuzzy random forecasting process $(\tilde{\vec{X}}_\tau)_{\tau\in\mathbf{T}}$ for future time points $\tau = N+h$. In order to characterize the fuzzy random variables $\tilde{\vec{X}}_{N+h}$ the fuzzy probability distribution functions form II ${}_{lr}F_{\tilde{\vec{X}}_\tau}(\tilde{x})$ and the characteristic moments are used.

Example 4.11. Based on the fuzzy time series of Example 4.10 it is intended to determine the fuzzy random variables $\tilde{\vec{X}}_\tau$ of the fuzzy random forecasting process $(\tilde{\vec{X}}_\tau)_{\tau\in\mathbf{T}}$ at time points $\tau = N+1$, $N+2$, $N+3$. For this purpose the simulated s sequences of future realizations $\tilde{\vec{x}}_{N+1}(k)$, $\tilde{\vec{x}}_{N+2}(k)$, $\tilde{\vec{x}}_{N+3}(k)$ ($k = 1, 2, \ldots, s$) of Example 4.10 are evaluated statistically. The last six values of the given fuzzy time series as well as three typical sequences of future realizations are shown in Fig. 4.6.

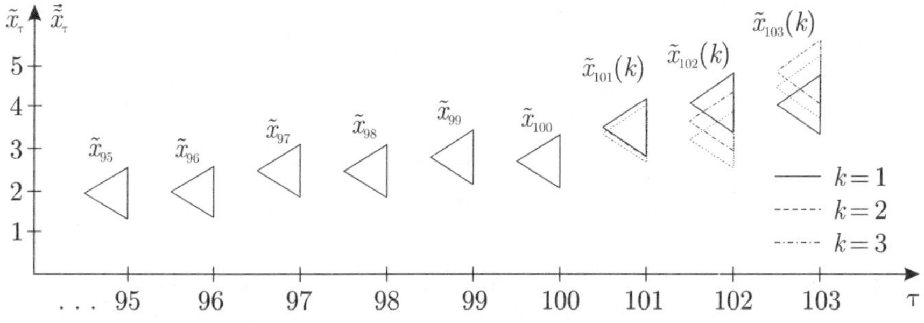

Fig. 4.6. Fuzzy time series and sequences of future realizations

This is demonstrated by considering the result of the statistical evaluation at time point $\tau = 102$ for an $l_\alpha r_\alpha$-discretization with $n = 2$ α-levels ($\alpha_1 = 0$, $\alpha_2 = 1$). The fuzzy expected value $E[\tilde{\vec{X}}_{N+2}]$ estimated according to Eq. (4.22) corresponds to the optimum forecast as follows:

$$\hat{E}[\tilde{\vec{X}}_{N+2}] = \tilde{\bar{x}}_{N+2} = \begin{bmatrix} 0.6 \\ 3.5 \\ 0 \\ 0.7 \end{bmatrix} . \qquad (4.26)$$

The empirical $l_\alpha r_\alpha$-correlation function ${}_{lr}\underline{\hat{R}}_{\tilde{\vec{X}}_\tau}(\tau_a, \tau_b)$ for $\tau_a = \tau_b = 102$ is given by Eq. (4.27).

$${}_{lr}\underline{\hat{R}}_{\tilde{\vec{X}}_\tau}(\tau_a = \tau_b = 102) = \begin{bmatrix} 1 & 0.21 & -0.84 \\ 0.21 & 1 & -0.17 \\ - & - & - & - \\ 0.84 & 0.17 & - & 1 \end{bmatrix} \qquad (4.27)$$

The empirical fuzzy probability distribution function form II ${}_{lr}\hat{F}_{\tilde{\vec{X}}_{N+2}}(\tilde{x})$ according to Eq. (2.115) is used as an estimator for the fuzzy probability distribution function form II ${}_{lr}F_{\tilde{\vec{X}}_{N+2}}(\tilde{x})$ of the fuzzy random variable $\tilde{\vec{X}}_{N+2}$ at time $\tau = N+2$. Selected functional values of the empirical fuzzy probability distribution function form II ${}_{lr}\hat{F}_{\tilde{\vec{X}}_{N+2}}(\tilde{x})$ are given in Table 4.3.

Table 4.3. Selected functional values of the empirical fuzzy probability distribution function form II ${}_{lr}\hat{F}_{\tilde{\vec{X}}_{N+2}}(\tilde{x})$ in the coordinate system of the increments

Coordinates				${}_{lr}\hat{F}_{\tilde{\vec{X}}_{N+2}}(\tilde{x})$
Δx_1	Δx_2	Δx_3	Δx_4	
0.53	3.44	0	0.64	0.45
0.64	3.58	0	0.71	0.53
0.63	3.72	0	0.69	0.64
0.55	3.41	0	0.68	0.42
⋮				⋮

◆

4.3 Forecasting on the Basis of Artificial Neural Networks

Forecasting on the basis of artificial neural networks is an alternative to forecasting on the basis of specific fuzzy random processes, as described in Sect. 4.2. Depending on the forecasting objective, a precondition for the latter is that an optimum forecasting or an optimum simulation network architecture has been trained. An optimum forecast, fuzzy forecast intervals or a fuzzy random forecast may be chosen as the forecasting objective.

4.3.1 Optimum Forecast

Optimum h-step forecast. For the optimum h-step forecast, i.e. for forecasting $\mathring{\tilde{x}}_{N+1}$, $\mathring{\tilde{x}}_{N+2}$, ..., $\mathring{\tilde{x}}_{N+h}$, an optimum forecasting network architecture (see Sect. 3.6.4) with $n_O = 1$ control variables must already have been trained. By means of this network architecture the optimum forecasts are determined step by step. Because the optimum forecasting network has n_I neurons in the input layer and $n_O = 1$ neurons in the output layer after training, the fuzzy input vector must always contain n_I fuzzy elements: $\underline{\tilde{x}} = (\tilde{x}_1, \tilde{x}_2, ..., \tilde{x}_{n_I})^T$.

In the first step the last n_I elements of the given fuzzy time series $\underline{\tilde{x}}_{N+1} = (\tilde{x}_{N-n_I+1}, \tilde{x}_{N-n_I+2}, ..., \tilde{x}_N)^T$ are assigned to the fuzzy input vector. The optimum forecasting network yields the optimum forecast $\mathring{\tilde{x}}_{N+1}$ via the neuron in the output layer.

In the second step the optimum forecast $\mathring{\tilde{x}}_{N+2}$ is computed. Using $\mathring{\tilde{x}}_{N+1}$, the new values $\underline{\tilde{x}}_{N+2} = (\tilde{x}_{N-n_I+2}, \tilde{x}_{N-n_I+3}, ..., \mathring{\tilde{x}}_{N+1})^T$ are assigned to the elements of the fuzzy input vector. The first three steps are presented graphically in Fig. 4.7. The values assigned to the elements of the fuzzy input vectors in subsequent steps are listed in Table 4.4. This serves to explain how the optimum h-step forecast is obtained with aid of an optimum forecasting network architecture.

Before being transferred to the artificial neural network each fuzzy input vector is subject to a conditioning procedure, as outlined in Sect. 3.6.5. The $\mathring{\tilde{x}}_{N+1}$, $\mathring{\tilde{x}}_{N+2}$, ..., $\mathring{\tilde{x}}_{N+h}$ are obtained in the original number range.

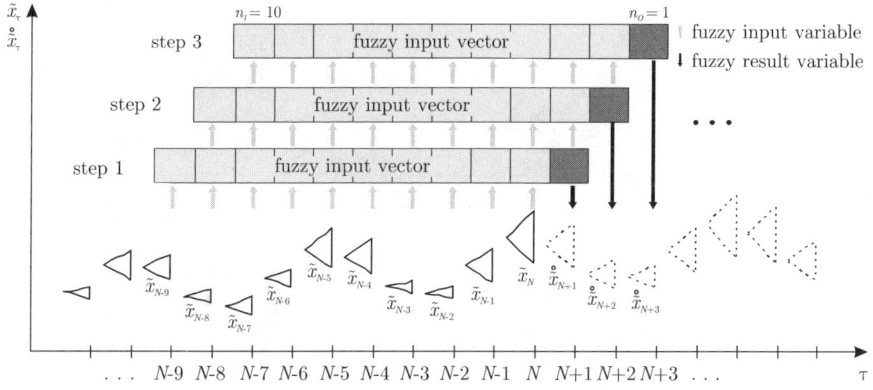

Fig. 4.7. Schematic representation of the optimum h-step forecast

Table 4.4. Elements of the fuzzy input vectors in the optimum h-step forecast

fuzzy input vector	optimum forecast
$\underline{\tilde{x}}_{N+1} = (\tilde{x}_{N-n_I+1}, \tilde{x}_{N-n_I+2}, ..., ..., ..., \tilde{x}_N)^T$	$\overset{\circ}{\tilde{x}}_{N+1}$
$\underline{\tilde{x}}_{N+2} = (\tilde{x}_{N-n_I+2}, \tilde{x}_{N-n_I+3}, ..., \tilde{x}_N, \overset{\circ}{\tilde{x}}_{N+1})^T$	$\overset{\circ}{\tilde{x}}_{N+2}$
$\underline{\tilde{x}}_{N+3} = (\tilde{x}_{N-n_I+3}, \tilde{x}_{N-n_I+4}, ..., \overset{\circ}{\tilde{x}}_{N+1}, \overset{\circ}{\tilde{x}}_{N+2})^T$	$\overset{\circ}{\tilde{x}}_{N+3}$
$\underline{\tilde{x}}_{N+4} = (\tilde{x}_{N-n_I+4}, \tilde{x}_{N-n_I+5}, ..., \overset{\circ}{\tilde{x}}_{N+2}, \overset{\circ}{\tilde{x}}_{N+3})^T$	$\overset{\circ}{\tilde{x}}_{N+4}$
$\vdots$	$\vdots$

Remark 4.12. Besides providing forecasts at time points $\tau = N+1, N+2, ..., N+h$, optimum h-step forecasts are also applied for validating the optimum forecasting network architecture (see Sect. 3.6.4). The evolution of the elements of the fuzzy input vectors is similar to that shown in Table 4.4, with N now replaced by N_T. The optimum h-step forecasts $\overset{\circ}{\tilde{x}}_{N_T+1}, \overset{\circ}{\tilde{x}}_{N_T+2}, ..., \overset{\circ}{\tilde{x}}_{N_T+h}$ are computed for the period of the fuzzy validation series and then compared with the given fuzzy variables of the validation series. The error computed from the differences serves as a means of assessing the quality of the optimum forecasting network architecture. ◆

Optimum single-step forecast. The optimum single-step forecast on the basis of artificial neural networks is again a special case of the optimum h-step forecast. The optimum forecast $\overset{\circ}{\tilde{x}}_{N+1}$ is computed using the observed values $\tilde{x}_\tau$ $(\tau \leqslant N)$ of the fuzzy time series. The fuzzy input vector contains the observed values $\tilde{x}_{N-n_I+1}, \tilde{x}_{N-n_I+2}, ..., \tilde{x}_N$ (see Fig. 4.8).

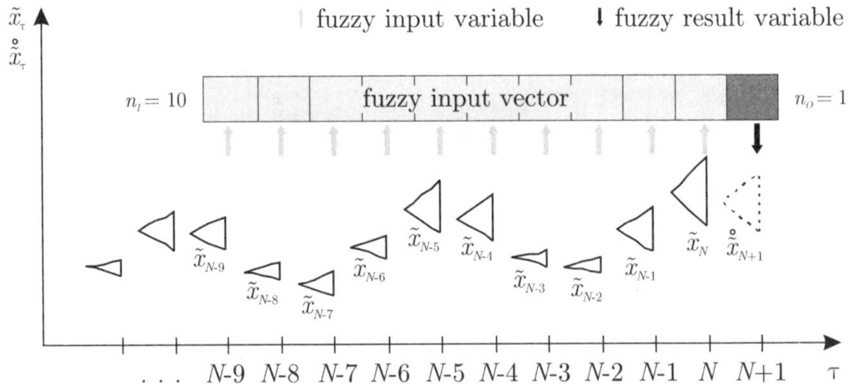

Fig. 4.8. Schematic representation of the optimum single-step forecast

Remark 4.13. Optimum single-step forecasts are applied in the determination of the optimum simulation network architecture (see Sect. 3.6.4). The optimum single-step forecasts $\overset{\circ}{\tilde{x}}_\tau$ are computed for the observation period

$n_I < \tau \leqslant N$ and subtracted from the observed values according to Eq. (3.183). The fuzzy input vector contains the observed values $\tilde{x}_{\tau-n_I}, \tilde{x}_{\tau-n_I+1}, ..., \tilde{x}_{\tau-1}$ for time points $\tau = n_I + 1, n_I + 2, ..., N$. In contrast to the optimum h-step forecast, only the observed values $\tilde{x}_\tau$ of the fuzzy time series are used for computing the optimum single-step forecasts. ◆

4.3.2 Fuzzy Forecast Intervals

According to Def. 4.3, a fuzzy interval $\tilde{x}_I$ is referred to as a fuzzy forecast interval $\tilde{x}^{\kappa}_{N+h}$ if the realization of the fuzzy random variable $\vec{\tilde{X}}_{N+h}$ of the fuzzy random forecasting process $(\vec{\tilde{X}}_\tau)_{\tau\in\mathbf{T}}$ will be completely included in the interval with a probability κ. The definition of complete inclusion of a fuzzy variable $\vec{\tilde{x}}_{N+h}$ in a fuzzy interval $\tilde{x}_I$ is given by Eq. (4.2). The determination of fuzzy forecast intervals $\tilde{x}^{\kappa}_{N+h}$ using artificial neural networks for fuzzy variables is presented in the following section.

Firstly, the optimum forecasting artificial neural network belonging to the given fuzzy time series is determined as outlined in Sect. 3.6.4. With the aid of the optimum forecasting network architecture optimum single-step forecasts $\mathring{\tilde{x}}_\tau$ are determined for time points $\tau = n_I^{(p)} + 1, n_I^{(p)} + 2, ..., N$. The term $n_I^{(p)}$ hereby denotes the number of artificial neurons in the input layer of the optimum forecasting multilayer perceptron. By subtracting the optimum single-step forecasts $\mathring{\tilde{x}}_\tau$ from the known observed values $\tilde{x}_\tau$ it is possible to compute the fuzzy variables $\tilde{e}_\tau$ for $\tau = n_I^{(p)} + 1, n_I^{(p)} + 2, ..., N$ according to Eq. (4.28).

$$\tilde{e}_\tau = \tilde{x}_\tau \ominus \mathring{\tilde{x}}_\tau \tag{4.28}$$

The fuzzy variables $\tilde{e}_\tau$ form a stationary fuzzy time series which reproduces the random components of the original fuzzy time series. These are fuzzy variables in the improper sense, however, as they do not fulfill the requirement expressed by Eq. (2.47). For this reason the computed fuzzy variables $\tilde{e}_\tau$ cannot be considered as a realization of a fuzzy white-noise process. Under the precondition that all relevant dependencies within the fuzzy time series are accounted for by the optimum forecasting artificial neural network, the $\tilde{e}_\tau$ are considered to be realizations of the fuzzy random process $(\tilde{\mathsf{E}}_\tau)_{\tau\in\mathbf{T}}$ exhibiting the following properties.

$$E[\tilde{\mathsf{E}}_\tau] = \tilde{m}_{\tilde{\mathsf{E}}_\tau} = 0 \quad \forall\, \tau \in \mathbf{T} \tag{4.29}$$

$$_{lr}Var[\tilde{\mathsf{E}}_\tau] = {}_{lr}\underline{\sigma}^2_{\tilde{\mathsf{E}}_\tau} \tag{4.30}$$

$$_{lr}K_{\tilde{\mathsf{E}}_\tau}(\tau_a, \tau_b) = \begin{cases} _{lr}K_{\tilde{\mathsf{E}}_\tau}(\tau_a, \tau_b) & \text{for} \quad \tau_a = \tau_b \\ \underline{0} & \text{for} \quad \tau_a \neq \tau_b \end{cases} \tag{4.31}$$

In the next step, an optimum simulation artificial neural network is trained according to Sect. 3.6.4 for the time series comprised of fuzzy variables $(\tilde{e}_\tau)_{\tau\in\mathbf{T}}$ in the improper sense.

By means of the optimum simulation artificial neural network it is possible to simulate several conditional realizations at each time point $\tau = N + h$ $(h = 1, 2, ...)$ which may be used to specify fuzzy forecast intervals. The realizations are simulated following a similar procedure to that of the optimum h-step forecast. In the first step, the last $n_I^{(s)}$ fuzzy variables of the time series $(\tilde{e}_\tau)_{\tau\in\mathbf{T}}$ obtained from Eq. (4.28) are transformed according to Eq. (3.198) and lumped together in the fuzzy input vector $\underline{\tilde{e}}^*_{N+1} = (\tilde{e}^*_{N-n_I^{(s)}+1}, \tilde{e}^*_{N-n_I^{(s)}+2}, ..., \tilde{e}^*_N)^T$. In the next step, the corresponding fuzzy output variable $\tilde{o}^{O*}_{N+1}$ of the optimum simulation artificial neural network is inserted into the fuzzy input vector $\underline{\tilde{e}}^*_{N+2} = (\tilde{e}^*_{N-n_I^{(s)}+2}, \tilde{e}^*_{N-n_I^{(s)}+3}, ..., \tilde{e}^*_N, \tilde{o}^{O*}_{N+1})^T$. The analogous continuation of this procedure for the subsequent fuzzy variables $\tilde{o}^{O*}_{N+h}$, and the transformation of these into the fuzzy variables $\tilde{o}^{O}_{N+h} = \vec{\tilde{x}}_{N+h}$ according to Eq. (3.203), yields a first sequence of generated realizations for $h = 1, 2, ...$ The stepwise modification of the fuzzy input vectors is presented in Table 4.5.

Table 4.5. Elements of the conditioned fuzzy input vectors and fuzzy results of the optimum simulation artificial neural network

fuzzy input vector	fuzzy result
$\underline{\tilde{e}}^*_{N+1} = (\tilde{e}^*_{N-n_I^{(s)}+1}, \tilde{e}^*_{N-n_I^{(s)}+2}, ..., ..., ..., \tilde{e}^*_N)^T$	$\tilde{o}^{O*}_{N+1}$
$\underline{\tilde{e}}^*_{N+2} = (\tilde{e}^*_{N-n_I^{(s)}+2}, \tilde{e}^*_{N-n_I^{(s)}+3}, ..., ..., \tilde{e}^*_N, \tilde{o}^{O*}_{N+1})^T$	$\tilde{o}^{O*}_{N+2}$
$\underline{\tilde{e}}^*_{N+3} = (\tilde{e}^*_{N-n_I^{(s)}+3}, \tilde{e}^*_{N-n_I^{(s)}+4}, ..., \tilde{e}^*_N, \tilde{o}^{O*}_{N+1}, \tilde{o}^{O*}_{N+2})^T$	$\tilde{o}^{O*}_{N+3}$
$\vdots$	$\vdots$

In order to simulate further sequences of realizations the fuzzy input vector $\underline{\tilde{e}}^*_{N+1} = (\tilde{e}^*_{N-n_I^{(s)}+1}, \tilde{e}^*_{N-n_I^{(s)}+2}, ..., \tilde{e}^*_N)^T$ must be varied according to the characteristics of the underlying fuzzy random process $(\tilde{E}_\tau)_{\tau\in\mathbf{T}}$. In accordance with Eqs. (4.29) to (4.31) the fuzzy random process $(\tilde{E}_\tau)_{\tau\in\mathbf{T}}$ is stationary. This means that the variation may be achieved by a permutation of the elements of the fuzzy input vector $\underline{\tilde{e}}^*_{N+1}$. This type of approach is permissible due to the fact no correlation exists according to Eq. (4.31) between the realizations $\tilde{e}_\tau$ at different time points τ. A permutation of the elements of the fuzzy input vector $\underline{\tilde{e}}^*_{N+1}$ permits the simulation of $n_I^{(s)}!$ realizations, i.e. of $n_I^{(s)}!$ potential future progressions of a fuzzy time series $(\tilde{x}_\tau)_{\tau\in\mathbf{T}}$.

Owing to the characteristics of the fuzzy random process $(\tilde{E}_\tau)_{\tau\in\mathbf{T}}$ the permutation may be applied to the fuzzy input vectors $\underline{\tilde{e}}^*_\tau$ at time points

$$\tau = \begin{cases} n_I^{(p)} + 1,\, n_I^{(p)} + 2,\, ...,\, N & \text{for} \quad n_I^{(p)} \geqslant n_I^{(s)} \\ n_I^{(s)} + 1,\, n_I^{(s)} + 2,\, ...,\, N & \text{for} \quad n_I^{(p)} < n_I^{(s)} \end{cases} . \tag{4.32}$$

New fuzzy input vectors at these time points are generated by permutation of the elements of the fuzzy input vector $\underline{\tilde{e}}_\tau^*$. Subsequent simulations analogous to the successive procedure outlined in Table 4.5 with starting times τ according to Eq. (4.32) yield new sequences of realizations. In accordance with the latter, the number of possible ways in which such sequences may be simulated is given by

$$\begin{aligned} & n_I^{(s)}! \cdot \left(N - n_I^{(p)}\right) \quad \text{for} \quad n_I^{(p)} \geqslant n_I^{(s)} \\ \text{and} & \\ & n_I^{(s)}! \cdot \left(N - n_I^{(s)}\right) \quad \text{for} \quad n_I^{(p)} < n_I^{(s)} . \end{aligned} \tag{4.33}$$

This means that it is possible to simulate the same number of potential future progressions of a fuzzy time series $(\tilde{x}_\tau)_{\tau \in \mathbf{T}}$.

Remark 4.14. A very long fuzzy time series containing the fuzzy variables $\tilde{o}_\tau^{O*}$ ($\tau = N+1,\, N+2,\, ...$) according to Table 4.5 may alternatively be simulated using the fuzzy input vector $\underline{\tilde{e}}_{N+1}^* = (\tilde{e}^*_{N-n_I^{(s)}+1},\, \tilde{e}^*_{N-n_I^{(s)}+2},\, ...,\, \tilde{e}_N^*)^T$. Owing to the stationarity of the fuzzy random process $(\tilde{\mathsf{E}}_\tau)_{\tau \in \mathbf{T}}$ it is also possible to permutate the simulated fuzzy variables $\tilde{o}_k^{O*}$ (multiplied by the corresponding transformation matrix $\underline{T}_k^*$). In other words, the term $\underline{T}_k^{*^{-1}} \odot \tilde{o}_k^{O*}$ in Eq. (3.203) may be arbitrarily replaced by simulated fuzzy variables $\underline{T}_{k+\Delta\tau}^{*^{-1}} \odot \tilde{o}_{k+\Delta\tau}^{O*}$. Each fuzzy variable $\tilde{o}_{k+\Delta\tau}^{O*}$ may thereby only be used once. This procedure permits the simulation of an arbitrary number s of potential future progressions of a fuzzy time series $(\tilde{x}_\tau)_{\tau \in \mathbf{T}}$. ◆

With the aid of s simulated potential future progressions of a fuzzy time series $(\tilde{x}_\tau)_{\tau \in \mathbf{T}}$ it is possible to estimate the corresponding fuzzy forecast intervals $\tilde{x}_{N+h}^\kappa$ for each time point $\tau = N+h$ in a similar manner to that described in Sect. 4.3. For this purpose the interval limits $x_{\alpha_i l}(N+h)$ and $x_{\alpha_i r}(N+h)]$ of the α-level sets $X_{\alpha_i}(N+h)$ of all s simulated fuzzy variables $\tilde{o}_{N+h}^O = \vec{\tilde{x}}_{N+h}$ are sorted and indexed from the smallest to the largest value according to the inequality given by Eq. (4.34).

$$\begin{aligned} & \vec{x}_{\alpha_i l}^1(N+h) \leqslant \vec{x}_{\alpha_i l}^2(N+h) \leqslant ... \leqslant \vec{x}_{\alpha_i l}^s(N+h) \\ \text{and} & \\ & \vec{x}_{\alpha_i r}^1(N+h) \leqslant \vec{x}_{\alpha_i r}^2(N+h) \leqslant ... \leqslant \vec{x}_{\alpha_i r}^s(N+h) \end{aligned} \tag{4.34}$$

For the confidence level κ the interval limits $x_{\alpha_i l}^\kappa(N+h)$ and $x_{\alpha_i r}^\kappa(N+h)$ of the α-level sets $X_{\alpha_i}^\kappa(N+h)$ of a fuzzy forecast interval $\tilde{x}_{N+h}^\kappa$ at time point

$\tau = N + h$ may be obtained according to Eq. (4.35). Eq. (4.35) only holds provided s is an even number.

$$x^{\kappa}_{\alpha_i l}(N+h) = \begin{cases} \leqslant \vec{x}^{1}_{\alpha_i l}(N+h) \text{ for } a = 0 \\ \vec{x}^{a}_{\alpha_i l}(N+h) \text{ for } 0 < a \leqslant \frac{s}{2} \end{cases} \quad \text{with } a = \text{int}\left[s \cdot \left(\frac{1}{2} - \frac{\kappa}{2}\right)\right]$$

and (4.35)

$$x^{\kappa}_{\alpha_i r}(N+h) = \begin{cases} \vec{x}^{b+1}_{\alpha_i r}(N+h) \text{ for } \frac{s}{2} \leqslant b < s \\ \geqslant \vec{x}^{s}_{\alpha_i r}(N+h) \text{ for } b = s \end{cases} \quad \text{with } b = \frac{s}{2} + \text{int}\left[s \cdot \left(\frac{\kappa}{2}\right)\right]$$

Analogous to Sect. 4.2.2, the interval limits of the α-level sets for the fuzzy forecast intervals $\tilde{x}^{\kappa}_{N+h}$ according to Eq. (4.35) correspond to the lower and upper quantile values of the empirical distributions of the interval limits for the simulated realizations $\vec{\tilde{x}}_{N+h}$. In other words, the future realizations of a fuzzy time series $(\tilde{x}_{\tau})_{\tau \in \mathbf{T}}$ at time point $\tau = N + h$ are completely included in the fuzzy forecast interval $\tilde{x}^{\kappa}_{N+h}$ with a probability κ.

As explained in Sect. 4.2.2, the fuzzy forecast intervals $\tilde{x}^{\kappa}_{N+h}$ only apply in each case to the fuzzy variables to be expected at the individual time points $\tau = N + h$. For a determination of the probability with which possible future progressions of a fuzzy time series lie completely within a prescribed forecast hose, the reader is referred to Sect. 4.2.2.

4.3.3 Fuzzy Random Forecast

In the case of a fuzzy random forecast the fuzzy random variables $\vec{\tilde{X}}_{\tau}$ of the fuzzy random forecasting process $(\vec{\tilde{X}}_{\tau})_{\tau \in \mathbf{T}}$ are determined for time points $\tau = N + h$ $(h = 1, 2, ...)$. This involves the simulation of s sequences of fuzzy realizations, which are subsequently evaluated statistically. The simulation of the s sequences of fuzzy realizations of the fuzzy random forecasting process $(\vec{\tilde{X}}_{\tau})_{\tau \in \mathbf{T}}$ by means of artificial neural networks has already been developed in Sect. 4.3.2 for computing fuzzy forecast intervals. This approach may also be applied to fuzzy random forecasts.

If s sequences of realizations are available, the fuzzy probability distribution functions form II $_{lr}F_{\vec{\tilde{X}}_{N+h}}(\tilde{x})$ (see Sect. 2.2.2) and the characteristic moments (see Sect. 2.2.3) may be estimated for the purpose of characterizing the fuzzy random variables $\vec{\tilde{X}}_{N+h}$.

With the aid of s simulated realizations $\vec{\tilde{x}}^{c}_{\tau}$ for $\tau = N + 1,\ N + 2,$ ... with $c = 1,\ 2,\ ...,\ s$, estimators for the characteristic values of the fuzzy random variables $\vec{\tilde{X}}_{N+h}$ may be obtained analogous to Sect. 4.3.3 for each time point $\tau = N + h$. The fuzzy expected value $E[\vec{\tilde{X}}_{N+h}]$ is estimated according to Eq. (4.22) as the fuzzy mean value $\bar{\tilde{x}}_{N+h}$ of the simulated fuzzy variables $\vec{\tilde{x}}^{c}_{N+h}$ at time point $\tau = N + h$. The estimator for the $l_{\alpha}r_{\alpha}$-covariance function

${}_{lr}\underline{K}_{\vec{\tilde{\mathrm{X}}}_\tau}(\tau_a, \tau_b)$ of the fuzzy random forecasting process $(\vec{\tilde{\mathrm{X}}}_\tau)_{\tau \in \mathbf{T}}$ is defined by Eq. (4.25) for $\tau_a, \tau_b = N+1,\ N+2,\ \ldots$. The leading diagonal elements of the estimated $l_\alpha r_\alpha$-covariance function ${}_{lr}\hat{\underline{K}}_{\vec{\tilde{\mathrm{X}}}_\tau}(\tau_a, \tau_b)$ correspond to the elements of the estimator for the $l_\alpha r_\alpha$-variance ${}_{lr}\underline{\sigma}^2_{\vec{\tilde{\mathrm{X}}}_\tau}$ for $\tau_a = \tau_b = \tau$.

The respective fuzzy probability distribution function form II ${}_{lr}F_{\vec{\tilde{\mathrm{X}}}_{N+h}}(\tilde{x})$ may also be estimated with the aid of the simulated fuzzy variables $\vec{\tilde{x}}^c_{N+h}$ with $c = 1,\ 2,\ \ldots,\ s$ for each time point $\tau = N+h$. From a statistical evaluation of the simulated realizations $\vec{\tilde{x}}^c_{N+h}$ the empirical fuzzy probability distribution functions form II ${}_{lr}\hat{F}_{\vec{\tilde{\mathrm{X}}}_{N+h}}(\tilde{x})$ are obtained as unbiased estimators for ${}_{lr}F_{\vec{\tilde{\mathrm{X}}}_{N+h}}(\tilde{x})$ according to Sect. 2.2.2. In this way it is also possible to derive theoretical fuzzy probability distribution functions form II.

Analogous to Sect. 4.3.3, the quality of the simulation may be checked in relation to the number s of realizations with the aid of the estimator $\hat{E}[\vec{\tilde{\mathrm{X}}}_{N+h}]$ of the fuzzy expected value. As already described, a minimum number s_m of realizations may be specified for which the absolute value of the empirical $l_\alpha r_\alpha$-variance ${}_{lr}Var$ of the $l_\alpha r_\alpha$-subtraction according to Eq. (4.23) does not exceed a prescribed maximum value.

Example 4.15. Given is a fuzzy random forecasting process $(\vec{\tilde{\mathrm{X}}}_\tau)_{\tau \in \mathbf{T}}$ with an optimum forecasting and an optimum simulation artificial neural network. The minimum number s_m of realizations $\vec{\tilde{x}}^s_{N+10}$ ($s = 1,\ 2,\ \ldots,\ s_m$) to be simulated for time point $\tau = N+10$ is sought, under the condition that $\eta = 0.05$ (see Eq. (4.24)) should be complied with. The optimum fuzzy forecast for this time point is $\mathring{\tilde{x}}_{N+10}$. In order to check the quality of the simulation the fuzzy mean value $\bar{\tilde{x}}_{N+10}(s)$ is computed in each case for $s = 1,\ 2,\ \ldots$ simulated fuzzy variables $\vec{\tilde{x}}^1_{N+10},\ \vec{\tilde{x}}^2_{N+10},\ \ldots,\ \vec{\tilde{x}}^s_{N+10}$ according to Eq. (4.22). In the next step, the differences between the computed fuzzy mean values $\bar{\tilde{x}}_{N+10}(s)$ and the optimum fuzzy forecast $\mathring{\tilde{x}}_{N+10}$ are calculated. The empirical $l_\alpha r_\alpha$-variance ${}_{lr}Var_{N+10}(s)$ is then computed according to Eq. (4.36) for $s = 2,\ 3,\ \ldots 180$ on the basis of the differences $\bar{\tilde{x}}_{N+10}(c) \ominus \mathring{\tilde{x}}_{N+10}$ with $c = 1,\ 2,\ \ldots,\ s$ for each s.

$$ {}_{lr}Var_{N+10}(s) = {}_{lr}Var\left[\bar{\tilde{x}}_{N+10}(c) \ominus \mathring{\tilde{x}}_{N+10} \mid c = 1,\ 2,\ \ldots,\ s\right] \qquad (4.36) $$

The result of the foregoing is presented graphically in Fig. 4.9. For $s = 112$ it is found that the specified quality of $\eta = 0.05$ is undershot, i.e. at least $s_m = 112$ realizations must be simulated. ◆

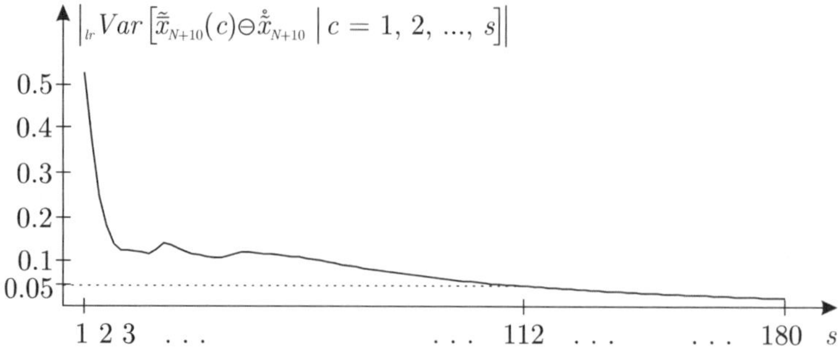

Fig. 4.9. Quality of the simulation versus the number of realizations s

5

Uncertain Forecasting in Engineering and Environmental Science

5.1 Model-Free Forecasting

Time series comprised of fuzzy data are frequently encountered in engineering and environmental science. These represent the results of regular observations and measurements, and contain information on measurable physical parameters. Time series in engineering either relate to measured actions such as settlement, displacements, loads, temperature, moisture and toxic substances, or to measured structural responses such as settlement, displacements, crack widths, concrete spalling, corrosion and carbonation (see Fig. 5.1). The reasons for the uncertainty of the measurements are manifold: individual measurements may be uncertain or fluctuate within an interval whereas a number of measurements may vary by different amounts.

The forecasting of fuzzy time series enables future actions and structural responses to be computed. For this purpose the given measured values are treated as a realization of a fuzzy random process. The unknown underlying process is modeled either as a specific fuzzy random process or as an artificial neural network. By means of the matched fuzzy random process or the trained artificial neural network it is possible to directly forecast future actions or structural responses. This type of forecasting is referred to as model-free forecasting (see Fig. 5.1). In the case of model-free forecasting only the variants of optimum forecasting, fuzzy forecast intervals and fuzzy random forecasting developed in Chap. 4 are applied without combining these with a computational model.

By combining model-free forecasting with a computational model it is possible to generate time series comprised of non-measurable physical parameters. Non-measurable parameters in engineering include, e.g. characteristic parameters for describing damage state, robustness or safety level. These are computed with the aid of a computational model. Computational models include among others finite element models or models for computing failure probability.

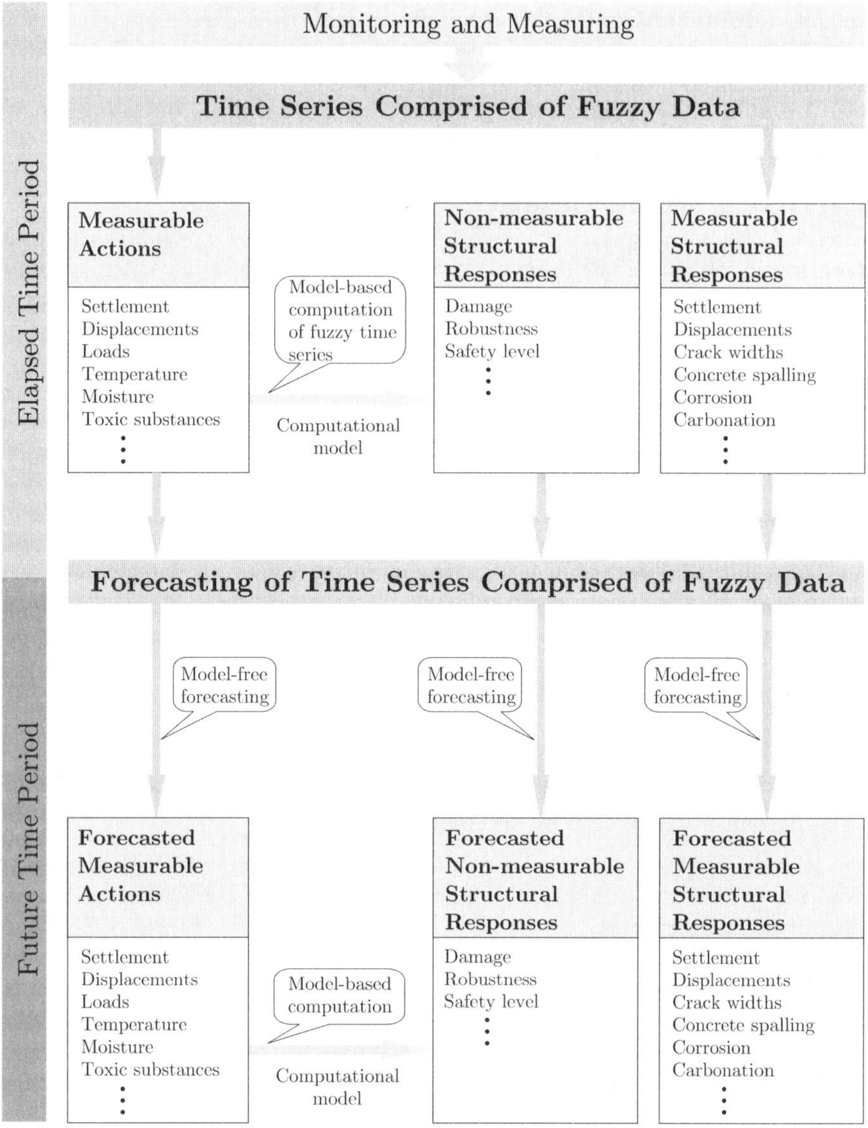

Fig. 5.1. Strategies for forecasting structural responses

The principle of model-free forecasting is shown in Fig. 5.2. Given is a sequence of measured uncertain data. The data describe measurable actions or measurable structural responses. These constitute a fuzzy time series. Each measurement date is treated as a convex fuzzy variable in $l_\alpha r_\alpha$-increment representation. Forecasting by means of a fuzzy forecasting process or an artificial neural network yields future values of the fuzzy time series directly. An optimum forecast, a fuzzy random forecast or fuzzy forecast intervals may be chosen as the forecasting variant.

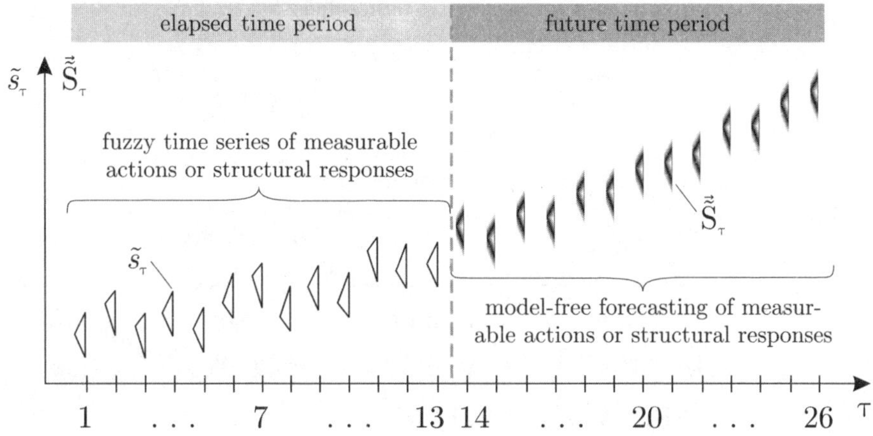

Fig. 5.2. Model-free forecasting

5.2 Model-Based Forecasting

A combination of model-free forecasting with a computational model results in model-based forecasting. In this respect, a distinction is made between two strategies.

In the case of Strategy I (see Fig. 5.3) future values are initially forecasted for a fuzzy time series comprised of measured actions by means of an optimum forecast, fuzzy forecast intervals or a fuzzy random forecast. These forecasted fuzzy variables or fuzzy random variables are the input data for a computational model, which may be used to compute non-measurable fuzzy structural responses. For example, the forecasted structural response might be the future damage state.

In the case of Strategy II (see Fig. 5.4) the starting point is also a fuzzy time series of measured actions. These fuzzy data are first transferred as input data to a computational model. Non-measurable fuzzy structural responses are computed using the computational model. For example, the result might be the simulated time history of the damage state in the form of a fuzzy

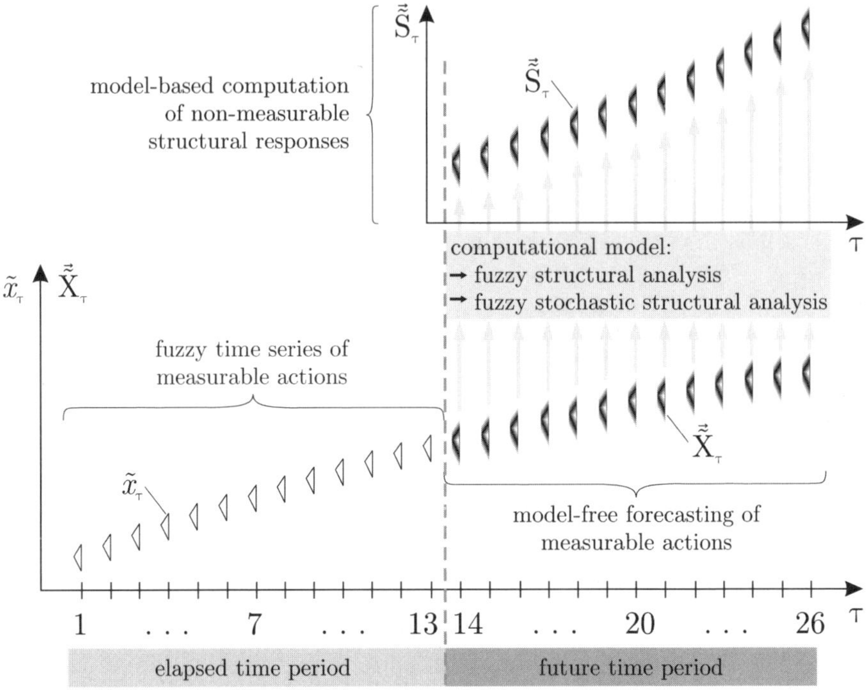

Fig. 5.3. Model-based forecasting – Strategy I

time series. This fuzzy time series of a non-measurable variable is treated as a realization of a fuzzy random process, and the underlying process is either modeled as a specific fuzzy random process or as an artificial neural network. Future realizations may then be forecasted by means of an optimum forecast, fuzzy forecast intervals or a fuzzy random forecast.

In the case of Strategy I the computational model must be capable of processing fuzzy variables or fuzzy random variables as input data. In the case of Strategy II, however, only fuzzy variables are transferred as input data to the computational model. Depending on the input data, i.e. fuzzy variables or fuzzy random variables, a different analysis algorithm is implemented in the computational model. In the case of fuzzy variables the required analysis algorithm is based on fuzzy structural analysis, whereas for fuzzy random variables, fuzzy stochastic structural analysis is applied.

Model-Based Forecasting and Fuzzy Structural Analysis

By means of fuzzy structural analysis it is possible to map the fuzzy input variables $\tilde{x}_1, \tilde{x}_2, ..., \tilde{x}_l$ onto the fuzzy result variables $\tilde{z}_1, \tilde{z}_2, ..., \tilde{z}_m$.

$$\underline{\tilde{z}} = (\tilde{z}_1, \tilde{z}_2, ..., \tilde{z}_m) = \tilde{f}(\tilde{x}_1, \tilde{x}_2, ..., \tilde{x}_l) \tag{5.1}$$

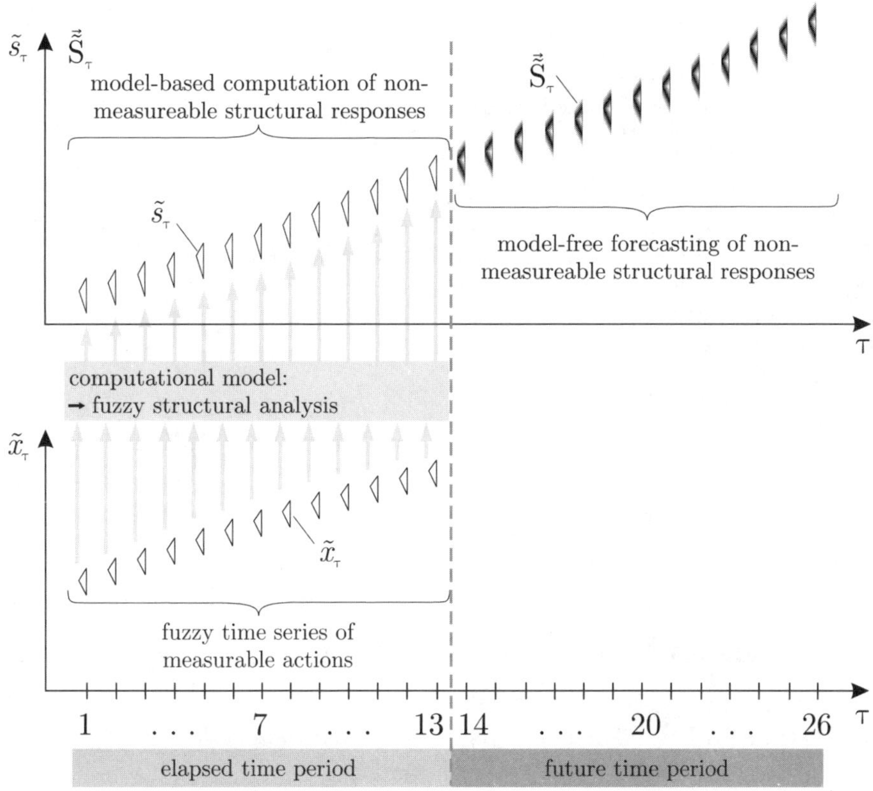

Fig. 5.4. Model-based forecasting – Strategy II

In Strategy I the $\tilde{x}_1$, $\tilde{x}_2$, ..., $\tilde{x}_l$ are given as the result of an optimum forecast or fuzzy forecast intervals, whereas in Strategy II, the measured fuzzy data of the fuzzy time series constitute the fuzzy input variables $\tilde{x}_1$, $\tilde{x}_2$, ..., $\tilde{x}_l$.

The solution of Eq. (5.1) may be found by applying the extension principle. Under the condition that the fuzzy variables $\tilde{x}$ are convex, however, α-level optimization is numerically more efficient [37]. The α-level optimization approach (see Fig. 5.5) is based on multiple discretization. All fuzzy variables $\tilde{x}$ and $\tilde{z}$ are discretized using the same number of α-levels α_i, $i = 1, 2, ..., n$. The α-level set $X_{k,\,\alpha_i}$ on the level α_i is then assigned to each fuzzy input variable $\tilde{x}_k$, $k = 1, 2, ..., l$.

The α-level sets $X_{k,\,\alpha_i}$, $k = 1, 2, ..., l$ form the l-dimensional crisp subspace $\underline{X}_{\alpha_i}$. A three-dimensional subspace $\underline{X}_{\alpha_i}$ is shown by way of example in Fig. 5.5. For $\alpha_i = 0$ the crisp support space $\underline{X}_{\alpha_i=0}$ is obtained. If no interaction exists between the fuzzy input variables, the subspace $\underline{X}_{\alpha_i}$ forms an l-dimensional hypercuboid. In the case of interaction, however, this hypercuboid forms the envelope curve. The crisp subspace $\underline{Z}_{\alpha_i}$ is assigned to the crisp subspace $\underline{X}_{\alpha_i}$

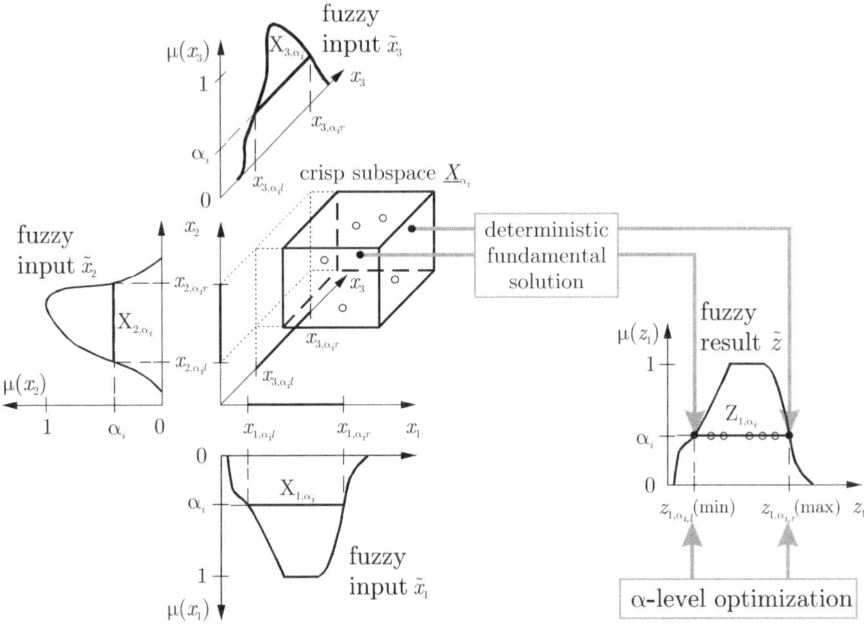

Fig. 5.5. Fuzzy structural analysis with α-level optimization

on the same α-level. These are constructed from the α-level sets Z_{j,α_i}, $j = 1, 2, ..., m$ of the fuzzy result variables $\tilde{z}_j$.

Each point of the hypercuboid $\underline{X}_{\alpha_i}$ is uniquely described by the coordinates $x_1, x_2, ..., x_l$. Each point in the subspace $\underline{Z}_{\alpha_i}$ may be computed by means of

$$\underline{z} = (z_1, z_2, ..., z_m) = f(x_1, x_2, ..., x_l) . \tag{5.2}$$

The mapping $f(\cdot)$ is referred to as the deterministic fundamental solution. This represents an arbitrary computational model, e.g. a finite element model.

Under the condition that the fuzzy variables are convex, it is sufficient to compute the largest element $z_{j,\alpha_i r}$ and the smallest element $z_{j,\alpha_i l}$ of Z_{j,α_i}. If these two elements are known for a sufficient number of α-levels known, the membership function $\mu_{\tilde{z}_j}(z_j)$ may be stated in discretized form.

The determination of $z_{j,\alpha_i r}$ and $z_{j,\alpha_i l}$ is an optimization problem with the objective functions

$$z_j = f(x_1, x_2, ..., x_l) \Rightarrow \max \tag{5.3}$$

$$z_j = f(x_1, x_2, ..., x_l) \Rightarrow \min \tag{5.4}$$

and the constraint

$$(x_1, x_2, ..., x_l) \in \underline{X}_{\alpha_i} . \tag{5.5}$$

Each of the two objective functions is satisfied by an optimum point in the subspace $\underline{X}_{\alpha_i}$. The optimization problem may be solved, e.g. by means of α-level optimization [36]. This replaces the Min-Max operator of the extension principle.

If p fuzzy model parameters $\tilde{m}$ are also present in addition to the fuzzy input variables $\tilde{x}$, the dimension of each l-dimensional crisp subspace $\underline{X}_{\alpha_i}$ increases to $l + p$ dimensions.

Model-Based Forecasting and Fuzzy Stochastic Structural Analysis

By means of fuzzy stochastic structural analysis it is possible to map the fuzzy random variables $\tilde{X}_1$, $\tilde{X}_2$, ..., $\tilde{X}_l$ onto the fuzzy random variables $\tilde{Z}_1$, $\tilde{Z}_2$, ..., $\tilde{Z}_m$.

$$\underline{\tilde{Z}} = (\tilde{Z}_1, \tilde{Z}_2, ..., \tilde{Z}_m) = \tilde{f}(\tilde{X}_1, \tilde{X}_2, ..., \tilde{X}_l) \tag{5.6}$$

In the case of Strategy I the $\tilde{X}_1$, $\tilde{X}_2$, ..., $\tilde{X}_l$ are given as the result of a fuzzy random forecast. In the case of Strategy II no fuzzy random variables are present.

The mapping problem given by Eq. (5.6) is solved by means of a three-step analysis algorithm. In this respect, a distinction is made between two variants.

Fuzzy stochastic structural analysis – Variant I. Variant I of the three-step algorithm is shown in Fig. 5.6. This includes fuzzy analysis as an outer loop, stochastic analysis as a middle loop, and the deterministic fundamental solution as an inner loop.

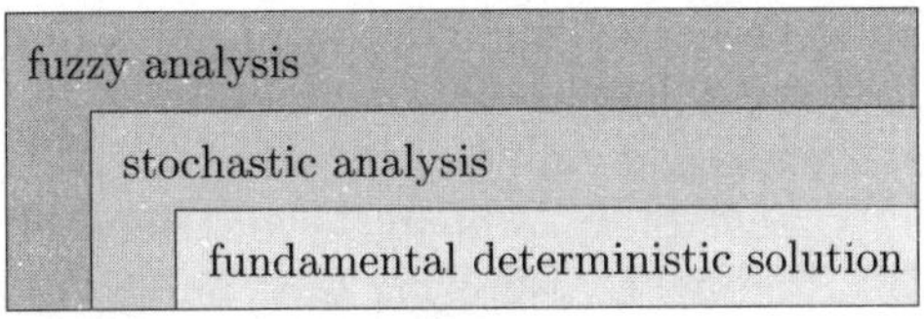

Fig. 5.6. Fuzzy stochastic structural analysis – Variant I

A precondition for this arrangement of the three loops is the representation of fuzzy random variables as bunch parameters according to Eq. (2.100) and a quantification of the fuzzy random variables by means of fuzzy probability distribution functions form I , i.e. in bunch parameter representation according to Eq. (2.101).

The bunch parameter representation $\tilde{X} = X(\underline{\tilde{s}})$ converts the fuzzy random variable $\tilde{X}$ into a family of originals $X_j \in \tilde{X}$ with $\mu(X_j) = \mu(X(\underline{s}_j)) = \mu(\underline{s}_j)$. The originals X_j are real-valued fuzzy variables which are described using real-valued probability distribution functions $F_{X_j}(\underline{s}_j, x)$. The fuzzy probability distribution function form I $\tilde{F}_{\tilde{X}}(x)$ of the fuzzy random variable $\tilde{X}$ may thus be expressed as a family of real-valued probability distribution functions also in bunch parameter representation $\tilde{F}_{\tilde{X}}(x) = F_{\tilde{X}}(\tilde{\underline{s}}, x)$. Typical fuzzy bunch parameters $\tilde{\underline{s}}$ are the first two moments of the fuzzy probability distribution functions form I.

In order to solve Eq. (5.6) the fuzzy random variables $\tilde{X}_k$, $k = 1, 2, ..., l$ are described by the fuzzy probability distribution functions form I $\tilde{F}_{\tilde{X}_k}(x_k) = F_{\tilde{X}_k}(\tilde{\underline{s}}_k, x_k)$, and the fuzzy random variables $\tilde{Z}_j$, $j = 1, 2, ..., m$ by the fuzzy probability distribution functions form I $\tilde{F}_{\tilde{Z}_j}(z_j) = F_{\tilde{Z}_j}(\tilde{\underline{\sigma}}_j, z_j)$, whereby the $F_{\tilde{Z}_j}(\tilde{\underline{\sigma}}_j, z_j)$ are sought. The $\tilde{\underline{s}}_k$, $k = 1, 2, ..., l$ are lumped together in the fuzzy bunch parameter vector $\tilde{\underline{s}}$ of all input variables $\tilde{X}_k$, and the $\tilde{\underline{\sigma}}_j$, $j = 1, 2, ..., m$ in the fuzzy bunch parameter vector $\tilde{\underline{\sigma}}$ of all result variables $\tilde{Z}_j$. The lengths of the vectors $\tilde{\underline{s}}$ and $\tilde{\underline{\sigma}}$ are denoted by t and u, respectively.

By means of the bunch parameter representation of the fuzzy probability distribution functions the mapping problem according to Eq. (5.6) reduces to the mapping problem

$$\tilde{\underline{\sigma}} = \tilde{f}_S(\tilde{\underline{s}}) . \tag{5.7}$$

Eq. (5.7) maps the fuzzy bunch parameters $\tilde{\underline{s}}$ onto the fuzzy bunch parameters $\tilde{\underline{\sigma}}$ by means of the fuzzy operator $\tilde{f}_S(\cdot)$. Eq. (5.7) may be solved by means of fuzzy structural analysis, as Eq. (5.7), analogous to Eq. (5.1), maps fuzzy variables onto fuzzy variables.

The fuzzy bunch parameters $\tilde{\underline{s}}$ and $\tilde{\underline{\sigma}}$ are subject to α-discretization. The α-level sets S_{1,α_i}, S_{2,α_i}, ..., S_{t,α_i} (Fig. 5.7, line a) obtained on the α-level α_i form the t-dimensional subspace $\underline{S}_{\alpha_i}$. A crisp point r in the subspace $\underline{S}_{\alpha_i}$ is defined by the elements $s_{1,r} \in S_{1,\alpha_i}$, ..., $s_{t,r} \in S_{t,\alpha_i}$. These form the vector $\underline{s}_r \in \underline{S}_{\alpha_i}$.

By means of the bunch parameters $s_{1,r} \in S_{1,\alpha_i}$, ..., $s_{t,r} \in S_{t,\alpha_i}$ a real-valued probability distribution function $F_{X_k}(\underline{s}_{k,r}, x_k)$ is specified in each case in the $k = 1, 2, ..., l$, fuzzy probability distribution functions form I $F_{\tilde{X}_k}(\tilde{\underline{s}}_k, x_k)$ (Fig. 5.7, line b). By selecting the point $\underline{s}_r$ in the subspace $\underline{S}_{\alpha_i}$ a real-valued probability distribution function is thus known for each fuzzy random variable $\tilde{X}_k$, and the mapping problem according to Eq. (5.7) reduces to a problem of stochastic analysis (Fig. 5.7, line c).

$$\left(F_{Z_j}(\underline{\sigma}_{j,r}, z_j) \mid j = 1, **2, ..., m\right) = g\left(F_{X_k}(\underline{s}_{k,r}, x_k) \mid k = 1, 2, ..., l\right) \tag{5.8}$$

The mapping $g(\cdot)$ in Eq. (5.8) represents the stochastic analysis. A Monte Carlo simulation is suitable for obtaining a solution, especially if the mapping operator $g(\cdot)$ is highly nonlinear. A deterministic fundamental solution is processed several times during the Monte Carlo simulation.

The Monte Carlo simulation yields m samples with real-valued realizations for the $\tilde{Z}_1$, $\tilde{Z}_2$, ..., $\tilde{Z}_m$ fuzzy random variables, on the basis of which the probability distribution functions (trajectories) $F_{\tilde{Z}_j}(\tilde{\underline{\sigma}}_j, z_j)$, $j = 1, 2, ..., m$

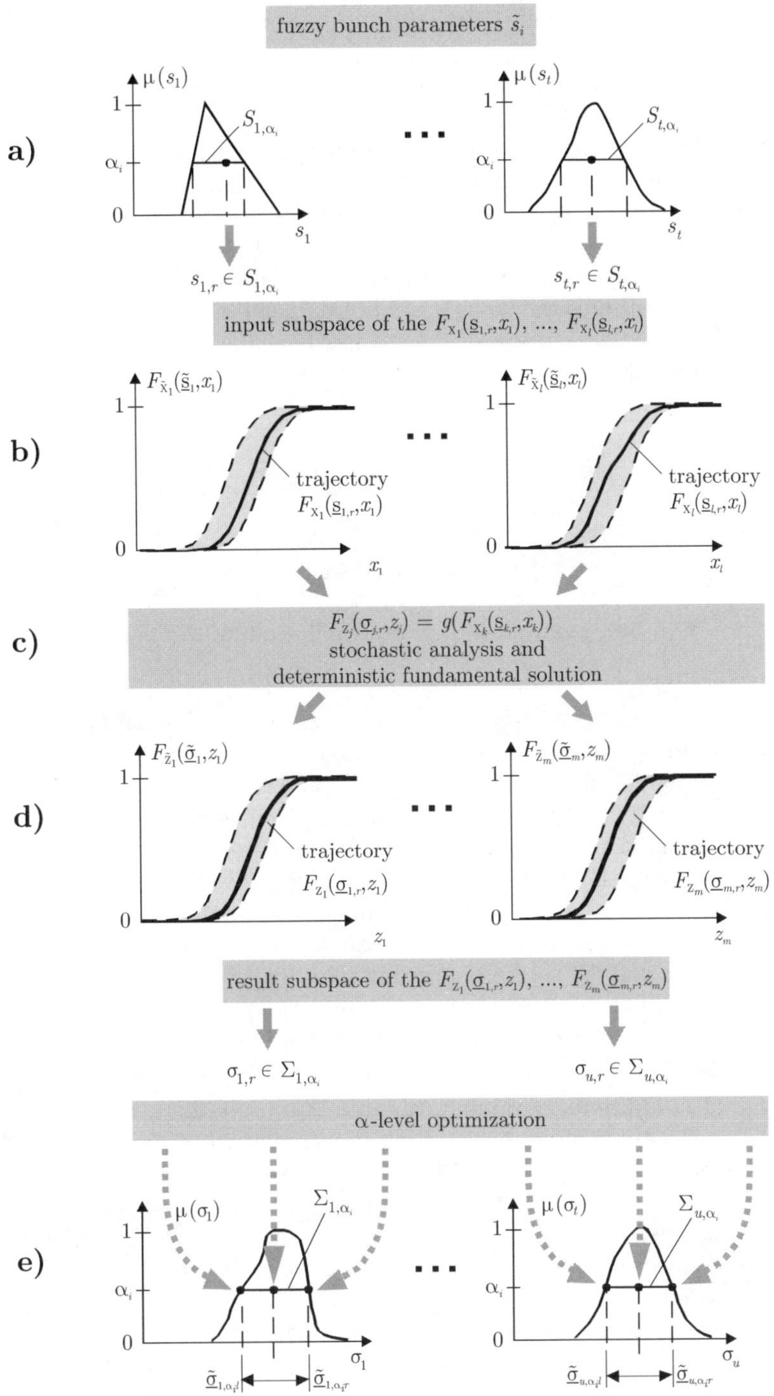

Fig. 5.7. Fuzzy stochastic structural analysis – Variant I

are estimated (Fig. 5.7, line d). From this it follows that the bunch parameters $\sigma_{1,r} \in \Sigma_{1,\alpha_i}$, ..., $\sigma_{u,r} \in \Sigma_{u,\alpha_i}$ are also known. With the aid of α-level optimization the largest values $\tilde{\underline{\sigma}}_{1,\alpha_i r}, ..., \tilde{\underline{\sigma}}_{m,\alpha_i r}$ and the smallest values $\tilde{\underline{\sigma}}_{1,\alpha_i l}, ..., \tilde{\underline{\sigma}}_{m,\alpha_i l}$ are computed for each α-level.

Once the fuzzy bunch parameter vector $\tilde{\underline{\sigma}}$ has been determined, the application of the fuzzy probability distribution functions means that the fuzzy random variables $\tilde{\mathrm{Z}}_1$, $\tilde{\mathrm{Z}}_2$, ..., $\tilde{\mathrm{Z}}_m$ in Eq. (5.6) are also known.

Fuzzy Stochastic Structural Analysis – Variant II. Variant II of the three-step algorithm is shown in Fig. 5.8. Compared with Variant I, the order of the three loops is rearranged: the stochastic analysis forms the outer loop, and the fuzzy analysis, the middle loop.

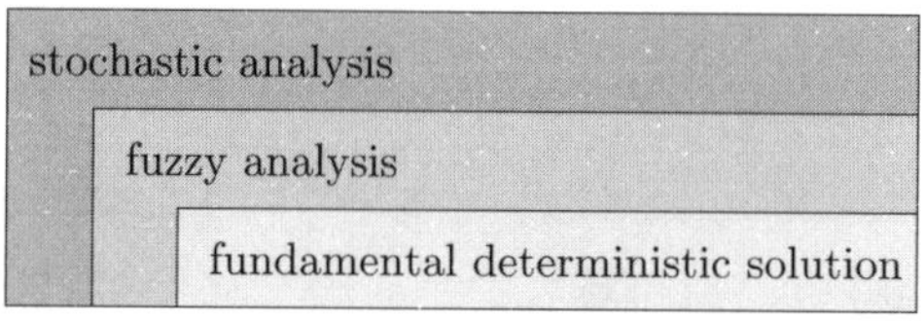

Fig. 5.8. Fuzzy stochastic structural analysis – Variant II

A precondition for this rearrangement is that the individual fuzzy random variables of Eq. (5.6) are represented in each case by a fuzzy probability distribution function form II (see Sect. 2.2.2). If the fuzzy probability distribution functions form II are known for the fuzzy random variables $\tilde{\mathrm{X}}_1$, $\tilde{\mathrm{X}}_2$, ..., $\tilde{\mathrm{X}}_l$, the fuzzy realizations $\tilde{x}_1$, $\tilde{x}_2$, ..., $\tilde{x}_l$ may be simulated. This marks a distinction between form II and form I fuzzy probability distribution functions. By means of the latter it is only possible to simulate real-valued realizations of the individual trajectories. Because fuzzy realizations are immediately available, however, these may be used as input variables for a fuzzy analysis.

In order to solve Eq. (5.6) the fuzzy random variables $\tilde{\mathrm{X}}_k$, $k = 1, 2, ..., l$ are described by their fuzzy probability distribution functions form II ${}_{lr}F_{\tilde{\mathrm{X}}_1}(\tilde{x}_1)$, ..., ${}_{lr}F_{\tilde{\mathrm{X}}_l}(\tilde{x}_l)$. The fuzzy probability distribution functions form II ${}_{lr}F_{\tilde{\mathrm{Z}}_1}(\tilde{z}_1)$, ..., ${}_{lr}F_{\tilde{\mathrm{Z}}_m}(\tilde{z}_m)$ of the fuzzy random variables $\tilde{\mathrm{Z}}_j$, $j = 1, 2, ..., m$ are sought.

The stochastic analysis begins with the simulation of s sequences of the fuzzy realizations $\tilde{x}_1$, $\tilde{x}_2$, ..., $\tilde{x}_l$ (Fig. 5.9). By means of fuzzy structural analysis the sequence of fuzzy result variables $\tilde{z}_1$, $\tilde{z}_2$, ..., $\tilde{z}_m$ corresponding to each sequence $\tilde{x}_1$, $\tilde{x}_2$, ..., $\tilde{x}_l$ may be computed analogous to Eq. (5.1). The algorithm developed for solving the mapping problem of Eq. (5.1) may also be applied in this case.

This results in s sequences of fuzzy result variables $\tilde{z}_1$, $\tilde{z}_2$, ..., $\tilde{z}_m$, i.e. a sample comprised of s fuzzy realizations is obtained for each fuzzy random variable $\tilde{\mathrm{Z}}_j$. A statistical evaluation of the samples by means of Eq. (2.115) yields an empirical fuzzy probability distribution function form II for each

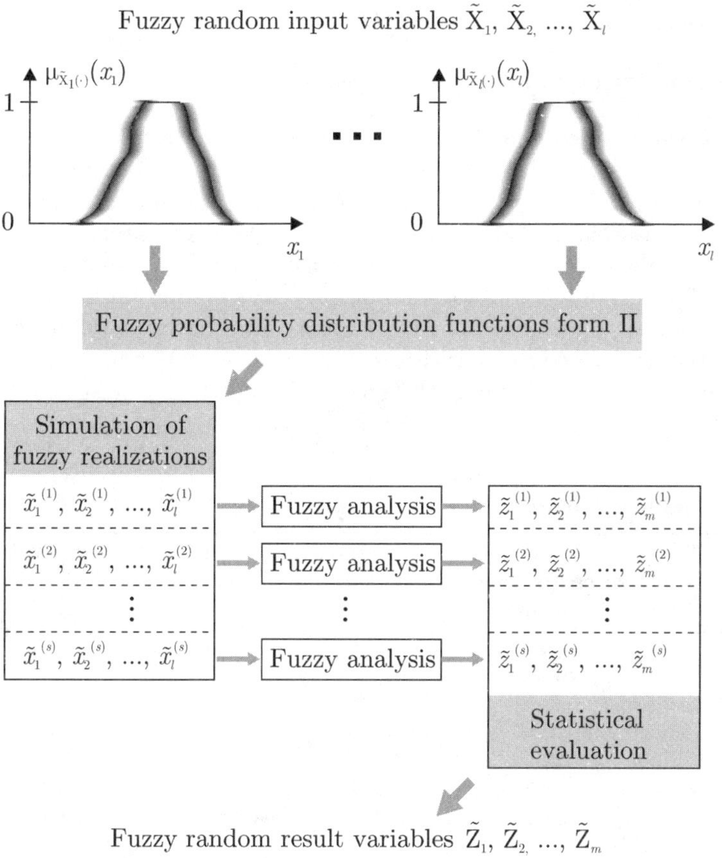

Fig. 5.9. Fuzzy stochastic structural analysis – Variant II

fuzzy random variable $\tilde{Z}_j$. The latter serve as unbiased estimators for the distributions of the fuzzy random result variables $\tilde{Z}_j$. Theoretical fuzzy probability distribution functions form II may also be derived from the latter as required.

5.3 Applications

The following examples demonstrate possible areas of application of the methods developed in Sects. 3 to 5 for analyzing and forecasting time series comprised of fuzzy data. These methods are applied for forecasting structural actions as well as structural responses.

5.3.1 Forecasting of Structural Actions

Forecasting of Foundation Soil Settlement

In the course of upgrading the B 172 main road, monthly extensometer measurements of incline movements were carried out between December 1998 and November 2002[1]. Three different measured values were obtained in each case for each measurement date and each extensometer measurement location[2], i.e. settlement at the respective measurement locations could not be measured unequivocally as real values, but only in uncertain terms. The normal conventional approach does not take account of this uncertainty, the uncertain information is reduced to an arithmetic mean. In order to realistically analyze and forecast the measured data, however, this uncertainty must be taken into consideration.

Table 5.1. Excerpt from a series of extensometer measurements

Date	1st measurement [mm]	2nd measurement [mm]	3rd measurement [mm]	mean value [mm]
⋮	⋮	⋮	⋮	⋮
30.05.2000	22.51	22.50	22.,52	22.510
27.06.2000	22.50	22.52	22.53	22.517
27.07.2000	22.40	22.40	22.41	22.403
30.08.2000	22.35	22.36	22.35	22.353
27.09.2000	21.72	21.80	21.77	21.763
⋮	⋮	⋮	⋮	⋮

Table 5.1 shows an excerpt of a series of extensometer measurements ([52]). The measured values lie in an interval which may be considered as a support of a fuzzy variable. It is thus appropriate to model the measured values as fuzzy variables. All values lying between the smallest and largest values measured on each measurement data are possible measurement results. These form the support of the corresponding fuzzy variable. The corresponding mean value is chosen in each case as the 'best possible crisped' measured value ($\mu = 1,0$). Using the support and the 'best possible crisped' measured value, fuzzy triangular numbers are constructed for each measurement date. The resulting time series comprised of fuzzy data is illustrated in Fig. 5.10. Due to limitations

[1] The measurements were carried out by the 'Gesellschaft für Geomechanik und Baumesstechnik mbH' under contract by the Dresden Road Construction Authority.

[2] The data were made available to the Institute of Structural Analysis by the 'Entwurfs- und Ingenieurbüro für Straßenwesen GmbH' in Dresden.

of scale the limits of the support of the fuzzy numbers are partly shown in magnified form.

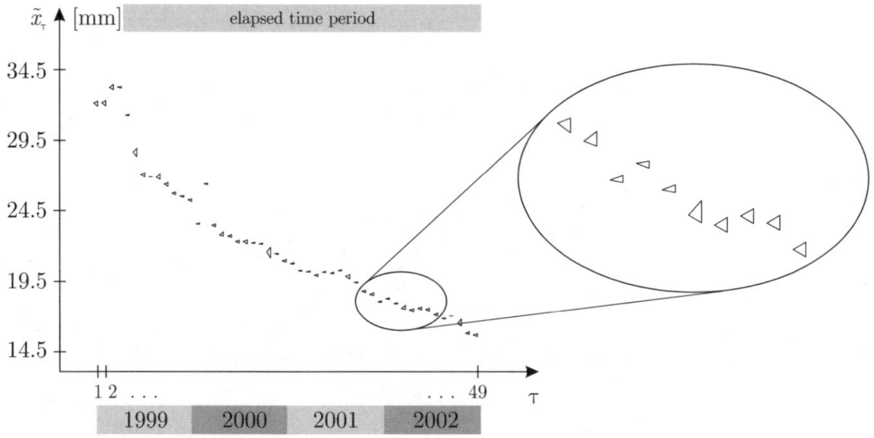

Fig. 5.10. Time series comprised of fuzzified extensometer measurements

For modeling the time series of fuzzy data constructed from the uncertain extensometer measurements a non-stationary fuzzy stochastic process must evidently be chosen as the underlying process. For this purpose the fuzzy stochastic process model is specified with the aid of the methods presented in Sect. 3.5.5. A fuzzy ARMA process of orders $p = 10$ and $q = 3$ is obtained. The parameters A_1, A_2, ..., A_{10} and B_1, B_2, B_3 are optimized by means of the gradient method presented in Sect. 3.5.6. In order to match the underlying fuzzy ARMA[10,3] process the fuzzy data of the fuzzy time series recorded between December 1998 and November 2002 are used. The now parameter-optimized fuzzy ARMA[10,3] process yields optimum single-step forecasts for the observation period $\tau \leqslant N$ with a minimized distance from the values of the given fuzzy time series. This result is shown in Fig. 5.11. The incongruence between the optimum single-step forecasts and the given fuzzy variables indicates the random properties of the fuzzy time series.

The parameter-optimized process permits the direct forecasting of future extensometer values. An optimum h-step forecast is carried out according to Sect. 4.2. In order to verify the result a second optimum h-step forecast is additionally performed using an artificial neural network for fuzzy variables The optimum forecasting multilayer perceptron used for this purpose consists of ten neurons in the input layer, three hidden layers with seven, five and six neurons, respectively, and one neuron in the output layer (10-7-5-6-1). The fuzzy forecast values are determined according to Sect. 4.3.1.

A comparison between the optimum h-step forecast obtained from the fuzzy ARMA process (FARMA) and the forecast obtained from the artificial

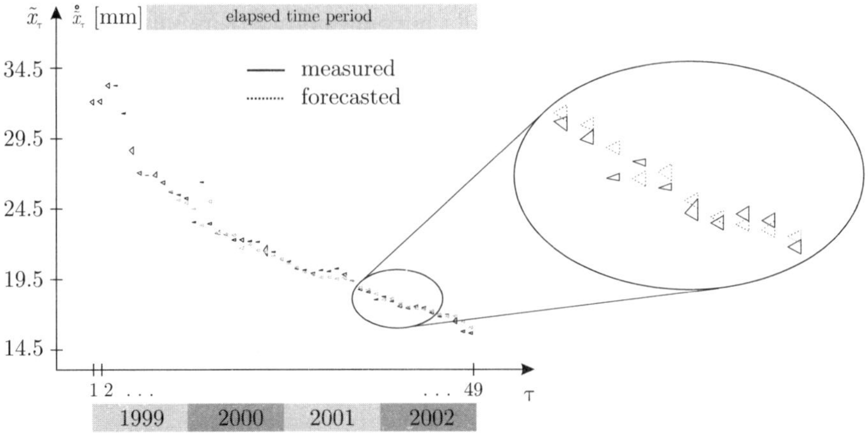

Fig. 5.11. Optimum single-step forecasts of the parameter-optimized fuzzy ARMA process

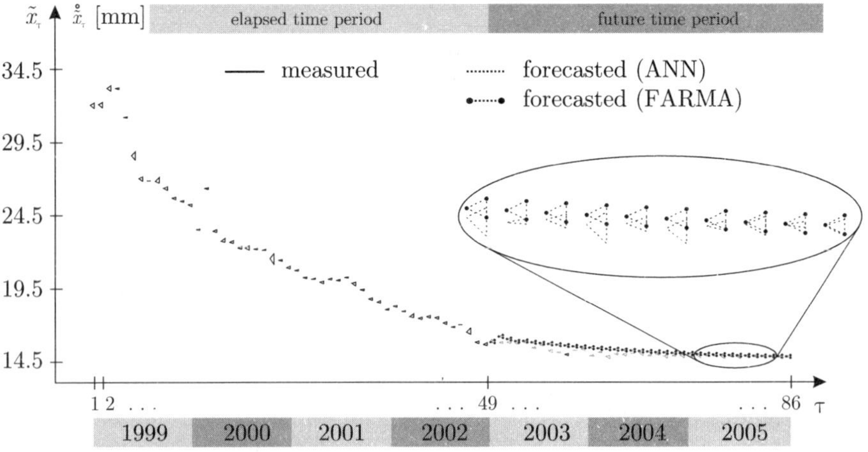

Fig. 5.12. Optimum h-step forecasts of the extensometer measurements

neural network (KNN) for 37 subsequent time steps is shown in Fig. 5.12. This represents a forecasting period of three years. Both of the optimum multistep forecasts closely follow an approximately nonlinear curve.

By means of the parameter-optimized fuzzy ARMA[10,3] process it is also possible to determine fuzzy forecast intervals according to the methods presented in Sect. 4.2.2. Fuzzy forecast intervals specify regions within which the expected extensometer measurements will lie with a confidence level of κ. The fuzzy forecast intervals for a 95% confidence level are shown by way of example in Fig. 5.13.

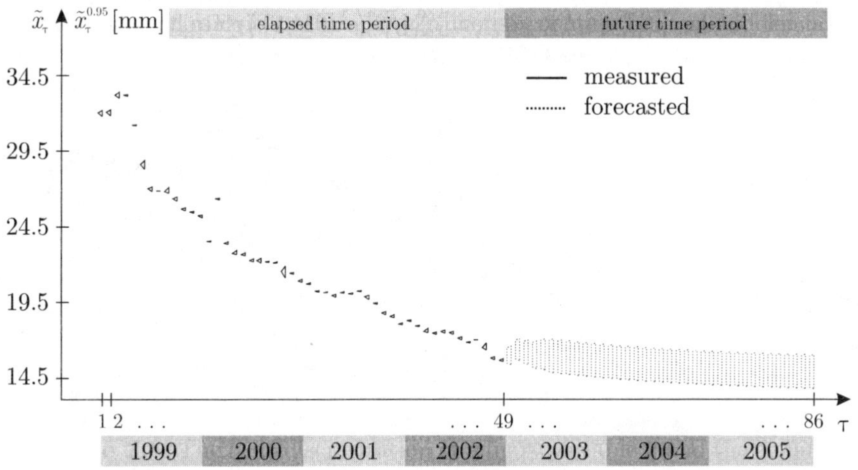

Fig. 5.13. Fuzzy forecast intervals of the extensometer measurements for a 95% confidence level

The random properties of the investigated fuzzy time series are found to be relatively slight. With regard to long-term forecasting, matching of the fuzzy trends alone (see Sect. 3.2) would yield reliable results in this case. In the following example, however, the random properties of the fuzzy time series are non-negligible and must therefore be taken into account in order to guarantee reliable long-term forecasts.

Forecasting of Fluvial Sediment Transport

The following example demonstrates the application of artificial neural networks for analyzing and forecasting time series comprised of fuzzy data. During the period 10.06.2003 to 27.08.2003 hydraulic engineering measurements were performed daily in the Kulfo river in southern Ethiopia[3][47]. Besides other measured parameters, the measurements also include information on sediment transport in the river. The sediment transport measurements form a basis for the design of sediment transport control structures. It is found, however, that the precise determination of sediment transport in a river is only possible to a limited extent. For this reason the given sediment transport data were subjectively fuzzified in relation to the river discharge. The fuzzified measured values are plotted in Fig. 5.14. Fig. 5.15 shows an alternative representation in which the interval limits for the α-levels $\alpha_1 = 0$ and $\alpha_2 = 1$ are connected polygonally. The observation period during which measurements were made is very short. As a sufficient number of time points with measured

[3] The data were made available to the Institute of Structural Analysis by the Institute of Hydraulic Engineering and Technical Hydromechanics of the Dresden University of Technology.

values are available, however, the time series is suitable for demonstrating a forecast with the aid of artificial neural networks.

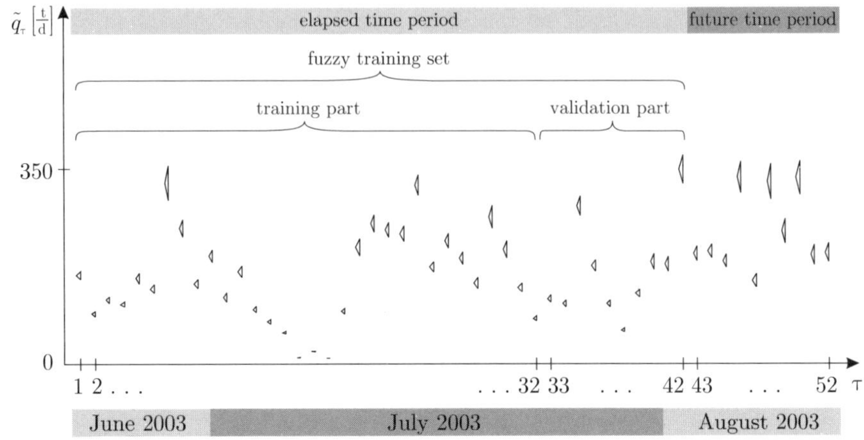

Fig. 5.14. Sediment transport $\tilde{q}_\tau$ in the Kulfo river at Arba Minch in tons per day [47]

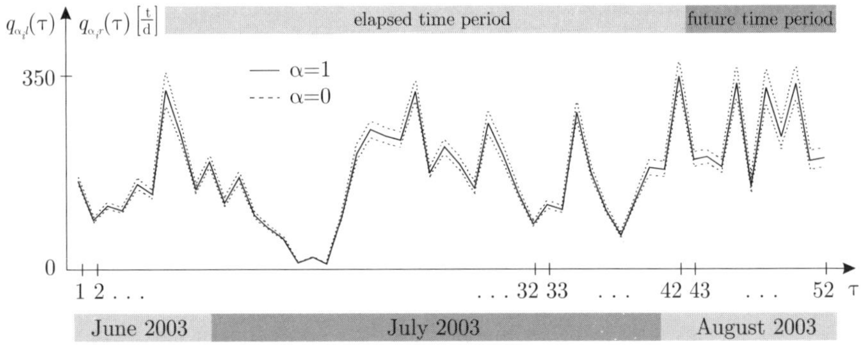

Fig. 5.15. Plot of the interval limits for $\alpha = 1$ and $\alpha = 0$

According to Sect. 3.1, a comprehensive plot of a fuzzy time series additionally requires a graphical representation of the variation of all $l_\alpha r_\alpha$-increments with time. For this purpose the $l_\alpha r_\alpha$-representation of the given fuzzy variables was carried out for the α-levels $\alpha_1 = 0$ and $\alpha_2 = 1$, i.e. for $n = 2$ α-level sets. The $l_\alpha r_\alpha$-increments $\Delta q_{\alpha_1 l}$ and $\Delta q_{\alpha_1 r}$ as well as the means values $q_{\alpha_2 l}$ are plotted in Fig. 5.16.

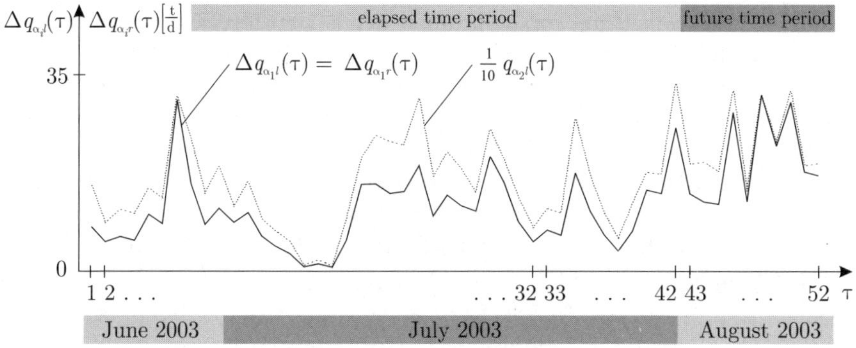

Fig. 5.16. Plot of the $l_\alpha r_\alpha$-increments

The plot of the $l_\alpha r_\alpha$-increments shows that the profiles of the $l_\alpha r_\alpha$-increments $\Delta q_{\alpha_1 l}$ and $\Delta q_{\alpha_1 r}$ are identical. The $l_\alpha r_\alpha$-increments $\Delta q_{\alpha_1 l}$ and $\Delta q_{\alpha_1 r}$ are thus completely positively correlated for $\Delta\tau = 0$. Moreover, the affinity of the plots of the mean value $q_{\alpha_2 l}$ and the $l_\alpha r_\alpha$-increments $\Delta q_{\alpha_1 l}$ and $\Delta q_{\alpha_1 r}$clearly reveals the mutual positive correlation for $\Delta\tau = 0$. This implies that for larger sediment transport rates the measured values are on average characterized by greater uncertainty, i.e. the support of the fuzzy numbers is wider. A numerical estimation confirms the visual impression, and yields the empirical $l_\alpha r_\alpha$-correlation function ${}_{lr}\hat{\underline{R}}_q(\Delta\tau)$ according to Eq. (5.9) for $\Delta\tau = 0$. A precondition for the determination of the empirical $l_\alpha r_\alpha$-correlation function is the (plausible) assumption of stationarity and ergodicity of the fuzzy time series.

$$ {}_{lr}\hat{\underline{R}}_q(\Delta\tau = 0) = \begin{bmatrix} 1 & 0.93 & - & 1 \\ 0.93 & 1 & - & 0.93 \\ - & - & - & - \\ 1 & 0.93 & - & 1 \end{bmatrix} \tag{5.9} $$

Using a multilayer perceptron for fuzzy variables, the sediment transport is investigated with the forecasting objectives optimum forecast, fuzzy forecast intervals and fuzzy random forecast. The forecasts do not rely on any assumptions regarding stationarity or ergodicity of the fuzzy time series. For the optimum h-step forecast an optimum forecasting artificial neural network is trained. For this purpose the fuzzy time series is subdivided into a fuzzy training series containing $N_T = 32$ fuzzy data values and a fuzzy validation series comprised of $N - N_T = 42 - 32 = 10$ fuzzy data values. The observed values between $\tau = 43$ and $\tau = 52$ are used for checking the forecasts. The training strategy according to Sect. 3.6.4 yields a network with the structure (3-9-10-3-1), i.e. the input layer contains three neurons, the three hidden layers contain nine, ten and three neurons, respectively, and the output layer contains one neuron. According to Eq. (3.182), the mean forecast error of this optimum forecasting network architecture is $MPF_{NN} = 51\frac{t}{d}$.

Using the trained network, optimum single-step forecasts are first computed for the time period $\tau = 1, 2, ..., N$ (see Sect. 4.3.1). The optimum single-step forecasts provide estimators for the conditional fuzzy expected values, as shown in Fig. 5.17. A comparison with the observed values also shown in Fig. 5.17 shows considerable differences. Information regarding the random components of the underlying fuzzy stochastic process may be deduced from the differences between the fuzzy expected values $\mathring{\tilde{x}}_4, \mathring{\tilde{x}}_5, ..., \mathring{\tilde{x}}_N$ and the observed values. In the case of this fuzzy time series the random components are obviously large.

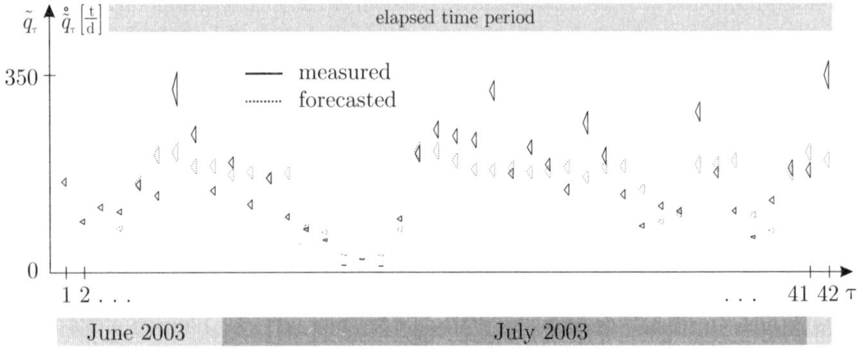

Fig. 5.17. Optimum single-step forecasts given by the optimum forecasting multilayer perceptron

Using the trained optimum forecasting network, an optimum 10-step forecast according to Sect. 4.3.1 is now performed for the time points $\tau = 43$ to $\tau = 52$. The result of the latter is shown in Fig. 5.18 (ANN). Independent of the forecast performed using the optimum forecasting network, a forecast was also obtained for the fuzzy time series by means of a fuzzy ARMA[3,1] process. The result of the optimum 10-step forecast given by this fuzzy ARMA process is also shown in Fig. 5.18 (FARMA). Good agreement is obtained in this case between the two optimum forecasts, i.e. between the conditional fuzzy expected values. The differences between the optimum forecasts and the observed values are explained by the large random component of the underlying fuzzy random process. The predictive capability of the optimum forecasts is hence limited. More reliable forecasts are given by fuzzy forecast intervals and fuzzy random forecasts, as these permit a determination of the probability of occurrence of future fuzzy variables.

For each of these forecasting variants an optimum simulation artificial neural network is trained with the aid of the training strategy developed in Sect. 3.6.4. The optimum forecasting network required for this purpose is already available. The training procedure yields an optimum simulation network with the structure (6-10-12-7-1), i.e. the input layer contains six neurons, the three

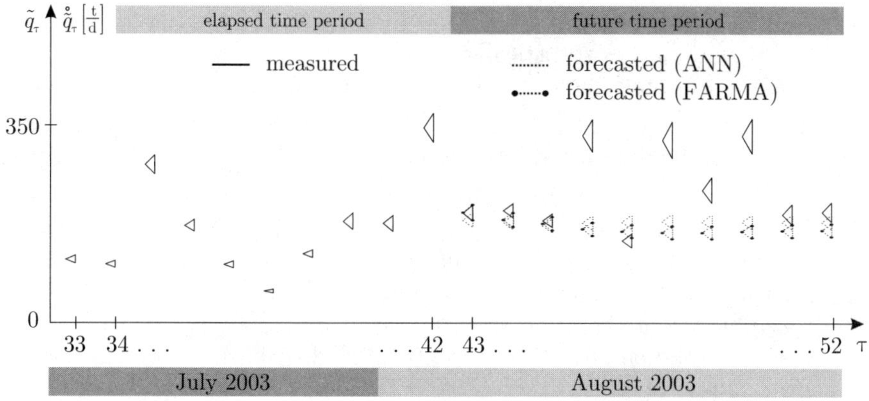

Fig. 5.18. Optimum 10-step forecasts given by the optimum forecasting multilayer perceptron and the fuzzy ARMA[3,1] process

hidden layers contain ten, twelve and seven neurons, respectively, and the output layer contains one neuron. The optimum structure was found by minimizing the forecast error PF_{so} in an optimization process (see Eq. (3.187)).

A sufficiently large number of sequences of potential future fuzzy realizations is simulated by means of the optimum simulation network. A simulated sample of fuzzy realizations is thus available at each time point $\tau > N$. Two sequences of simulated fuzzy realizations for $\tau = 43, ..., 52$ are shown in Fig. 5.19.

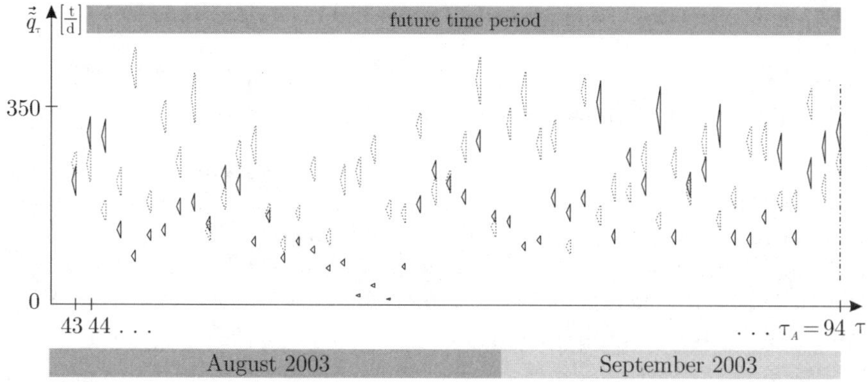

Fig. 5.19. Examples of possible future progressions of the fuzzy time series

In order to estimate fuzzy forecast intervals the simulated fuzzy realizations are evaluated by means of Eq. (4.35) for a prescribed confidence level. In order to estimate the fuzzy random variables $\vec{\tilde{Q}}_\tau$ of the sediment transport for $\tau = 43, ..., 52$ for the fuzzy random forecast, it is necessary to evaluate the simulated sequences statistically.

By way of example, the fuzzy random variable for the sediment transport $\vec{\tilde{Q}}_{\tau_A}$ at time point $\tau_A = 94$ is determined in the following by means of the empirical fuzzy probability distribution function form II $\hat{\tilde{F}}_{\vec{\tilde{Q}}_{\tau_A}}(\tilde{q})$. On the basis of the $s = 100$ simulated fuzzy realizations it is possible to estimate the empirical fuzzy probability distribution function form II according to Eq. 2.115. According to Remark 2.39, the result may be represented graphically in a simplified way with the aid of the marginal distribution functions. The simplified graphical representation given by the marginal distribution functions corresponds to the empirical fuzzy probability distribution function form I $\hat{\tilde{F}}_{\vec{\tilde{Q}}_{\tau_A}}(q)$ (see Fig. (5.20)).

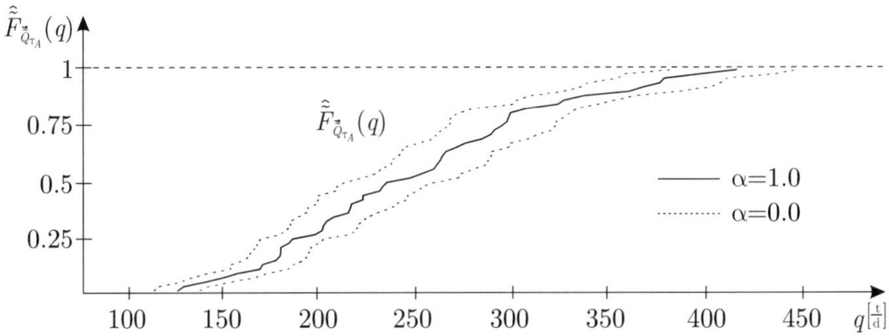

Fig. 5.20. Empirical fuzzy probability distribution function Type I $\hat{\tilde{F}}_{\vec{\tilde{Q}}_{\tau_A}}(q)$

The fuzzy probability distribution function form I $\hat{\tilde{F}}_{\vec{\tilde{Q}}_{\tau_A}}(q)$ is *not* sufficient for a full description of the forecasted fuzzy random variable $\vec{\tilde{Q}}_{\tau_A}$ at time time point τ_A. For this reason a tabular representation of all simulated fuzzy variables $\vec{\tilde{q}}_{\tau_A}$ at time point τ_A is chosen. The corresponding realization $\vec{\tilde{q}}_{\tau_A}$ is hereby assigned to each simulation line-by-line. The tabular representation automatically includes all interaction relationships for the fuzzy random variable $\vec{\tilde{Q}}_{\tau_A}$. An excerpt from the tabular representation is given in Table 5.2. Because the simulated fuzzy variables $\vec{\tilde{q}}_{\tau_A}$ are fuzzy triangular numbers, the abbreviated notation $\vec{\tilde{q}}_{\tau_A} = (\vec{q}_l; l; r)_{LR}$ according to Sect. 2.1 is used in the tabular representation.

The tabular representation is of advantage if the forecasted fuzzy random variable $\vec{\tilde{Q}}_{\tau_A}$ is the input variable of a fuzzy stochastic analysis Variant II. The Monte Carlo simulation performed during the fuzzy stochastic analysis is reduced to the extraction of elements from the table.

Table 5.2. Realizations $\vec{\tilde{q}}_{\tau_A}$ of the fuzzy random variable $\vec{\tilde{Q}}_{\tau_A}$ (excerpt)

Simulation	Fuzzy realization
⋮	⋮
k	$\vec{\tilde{q}}^{\,k}_{\tau_A} = (169.9;\ 186.4;\ 202.9)_{LR}\ \frac{\text{t}}{\text{d}}$
$k+1$	$\vec{\tilde{q}}^{\,k+1}_{\tau_A} = (305.5;\ 327.1;\ 348.5)_{LR}\ \frac{\text{t}}{\text{d}}$
$k+2$	$\vec{\tilde{q}}^{\,k+2}_{\tau_A} = (113.0;\ 125.6;\ 137.8)_{LR}\ \frac{\text{t}}{\text{d}}$
$k+3$	$\vec{\tilde{q}}^{\,k+3}_{\tau_A} = (239.2;\ 265.8;\ 292.1)_{LR}\ \frac{\text{t}}{\text{d}}$
⋮	⋮

Forecasting of Heavy Goods Vehicle Loading

The analysis and forecasting of a time series comprised of fuzzy data is demonstrated in the following by the example of heavy goods vehicle loading of the Loschwitzer Bridge in Dresden (referred to hereafter as the 'Blaues Wunder'). Since October 1999 all vehicles crossing the 'Blaues Wunder' from the left side of the Elbe River have been registered by a *weight in motion* measuring device, and their specific data[4] such as vehicle type, speed and weight have been archived.

For the time series analysis of the heavy goods vehicle loading the measured weights of heavy goods vehicles and articulated trucks were extracted from the database. An excerpt of the measured values is given in Table 5.3.

This information was compiled on a daily basis in the form of a histogram and fuzzified [36]. The daily loading due to heavy vehicle traffic was subjectively assessed with the aid of a polygonal membership function. A typical histogram of the weight measurements during one day using a non-normalized and normalized membership function is shown in Fig. 5.21.

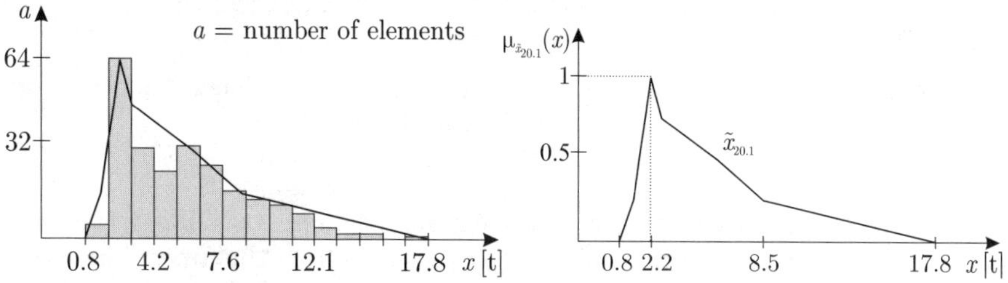

Fig. 5.21. Histogram and fuzzification of the weight measurements of 20.01.03; non-normalized and normalized membership function

Pure stochastic modeling of the available data by means of multivariate analysis and forecasting methods for real-valued time series is also possible and

[4] The data were made available to the Institute of Structural Analysis by the Dresden Road Construction and Public Works Authority, Dept. of Traffic Engineering.

Table 5.3. Excerpt from the weight measurements

Date	Time	Vehicle type	Total mass [kg]
⋮	⋮	⋮	⋮
01.08.2002	10:45:41	Heavy goods vehicle	4 228
01.08.2002	10:46:07	Heavy goods vehicle	5 648
01.08.2002	10:50:07	Heavy goods vehicle	3 562
01.08.2002	10:50:47	Heavy goods vehicle	1 924
01.08.2002	10:51:01		6 038
01.08.2002	10:52:07	Heavy goods vehicle	10 370
⋮	⋮	⋮	⋮

would also seem to be appropriate. The fuzzy stochastic modeling performed here is a suitable alternative method of solution. The application of fuzzy stochastic methods is especially recommended for very small sample sizes (a very small).

The fuzzy time series obtained by this means for the month of January 2003 is shown in Fig. 5.22. The plot indicates clearly recognizable regularities during weekends and public holidays.

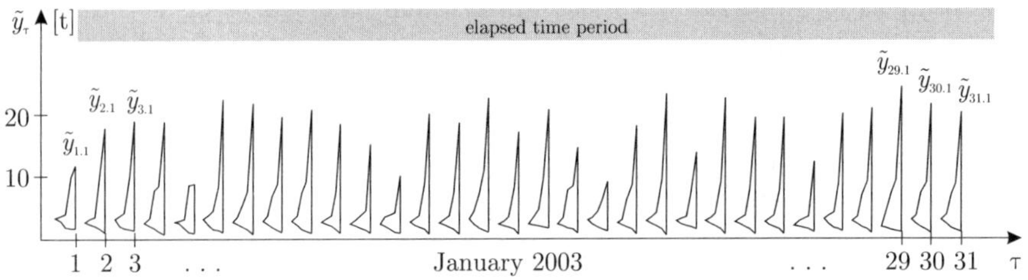

Fig. 5.22. Time series of the heavy-load traffic crossing the 'Blaues Wunder' bridge in January 2003 (including weekends and public holidays)

In order to approximately satisfy the assumption of stationarity the fuzzy data for weekends and public holidays are excluded from the fuzzy time series. A numerical representation of the fuzzy variables is achieved with the aid of $l_\alpha r_\alpha$-discretization for the α-levels $\alpha_1 = 0$, $\alpha_2 = 0.25$, $\alpha_3 = 0.5$, $\alpha_4 = 0.75$ and $\alpha_5 = 1$, i.e. for $n = 5$ α-level sets. The fuzzy numbers and selected $l_\alpha r_\alpha$-increments of the modified time series are plotted by way of example for the month of September 2002 in Figs. 5.23 and 5.24, respectively.

An analysis of the fuzzy time series is carried out for the period July 2002 to April 2003. Assuming stationarity and ergodicity, the empirical moments are determined according to Sect. 3.3.

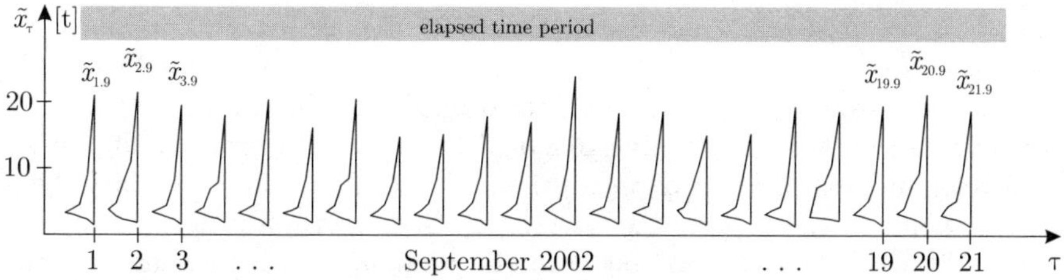

Fig. 5.23. Modified time series of the heavy-load traffic crossing the 'Blaues Wunder' bridge in September 2002 (excluding weekends and public holidays)

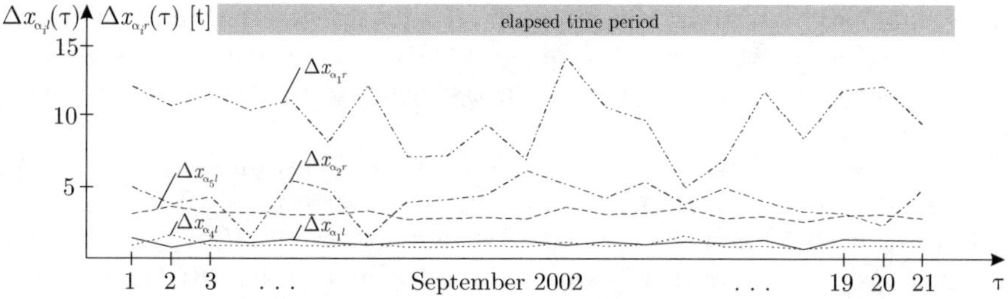

Fig. 5.24. Plot of selected $l_\alpha r_\alpha$-increments for September 2002

In order to proceed further, the fuzzy mean $\tilde{\bar{x}}$, the empirical $l_\alpha r_\alpha$-covariance function ${}_{lr}\hat{K}_{\tilde{x}_\tau}(\Delta\tau)$, and the empirical partial $l_\alpha r_\alpha$-correlation function ${}_{lr}\hat{P}_{\tilde{x}_\tau}(\Delta\tau)$ are required. The fuzzy mean $\tilde{\bar{x}}$ is shown by way of example in Fig. 5.25.

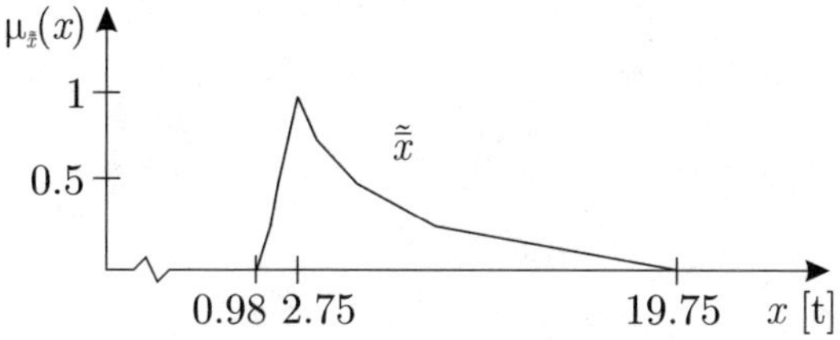

Fig. 5.25. Fuzzy mean value $\tilde{\bar{x}}$

A fuzzy AR process of order $p = 10$ is specified as the fuzzy stochastic process model (see Sect. 3.5.5). For this purpose the empirical partial $l_\alpha r_\alpha$-correlation function ${}_{lr}\hat{P}_{\tilde{x}_\tau}(\Delta\tau)$ is used. The process parameters A_1, A_2, ..., A_{10} are determined with the aid of the characteristic parameter method developed in Sect. 3.5.6. The modified evolution strategy after [36] is applied for solving the corresponding optimization problem (Eq. (3.112). The constraint of the minimization problem is the requirement of non-negativity of the $l_\alpha r_\alpha$-increments according to Eq. (2.47) for all fuzzy variables to be generated.

The values of the empirical $l_\alpha r_\alpha$-covariance function ${}_{lr}\underline{\hat{K}}_{\tilde{x}_\tau}(\Delta\tau)$ are relatively small outside the leading diagonal for $\tau > 0$. In order to obtain an efficient numerical solution of the optimization problem the secondary diagonal elements of the process parameters A_1, A_2, ..., A_{10} are thus set to zero a priori. Linear dependencies between the random $l_\alpha r_\alpha$-increments at time point τ are accounted for by the $l_\alpha r_\alpha$-covariance function ${}_{lr}\underline{K}_{\tilde{\mathcal{E}}_\tau}(\Delta\tau)$ of the included fuzzy white-noise process $(\tilde{\mathcal{E}}_\tau)_{\tau\in\mathbf{T}}$ for $\Delta\tau = 0$. In order to reduce the numerical effort, only the elements of the leading diagonal and the first upper and lower secondary diagonals of ${}_{lr}\underline{K}_{\tilde{\mathcal{E}}_\tau}(0)$ are determined. Although it is not possible to determine all (unknown) GRANGER-causal dependencies by this means, it is possible to adequately map selected (known) correlative relationships.

The solution of the optimization problem yields estimated values of the process parameters A_1, A_2, ..., A_{10}, the fuzzy expected value $E[\tilde{\mathcal{E}}_\tau]$, the $l_\alpha r_\alpha$-variance ${}_{lr}Var[\tilde{\mathcal{E}}_\tau]$ and the $l_\alpha r_\alpha$-covariance function ${}_{lr}\underline{K}_{\tilde{\mathcal{E}}_\tau}(\Delta\tau)$ of the fuzzy white-noise process. With the aid of the estimated fuzzy AR[10] process it is possible to forecast future values of the time series. The fuzzy data for the months July 2002 to April 2003 form the basis for modeling the fuzzy stochastic process. The forecast is made for May 2003, and compared with the values measured during the same month. The optimum single-step forecast for the fuzzy AR[10] process is given by Eq. (5.10). This is compared with the measured value in Fig. 5.26.

$$\mathring{\tilde{x}}_{N+1} = \underline{A}_1 \odot \tilde{x}_N \oplus ... \oplus \underline{A}_{10} \odot \tilde{x}_{N-9} \oplus E\left[\tilde{\mathcal{E}}_\tau\right] \tag{5.10}$$

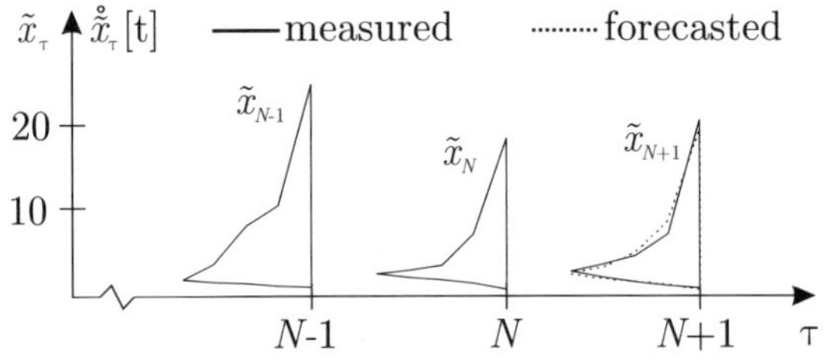

Fig. 5.26. Optimum single-step forecast and the given time series value

In contrast to the fluvial sediment transport investigated in Sect. 5.3.1, the underlying fuzzy AR[10] process has a less pronounced random component. The effect of the included fuzzy white-noise process is hence relatively small. This is also indicated by the comparison between the optimum h-step forecast and the measured values. The optimum 12-step forecast shown in Fig. 5.27 is seen to deviate far less from the measured values than the optimum forecast determined in Sect. 5.3.1.

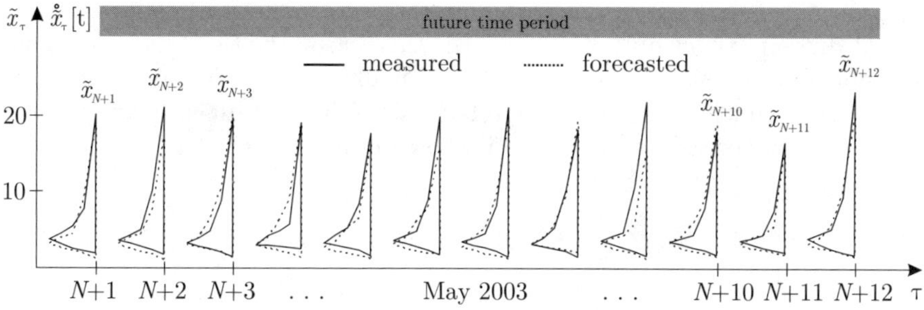

Fig. 5.27. Comparison of the optimum 12-step forecast with measured values

This result is confirmed by the optimum 12-step forecast determined using an artificial neural network for fuzzy variables, as shown in Fig. 5.28. The optimum forecasting multilayer perceptron used in this case consists of ten neurons in the input layer, three hidden layers with eleven, three and two neurons, respectively, and one neuron in the output layer (10-11-3-2-1).

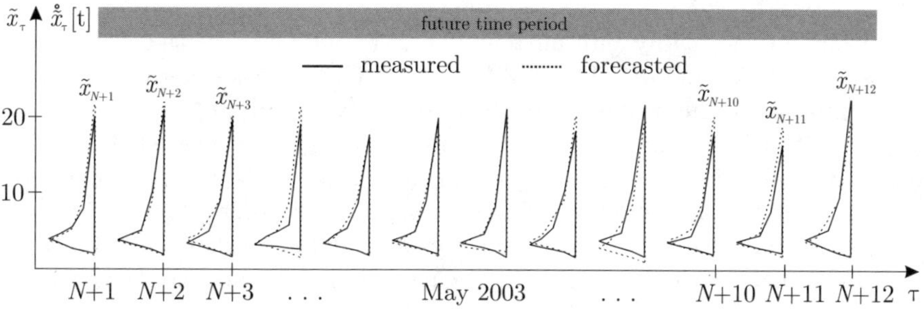

Fig. 5.28. Optimum 12-step forecast given by the multilayer perceptron

The optimum 12-step forecast yields the conditional fuzzy expected values of the future fuzzy random variables at corresponding points in time. In order to obtain probability information on the fuzzy time series during the forecasting period it is necessary to determine fuzzy forecast intervals or fuzzy random forecasts according to Sects. 4.2.2 and 4.3.3, respectively.

5.3.2 Forecasting of Structural Responses

Forecasting of Bearing Movements

The model-free forecasting of structural responses is demonstrated by the example of the movements of the pylon bearing of the Loschwitzer Bridge in

Dresden ('Blaues Wunder'). Measurements[5] of the horizontal movements of the pylon bearing on the right side of the Elbe River in the longitudinal direction of the bridge were made between January 1998 and November 2002 with the aid of slide gages. The measurements were made at irregular intervals and at different times during the day. For this reason the (few) measurements available for each month were treated as a small sample with non-constant reproduction conditions, and subjectively fuzzified. The obtained (fragmentary) time series comprised of fuzzy data is shown in Fig. 5.29.

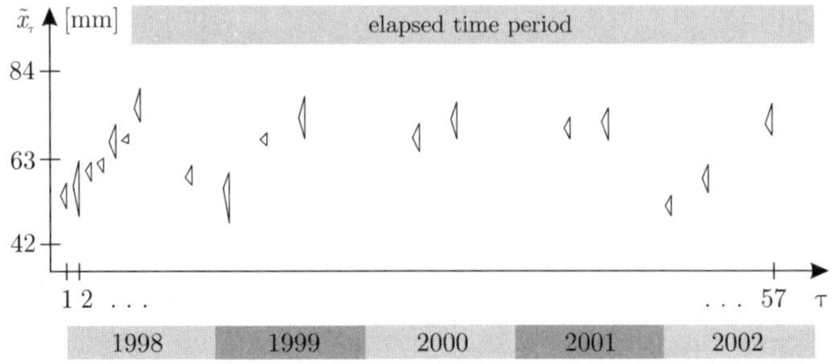

Fig. 5.29. Time series of the pylon bearing movements of the 'Blaues Wunder' bridge

Owing to the highly intermittent data, the modeling of this fuzzy time series as a realization of a fuzzy MA, AR or ARMA process is problematic. An estimation of the properties of the potentially underlying fuzzy stochastic process with the aid of an artificial neural network is also pointless. A suitable means of analyzing and forecasting the fuzzy time series, however, is offered by the fuzzy component model introduced in Sect. 3.2. In order to apply the fuzzy component model it is necessary to select and match a fuzzy trend function $\tilde{t}(\tau)$ and a fuzzy cycle function $\tilde{z}(\tau)$. Owing to the direct dependency between bearing movements and seasonal temperature variations of the structure a constant fuzzy trend function and a sinusoidal fuzzy cycle function are chosen. The fuzzy functions are matched with the aid of trend auxiliary functions and cycle auxiliary functions following the method presented in Sect. 3.2. The matched fuzzy trend function $\tilde{t}(\tau)$ is given by Eq. (5.11).

$$\tilde{t}(\tau) = \tilde{t}_0 \quad \text{with} \quad \tilde{t}_0 = (52.2;\, 52.8;\, 54.1)_{LR}\text{mm} \tag{5.11}$$

The matched fuzzy cycle function $\tilde{z}(\tau)$ is formulated in Eq. (5.12).

$$\tilde{z}(\tau) = \left[\sin\left(\frac{2\pi\tau}{12}\right) + 1\right] \tilde{z}_0 \quad \text{with} \quad \tilde{z}_0 = (22.0;\, 22.0;\, 22.0)_{LR}\text{mm} \tag{5.12}$$

[5] The measurements were carried out under contract by the Dresden Road Construction and Public Works Authority and were made available to the Institute of Structural Analysis by the 'GMG Ingenieurpartnerschaft Dresden'.

The fuzzy residual component $\tilde{r}_\tau$to be determined using Eq. (5.13) is modeled as a realization of a fuzzy white-noise process $(\tilde{\mathcal{E}}_\tau)_{\tau \in \mathbf{T}}$ according to Sect. 3.5.1.

$$\tilde{r}_\tau = \tilde{x}_\tau \ominus \tilde{t}_\tau \ominus \tilde{z}_\tau \tag{5.13}$$

The characteristic parameters of the fuzzy white-noise process $(\tilde{\mathcal{E}}_\tau)_{\tau \in \mathbf{T}}$ are estimated according to Sect. 3.3 from the realizations $\tilde{r}_\tau$. The fuzzy expected value $E[\tilde{\mathcal{E}}_\tau]$ of the fuzzy white-noise process $(\tilde{\mathcal{E}}_\tau)_{\tau \in \mathbf{T}}$ is given in Eq. (5.14).

$$E[\tilde{\mathcal{E}}_\tau] = (-3.51;\ 0;\ 3.45)_{LR}\text{mm} \tag{5.14}$$

By this means it is possible to formulate the optimum forecasts $\mathring{\tilde{x}}_\tau$ for the given fuzzy time series for time points $\tau = 1, 2, ..., N, N+1, N+2, ...$ by means of Eq. (5.15).

$$\mathring{\tilde{x}}_\tau = \tilde{t}_\tau \oplus \tilde{z}_\tau \oplus E[\tilde{\mathcal{E}}_\tau] \tag{5.15}$$

A plot of the optimum forecasts $\mathring{\tilde{x}}_\tau$ given by the fuzzy component model is presented in Fig. 5.30.

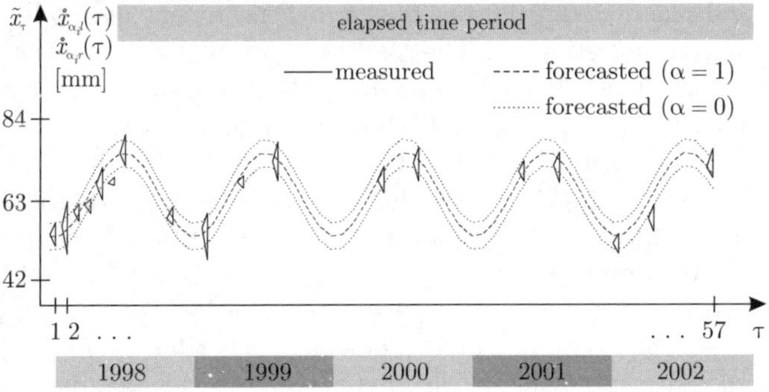

Fig. 5.30. Optimum forecasts $\mathring{\tilde{x}}_\tau$ given by the fuzzy component model

With the aid of the matched fuzzy component model it is also possible to determine fuzzy forecast intervals or fuzzy random forecasts according to Sect. 4.1. The fuzzy stochastic forecasting process $(\vec{\tilde{X}}_\tau)_{\tau \in \mathbf{T}}$ required for this purpose is described by Eq. (5.16).

$$\vec{\tilde{X}}_\tau = \tilde{t}_\tau \oplus \tilde{z}_\tau \oplus \tilde{\mathcal{E}}_\tau \tag{5.16}$$

By performing a Monte Carlo simulation of the realizations of the fuzzy white-noise process $(\tilde{\mathcal{E}}_\tau)_{\tau \in \mathbf{T}}$ potential future progressions of the fuzzy time series are obtained with the aid of Eq. (5.16). An evaluation of the simulated progressions according to Sects. 4.2.2 and 4.3.3 yields fuzzy forecast intervals and fuzzy random forecasts, respectively. The fuzzy forecast intervals for a 95% confidence level are shown in Fig. 5.31.

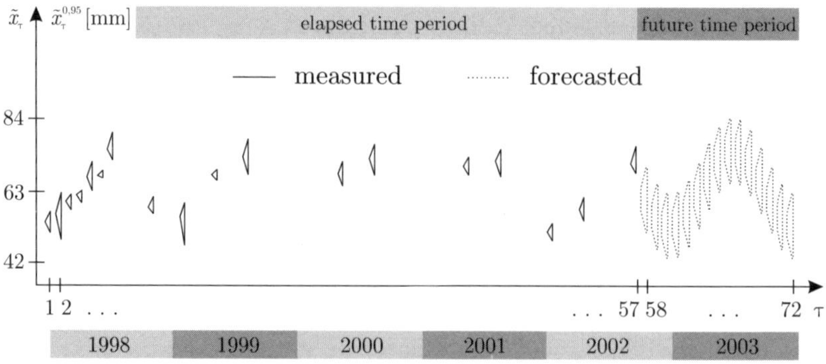

Fig. 5.31. Fuzzy forecast intervals of the pylon bearing movements for a 95% confidence level

Forecasting of Asphalt Deformations

In the following the model-based forecasting of structural responses is demonstrated for the example of a road pavement. In order to illustrate the method the elastic and plastic deformations of the road pavement due to heavy goods traffic loading are investigated. The loading process is given in the form of a time series of fuzzy load alternation numbers. Future load alternation numbers are forecasted directly. These directly forecasted values of the loading process are inputted to a computational model for computing a model-based forecast of asphalt deformations. The processing sequence corresponds to Strategy I.

For verification purposes the results of the model-based forecast according to Strategy I are compared with those of the model-based forecast according to Strategy II. For this purpose the fuzzy deformation of the road pavement is computed by the computational model at each time point $\tau \leqslant N$. The time series of the fuzzy deformations determined in this way is forecasted in a model-free manner.

The investigated road pavement consists of a 25 mm thick asphalt layer and a 275 mm thick base layer without binding agents. This special method of construction is the standard method adopted in New Zealand. The soil subgrade is located below the base layer without binding agents. In order to determine the structural responses of the road pavement numerically a three-dimensional finite element computational model is applied. By means of the finite element program FALT-FEM [43, 23, 48] it is possible to determine the elastic and plastic deformations resulting from heavy goods traffic loading. The heavy goods traffic loading is hereby simulated by two wheel loads of 40 kN and 50 kN which repeatedly roll over the structure with the given load alternation number.

The applied finite element model was developed within the framework of numerical investigations of accelerated testing methods[6] and validated on the basis of large-scale laboratory experiments. [67].

The plan dimensions of the construction segment considered in the computations are 4000 mm × 2400 mm. Making use of symmetry, it is only necessary to generate a mesh for half of this segment. The applied finite element mesh is shown in Fig. 5.32. The mesh consists of 1 962 elements with a total of 25 804 displacement degrees of freedom.

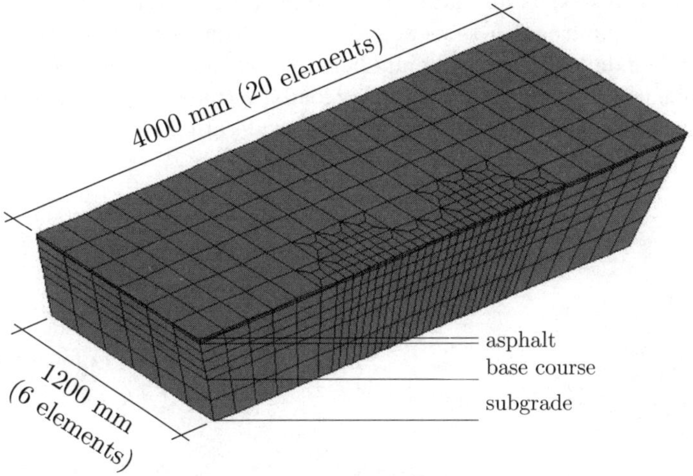

Fig. 5.32. Finite element mesh for the road pavement

The subgrade is modeled by two 150 mm thick sublayers, the base layer is subdivided into four 68.75 mm thick sublayers, and the asphalt layer is represented by two 12.5 mm thick sublayers. More finely meshed elements (66.7 mm × 66.7 mm × Höhe) are adopted in the areas of load application.

Owing to the relatively small thickness of the asphalt layer, the contribution of this layer to the overall deformation of the road pavement is very slight. For this reason a linear elastic material model is adopted for the asphalt layer with a Young's modulus of $E = 5000 \frac{\mathrm{N}}{\mathrm{mm}^2}$ and a Poisson's ratio of $\mu = 0.35$. The overall deformation of the road pavement is predominantly governed by the non-cohesive base layer. For this base layer without binding agents an empirical, nonlinear elastic-plastic material model after [48] is applied. The parameter values of the material model correspond to the characteristic values determined in [67]. The subgrade is modeled as a linear elastic medium with a Young's modulus of $E = 25 \frac{\mathrm{N}}{\mathrm{mm}^2}$ and a Poisson's ratio of $\mu = 0.4$.

[6] The accelerated tests were carried out by the Canterbury Accelerated Pavement Testing Indoor Facility (CAPTIF) in Christchurch, New Zealand.

The load alternation number is the input variable for the finite element computational model. Measured data[7] are used to construct a sample fuzzy time series of load alternations. The measurements were recorded between January 2003 and April 2004 within the scope of an automatic traffic census in Dresden. All vehicles passing the measuring point were registered and classified according to vehicle type.

In order to model the load alternation sequence the number of axle crossings and the corresponding axle loads are required. The measured data only contain information regarding vehicle type, however (e.g. cars, trucks, trucks with trailers, buses). This data uncertainty is taken into account by modeling the load alternation sequence as a fuzzy time series. For this purpose the possible axle loads and axle numbers are assumed for the different heavy goods vehicles, and corresponding fuzzy load alternation numbers $\tilde{l}_\tau$ are determined for each week of the period under consideration. The fuzzy time series obtained by this means is shown in Fig. 5.33.

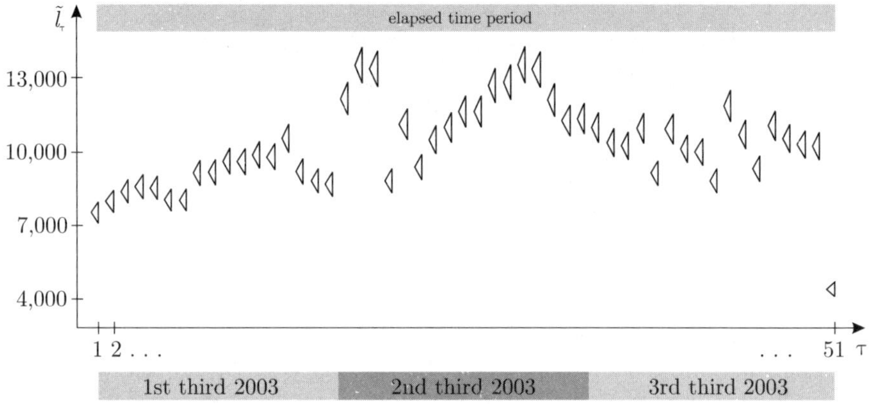

Fig. 5.33. Time series of the fuzzy load alternation numbers

In order to determine the elastic and plastic deformations of the road pavement the cumulative fuzzy load alternation numbers $\tilde{k}_\tau$ are required. The corresponding fuzzy time series is shown in Fig. 5.34.

For the purpose of analyzing and forecasting the cumulative fuzzy load alternation time series a multilayer perceptron for fuzzy variables was developed. By means of the method outlined in Sect. 3.6.4 a multilayer perceptron with four neurons in the input layer, eight, two and nine neurons in the three hidden layers, respectively, as well as one neuron in the output layer (4-8-2-9-1) was determined as the optimum forecasting network architecture. The

[7] The data were made available to the Institute of Structural Analysis by the Dresden Road Construction and Public Works Authority, Dept. of Traffic Engineering.

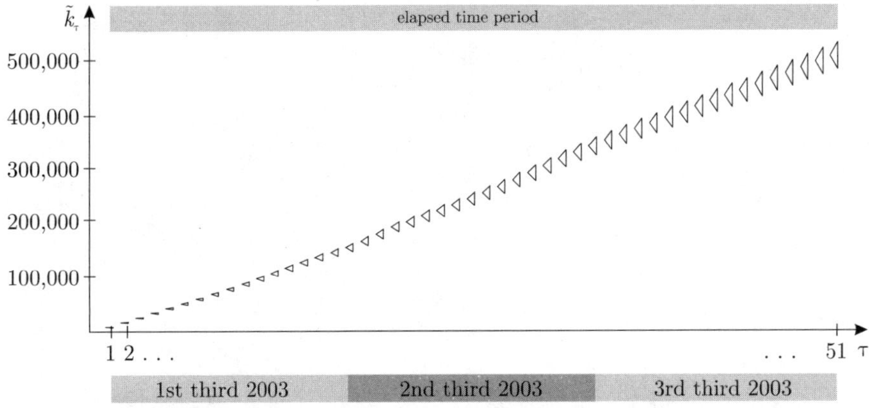

Fig. 5.34. Time series of the cumulative fuzzy load alternation numbers

optimum h-step forecast computed for the first third of 2004 using this artificial neural network is shown in Fig. 5.35.

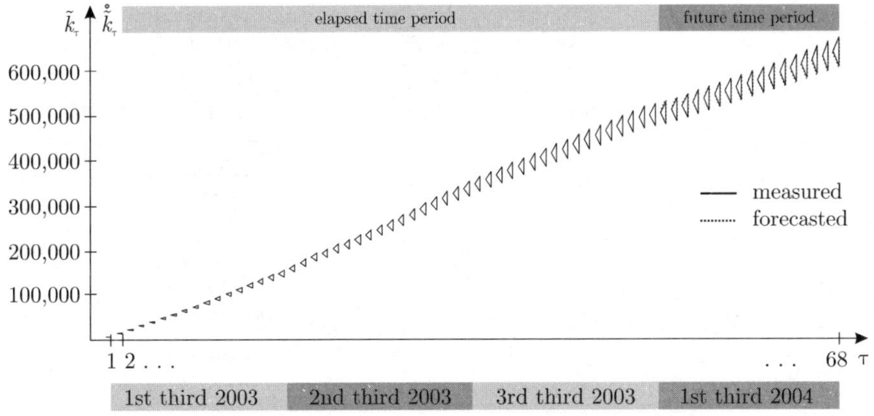

Fig. 5.35. Optimum 17-step forecast of the cumulative fuzzy load alternation numbers

A visual plausibility check of the optimum h-step forecast may easily be obtained on the basis of a representation of the fuzzy load alternation numbers per week (see Fig. 5.36).

The results of the model-free h-step forecast of the fuzzy load alternation numbers are used for the model-based forecast (Strategy I) of the elastic-plastic deformations $\tilde{v}_\tau$ of the road pavement. The deformations are determined by means of the finite element computational model, which uses the given and forecasted fuzzy load alternation numbers as input data (see Fig. 5.37).

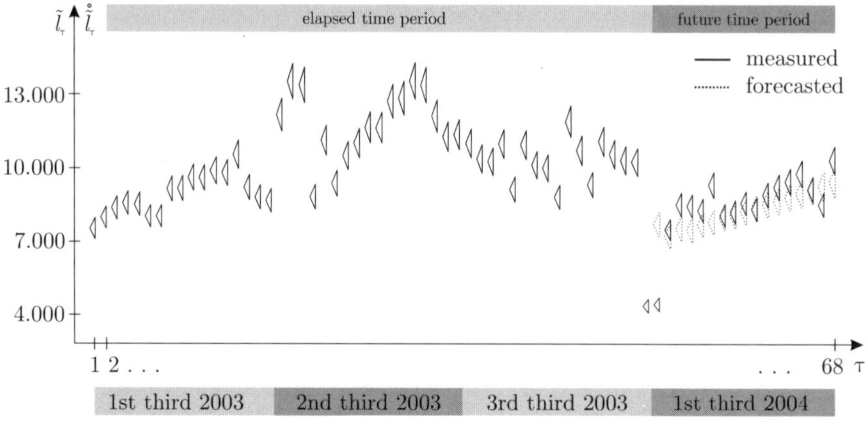

Fig. 5.36. Optimum 17-step forecast of the fuzzy load alternation numbers

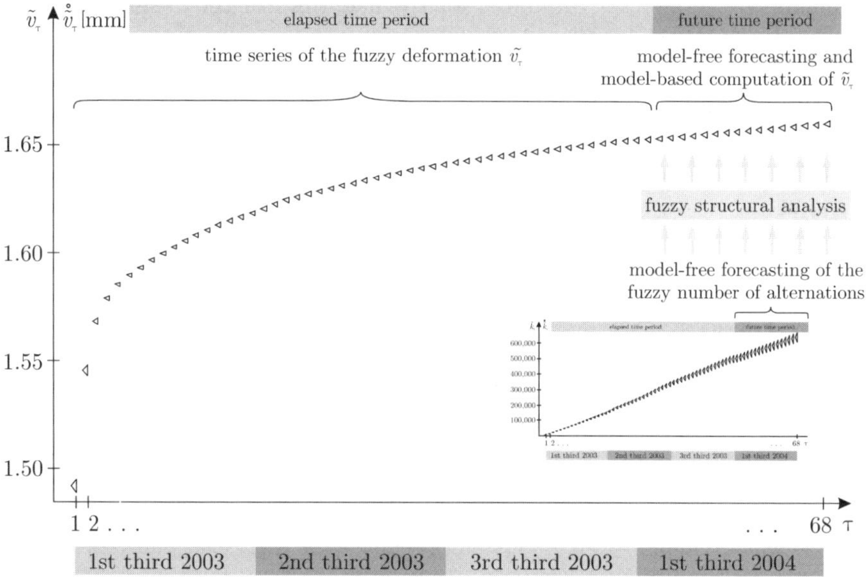

Fig. 5.37. Model-based optimum h-step forecast of the fuzzy asphalt deformations; comparison between Strategy I and Strategy II

If the fuzzy load alternation numbers are available as an optimum forecast or a fuzzy forecast interval, the computational model represents the deterministic fundamental solution of the fuzzy structural analysis. If the fuzzy load alternation numbers during the forecasting period are available as a fuzzy random forecast, the computational model represents the deterministic fundamental solution of the fuzzy stochastic structural analysis.

In order to validate the model-based forecast (Strategy I) the fuzzy deformations $\tilde{v}_\tau$ for the 1st third of 2004 were forecasted according to Strategy II. With fuzzy load alternation numbers as input data, the fuzzy deformations $\tilde{v}_\tau$ computed for 2003 by fuzzy structural analysis are shown in Fig. 5.37. These constitute a fuzzy time series, which serves as a basis for forecasting the deformations during the 1st third of 2004. The direct forecast was performed with the aid of an artificial neural network for fuzzy variables. The values forecasted according to Strategy I as well as Strategy II are in close agreement with the measured fuzzy deformations (see Fig. 5.37).

Forecasting of Structural Damage

The following example demonstrates a model-based forecast (Strategy I) of damage to a structure. The structural damage of the plate-beam floor shown in Fig. 5.38 is assessed by means of a global fuzzy damage indicator $\tilde{D}$.

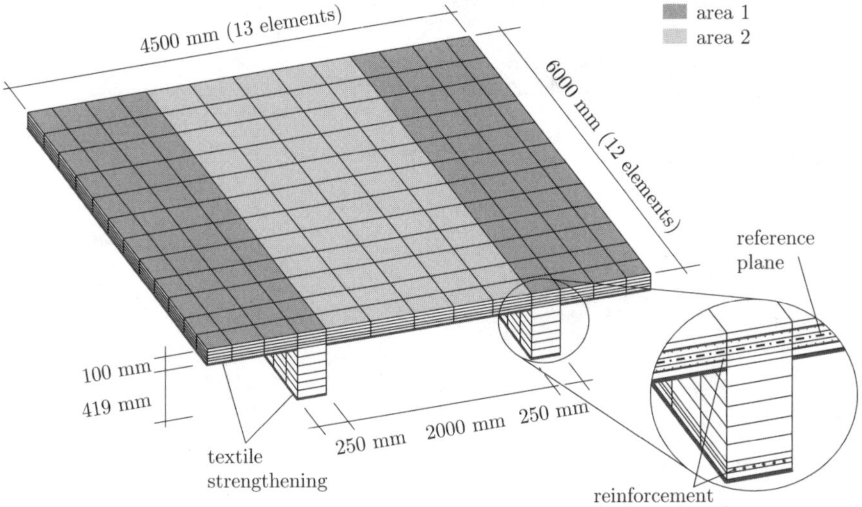

Fig. 5.38. Finite element model of the plate-beam floor

The global fuzzy damage indicator $\tilde{D}$ is defined by Eq. (5.17) (see e.g. [45] and [40, 26, 51]).

$$\tilde{D} = 1 - \frac{\text{stiffness of the damaged structure}}{\text{stiffness of the undamaged structure}} \tag{5.17}$$

The stiffness of the structure is assessed with the aid of the fuzzy determinant of the global tangential fuzzy system stiffness matrix $\underline{\tilde{K}}_T(\tau, \underline{\tilde{v}}, \tilde{s})$. The fuzzy system stiffness matrix $\underline{\tilde{K}}_T(\tau, \underline{\tilde{v}}, \tilde{s})$ is dependent on time τ, the fuzzy displacement state $\underline{\tilde{v}}$ and the fuzzy damage $\tilde{s}$. The global fuzzy damage indicator according to Eq. (5.17) is thus a fuzzy damage indicator. This indicator is given by Eq. (5.18). The term $\underline{\tilde{K}}_T(\tau_0, \underline{\tilde{v}}_0, \tilde{s}_0)$ is hereby the fuzzy system stiffness matrix at the reference time $\tau = \tau_0$.

$$\tilde{D}_{\underline{K}} = 1 - \frac{\det\left[\underline{\tilde{K}}_T(\tau, \underline{\tilde{v}}, \tilde{s})\right]}{\det\left[\underline{\tilde{K}}_T(\tau_0, \underline{\tilde{v}}_0, \tilde{s}_0)\right]} \tag{5.18}$$

In the following the fuzzy damage indicator $\tilde{D}_{\underline{K}}$ is forecasted for a plate-beam floor of an existing building. It is proposed to refurbish the existing building and use it as a warehouse from December 2006 onwards. Strengthening of the plate-beam floor by means of a textile-reinforced fine concrete layer on the underside of the floor is planned as a refurbishment measure. The future live load on the plate-beam floor is determined by future storage requirements. The storage requirement in the past is given in the form of a non-stationary fuzzy time series. The fuzzy time series of the live load $\tilde{p}$ corresponding to the monthly storage requirement is shown in Fig. 5.39.

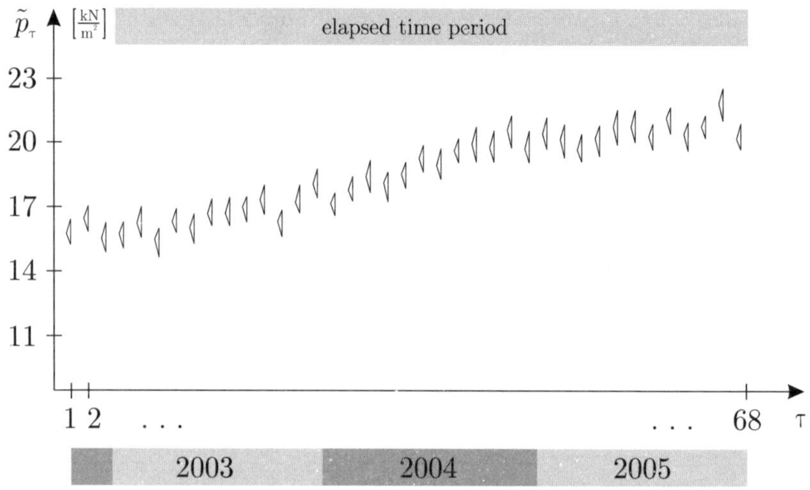

Fig. 5.39. Time series of the fuzzy live load $\tilde{p}$

In order to forecast the fuzzy live load a fuzzy ARMA[4, 4] process is taken to be the underlying process of the fuzzy time series shown in Fig. 5.39. The corresponding parameters are estimated by means of the gradient method outlined in Sect. 3.5.6. With the aid of the matched fuzzy ARMA[4, 4] process it is possible to determine optimum forecasts $\mathring{\tilde{p}}_{N+h}$ according to Sect. 4.2.1, fuzzy forecast intervals $\tilde{p}^{\kappa}_{N+h}$ according to Sect. 4.2.2, and fuzzy random forecasts $\vec{\tilde{\mathrm{P}}}_{N+h}$ according to Sect. 4.3.3 for the live load $\tilde{p}$. The optimum 12-step forecast is shown in Fig. 5.41. For the other two forecast objectives,

$s = 100$ potential realizations of the future progression of the fuzzy time series up to December 2006 are simulated and represented in tabular form. The tabular representation consists of twelve columns and 100 rows. One column is assigned to each month of 2006. The 100 potential realizations of the corresponding fuzzy random variable $\vec{\tilde{\mathrm{P}}}_{N+h}$ are listed in rows for each month. Because the simulated fuzzy variables $\vec{\tilde{p}}_{N+h}$ are fuzzy triangular numbers, the abbreviated notation $\vec{\tilde{p}}_{N+h} = (\vec{p}_l; l; r)_{LR}$ according to Sect. 2.1 is used in the tabular representation. An excerpt of the realizations is given in Table 5.4. Ten fuzzy realizations $\vec{\tilde{p}}_{N+12}$ of the fuzzy random live load $\vec{\tilde{\mathrm{P}}}_{N+12}$ for December 2006 are shown by way of example in Fig. 5.40.

Table 5.4. Excerpt of simulated future progressions of the fuzzy live load $\tilde{p}$

...	September 2006	October 2006	...
	⋮	⋮	
...	$\vec{\tilde{p}}^{k}_{N+9} = (19.8; 20.3; 20.8)_{LR}\ \frac{\mathrm{kN}}{\mathrm{m}^2}$	$\vec{\tilde{p}}^{k}_{N+10} = (20.5; 20.9; 21.6)_{LR}\ \frac{\mathrm{kN}}{\mathrm{m}^2}$	...
...	$\vec{\tilde{p}}^{k+1}_{N+9} = (19.4; 19.9; 20.3)_{LR}\ \frac{\mathrm{kN}}{\mathrm{m}^2}$	$\vec{\tilde{p}}^{k+1}_{N+10} = (19.8; 20.4; 20.8)_{LR}\ \frac{\mathrm{kN}}{\mathrm{m}^2}$	...
...	$\vec{\tilde{p}}^{k+2}_{N+9} = (21.1; 21.5; 22.3)_{LR}\ \frac{\mathrm{kN}}{\mathrm{m}^2}$	$\vec{\tilde{p}}^{k+2}_{N+10} = (19.9; 20.4; 21.1)_{LR}\ \frac{\mathrm{kN}}{\mathrm{m}^2}$	...
	⋮	⋮	

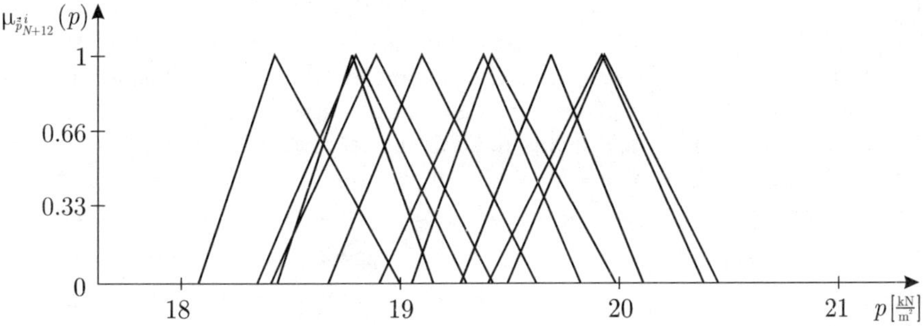

Fig. 5.40. Realizations $\vec{\tilde{p}}^{\,i}_{N+12}$ $(i = 1, 2, ..., 10)$ of the model-free forecast of the fuzzy random live load $\vec{\tilde{\mathrm{P}}}_{N+12}$

With the aid of the 100 simulated future progressions the fuzzy forecast intervals are estimated according to Sect. 4.2.2. These are shown for 2006 for a confidence level of $\kappa = 0.95$ in Fig. 5.41.

The 100 simulated future progressions also provide the basis for a fuzzy random forecast of the live load. It is not necessary, however, to express the fuzzy random variables $\vec{\tilde{\mathrm{P}}}_{N+h}$ of the live load by fuzzy probability distribution

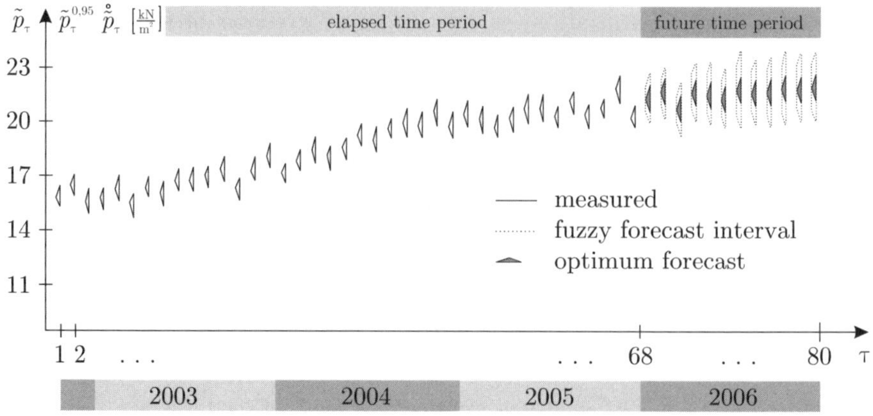

Fig. 5.41. Optimum 12-step forecast and fuzzy forecast intervals of the fuzzy live load

functions (form II). The simulated fuzzy values presented in Table 5.4 may be directly applied for the model-based forecast according to Strategy I.

Once the live load has been forecasted as a measurable action for different forecast objectives, the non-measurable system response 'damage indicator' may be computed according to Strategy I (see Fig. 5.3) by means of a computational model. It is intended to forecast the damage indicator as a fuzzy random variable according to Sect. 5.2. In order to assess the damage to the plate-beam floor (Fig. 5.38) due to the action of the forecasted fuzzy random live load $\vec{\tilde{\mathrm{P}}}_{N+12}$ in December, a model-based fuzzy random forecast (Strategy I) is performed. Variant II is chosen for the fuzzy stochastic structural analysis. The outer loop of the fuzzy stochastic structural analysis (Variant II), which constitutes the stochastic analysis, requires the simulation of s sequences of fuzzy realizations of the input variables. A total of $s = 100$ fuzzy realizations $\vec{\tilde{p}}_{N+12}$ are already available in tabular form (Table 5.4) for the fuzzy random input variable $\vec{\tilde{\mathrm{P}}}_{N+12}$. The Monte Carlo simulation of the fuzzy random live load $\vec{\tilde{\mathrm{P}}}_{N+12}$ is hence reduced to the extraction of elements from the table.

The middle loop of the fuzzy stochastic structural analysis Variant II (see Fig. 5.8) constitutes the fuzzy structural analysis. The input variable for the fuzzy structural analysis (which must be performed s-times) is a fuzzy realization $\vec{\tilde{p}}_{N+12}$ of the fuzzy random live load $\vec{\tilde{\mathrm{P}}}_{N+12}$ in each case. In order to take account of different loading situations the plate-beam floor is subdivided into two domains (see Fig. 5.38) in which stochastically independent loads occur. The live load is modeled in each case by a different fuzzy realization $\vec{\tilde{p}}_{N+12}$ for each domain. Moreover, the concrete compressive strength of the existing structure is modeled as a fuzzy triangular number

$\tilde{\beta}_C = (21.0; 30.0; 31.5)_{LR} \frac{\text{N}}{\text{mm}^2}$. The input space of the fuzzy structural analysis is thus formed by three fuzzy variables.

The deterministic fundamental solution is the finite element program FALT-FEM [43]. The finite element modeling is carried out using 156 layered hybrid elements with assumed stress distribution. This represents the modeling of a multilayered continuum, which is subsequently analyzed using the multi-reference-plane model after [42]. The plate is modeled by five concrete layers, and the beam, by twelve concrete layers. The steel reinforcement is specified as a uniaxial smeared layer. The textile strengthening on the underside of the floor is modeled as an additional fine concrete layer with textile reinforcement. The textile reinforcement is also specified as a uniaxial smeared layer. The physical nonlinear analysis is performed on the basis of endochronic material laws for concrete and steel (see [23]). Dead weight, crack formation, tension stiffening and steel yielding are also taken into consideration. The finite element model of the plate-beam floor is shown in Fig. 5.38. The material parameters adopted in the analysis are listed in Table 5.5.

Table 5.5. Material parameters for the reinforcement layers

Steel layer	**Textile layer**
Thickness	
$d_{S_{Pl}} =$ 0.221 mm (plate, longitudinal, top/bottom)	$d_G =$ 0.2 mm
$d_{S_{Pq}} =$ 0.050 mm (plate, transverse, top/bottom)	
$d_{S_B} =$ 4.020 mm (beam, bottom)	
Young's modulus	
$E_S = 210\,000 \frac{\text{N}}{\text{mm}^2}$	$E_G = 74\,000 \frac{\text{N}}{\text{mm}^2}$
Tensile strength	
$R_S = 500 \frac{\text{N}}{\text{mm}^2}$	$R_G = 1\,400 \frac{\text{N}}{\text{mm}^2}$

As a result of the fuzzy stochastic structural analysis (Variant II), a set of 100 fuzzy damage indicators are obtained in each case for the non-strengthened and strengthened plate-beam floor. A statistical evaluation of the latter yields a fuzzy random variable $\tilde{\vec{\text{D}}}$ for the damage indicator. Eight fuzzy realizations of the damage indicator for the non-strengthened and strengthened plate are shown by way of example in each case in Fig. 5.42.

Based on the samples of the 100 computed fuzzy damage indicators in each case for the non-strengthened and strengthened plate the fuzzy random variables $\tilde{\vec{\text{D}}}_u$ and $\tilde{\vec{\text{D}}}_v$ for the non-strengthened and strengthened plate-beam floor may be estimated for the month of December 2006. The respective empirical fuzzy probability distribution functions (form II) are determined with the aid of the statistical methods outlined in Sect. 2.2.2. With the aid of the two em-

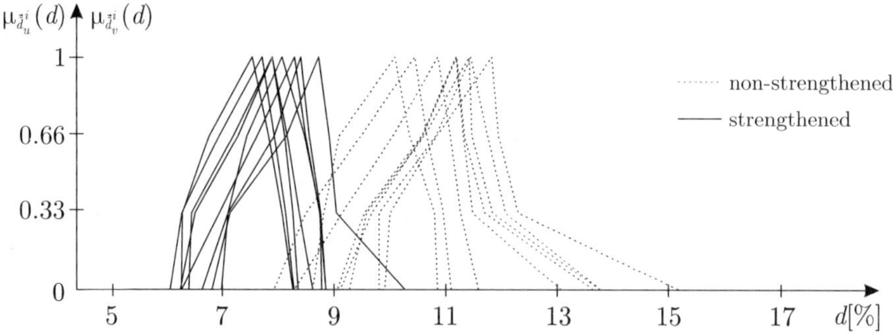

Fig. 5.42. Fuzzy realizations $\vec{\tilde{d}}_u^{\,i}$ and $\vec{\tilde{d}}_v^{\,i}$ ($i = 1, 2, ..., 8$) of the fuzzy random variables $\vec{\tilde{D}}_u$ and $\vec{\tilde{D}}_v$ of the non-strengthened and strengthened structure, respectively

pirical fuzzy probability distribution functions (form I) $\hat{\tilde{F}}_{\vec{\tilde{D}}_u}(d)$ and $\hat{\tilde{F}}_{\vec{\tilde{D}}_v}(d)$, the simplified graphical representation of the result is given in Fig. 5.43.

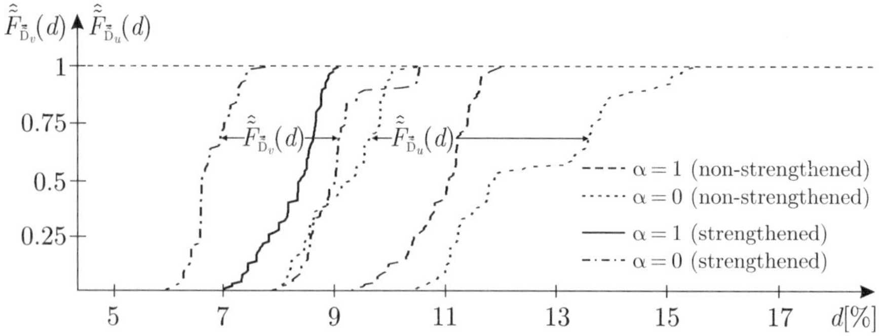

Fig. 5.43. $\hat{\tilde{F}}_{\vec{\tilde{D}}_u}(d)$ and $\hat{\tilde{F}}_{\vec{\tilde{D}}_v}(d)$ as empirical fuzzy probability distribution functions form I for the non-strengthened and strengthened structure, respectively

A comparison between the two fuzzy probability distribution functions (form I) $\hat{\tilde{F}}_{\vec{\tilde{D}}_u}(d)$ and $\hat{\tilde{F}}_{\vec{\tilde{D}}_v}(d)$ shows that the forecasted damage of the strengthened plate-beam floor is much less. The forecasted damage of the strengthened structure lies between 6 and 10 percent. The forecasted damage without strengthening lies between 8 and 15 percent. A comparison between the fuzzy probability distribution functions (form I) $\hat{\tilde{F}}_{\vec{\tilde{D}}_u}(d)$ and $\hat{\tilde{F}}_{\vec{\tilde{D}}_v}(d)$ in the upper range of the functional values clearly indicates a much larger distance between the right-hand and left-hand boundary function of $\hat{\tilde{F}}_{\vec{\tilde{D}}_u}(d)$. This means that the forecasted fuzzy random damage $\vec{\tilde{D}}_u$ of the non-strengthened structure exhibits a much greater uncertainty. This result is plausible. The plate-beam floor without textile strengthening has a lower system resistance. Owing to nonlinear material properties (e.g. progressive crack formation in the concrete) under comparable loading, this results in overproportional damage. The decisive kink in the plot of the right-hand boundary function corresponds to a functional value of about 0.5, i.e. the probability of occurrence of the overproportional damage is about 50 percent. The effect of the overproportional damage of the strengthened structure is far slighter. The decisive kink in the plot of the right-hand boundary function of $\tilde{\tilde{F}}_{\tilde{\vec{\mathrm{D}}}_v}(d)$ first occurs for a functional value of 0.9. Moreover, the distance between the left-hand and right-hand boundary function in this region is significantly smaller. This result clearly demonstrates the strengthening effect of the textile-reinforced fine concrete layer.

References

[1] Abidi SSR, Goh A (1998) Neural network based forecasting of bacteria-antibiotic interactions for infectious disease control. In: Cesnik B, McCray A, Scherrer J (eds) 9th World Congress on Medical Informatics (MedInfo 1998), Seoul, IOS Press, Amsterdam, pp 741–751

[2] Akaike H (1977) On entropy maximization principle. In: Krishnaiah PR (ed) Applications of Statistics, North Holland, Amsterdam, pp 27–41

[3] Bandemer H, Näther W (1992) Fuzzy Data Analysis. Kluwer Academic Publishers, Boston London

[4] Belew R, McInerney J, Schraudolph N (1992) Evolving networks: Using the genetic algorithm with connectionist learning. In: Langton C, Taylor C, Farmer J, Rasmussen S (eds) Artificial Life II, Addison Wesley, Redwood City, pp 511–547

[5] Bothe HH (1993) Fuzzy Logic. Springer, Berlin Heidelberg New York

[6] Box GEP, Jenkins GM, Reinsel GC (1994) Time Series Analysis, Forecasting and Control, 3rd edn. Prentice-Hall, Englewood Cliffs

[7] Brockwell PJ, Davis RA (1991) Time Series: Theory and Methods, 2nd edn. Springer, Berlin Heidelberg New York

[8] Brockwell PJ, Davis RA (1996) Introduction to Time Series and Forecasting. Springer, New York Berlin Heidelberg

[9] Chakraborty K, Mehrotra K, Mohan CK, Ranka S (1992) Forecasting the behaviour of multivariate time series using neural networks. Neural Networks (5):961–970

[10] Dallmann H, Elster KH (1991) Einführung in die höhere Mathematik Mathematik Band III, 2nd edn. Gustav Fischer, Jena

[11] Dubois D, Prade H (1980) Fuzzy Sets and Systems: Theory and Applications. Academic Press, New York London

[12] Durbin J (1960) The fitting of time-series models. Revue de l'Institut International de Statistique (28):233–243

[13] Feuring T (1995) Fuzzy-Neuronale Netze - Von kooperativen über hybride zu fusionierten vage-konnektionistischen Systemen. Dissertation, Westfälische Wilhelms-Universität Münster, Münster

[14] Franke D, Arnold M, Bartl U, Vogt L (2003) Erddruckmessungen an der Kelleraußenwand eines mehrgeschossigen Massivbaus. Bauingenieur 78(3):125–130

[15] Gioffre M, Gusellav V, Grigoriu M (2000) Simulation of non-gaussian field applied to wind pressure fluctuations. Probabilistic Engineering Mechanics 15:339–345

[16] Goldberg DE (1989) Genetic Algorithms in Search, Optimization, and Machine Learning. Addison Wesley, Massachusetts

[17] Granger CWJ (1969) Investigating causal relations by econometric models and cross-spectral methods. Econometrica (37):24–36

[18] Granger CWJ, Morris MJ (1976) Time series modelling and interpretation. Journal of the Royal Statistical Society A (139):246–257

[19] Gupta MM, Rao DH (1994) On the principles of fuzzy neural networks. Fuzzy Sets and Systems 61(1):1–18

[20] Hareter D (2003) Zeitreihenanalyse mit unscharfen Daten. Dissertation, TU Wien, Wien

[21] Hausdorff F (1949) Grundzüge der Mengenlehre. Chelsea Publishing, New York

[22] Haykin S (1999) Neural Networks – A Comprehensive Foundation, 2nd edn. Prentice Hall, New Jersey

[23] Kluger J (1999) Numerische Modelle zur physikalisch nichtlinearen Analyse von Stahlbeton-Faltwerken. Dissertation, TU Dresden, Veröffentlichungen des Lehrstuhls für Statik, Heft 1, Dresden

[24] Körner R (1997) Linear Models with Random Fuzzy Variables. Disseration, Technische Universität Bergakademie Freiberg, Freiberg

[25] Koza J, Rice J (1991) Genetic generation of both the weights and architecture for a neural network. In: IEEE – International Joint Conference on Neural Networks, Seattle, vol II, pp 397–404

[26] Krätzig WB, Noh SY (1998) Computersimulation progressiver Schädigungsprozesse von Stahlbetonkonstruktionen. In: Wriggers P, Meißner U, Stein E, Wunderlich W (eds) FEM' 98 – Finite Elemente in der Baupraxis, Ernst & Sohn, Berlin, pp 123–132

[27] Kwakernaak H (1978) Fuzzy random variables – i. definitions and theorems. Information Sciences (15):1–29

[28] Kwakernaak H (1979) Fuzzy random variables – ii. algorithms and examples for the discrete case. Information Sciences (17):253–278

[29] Lau J (1991) Point Process Analysis Techniques: Theory and Application to Complex Neurophysiological Systems. Dissertation, University of Glasgow, Glasgow

[30] Lee A, Ulbricht C, Dorffner G (1999) Application of artificial neural networks for detection of abnormal fetal heart rate pattern: A comparison with conventional algorithms. Journal of Obstetrics and Gynaecology 19(5):482–485

[31] Loredana Z (2001) Use of neural networks in a model-based predictive control system for anaerobic digestion optimization. Dissertation, Universität für Bodenkultur Wien, Wien

[32] Lütkepohl H (1993) Introduction to Multiple Time Series Analysis, 2nd edn. Springer, Berlin Heidelberg New York

[33] Lütkepohl H (2005) New Introduction to Multiple Time Series Analysis. Springer, Berlin Heidelberg New York

[34] Masters T (1995) Neural, Novel & Hybrid Algorithms for Time Series Prediction. John Wiley & Sons Ltd., Chinchester New York Weinheim

[35] Möller B (2004) Fuzzy-randomness – a contribution to imprecise probability. ZAMM 84:754–764

[36] Möller B, Beer M (2004) Fuzzy Randomness – Uncertainty in Civil Engineering and Computational Mechanics, 1st edn. Springer, Berlin Heidelberg New York

[37] Möller B, Graf W, Beer M (2000) Fuzzy structural analysis using α-level optimization. Computational Mechanics (26):547–565

[38] Möller B, Sickert JU, Graf W (2001) Berücksichtigung räumlich verteilter unschärfe bei der numerischen simulation von textilbeton. In: Curbach M (ed) 2nd Colloquium on Textile Reinforced Structures (CTRS2), TU Dresden, Dresden, pp 435–446

[39] Möller B, Graf W, Beer M (2003) Safety assessment of structures in view of fuzzy randomness. Computers & Structures 81(15):1567–1582

[40] Möller B, Graf W, Hoffmann A, Steinigen F (2004) Numerical models concerning structures with multi-layered textile strengthening. In: Topping BHV, Mota Soares CA (eds) 7th International Conference on Computational Structures Technology, Civil-Comp Press, Stirling, Lissabon, pp 97–98, Abstract, Volltext (21 Seiten) CD–ROM

[41] Möller B, Beer M, Reuter U (2005) Theoretical basics of fuzzy randomness – application to time series with fuzzy data. In: Augusti G, Schueller GI, Ciampoli M (eds) Safety and Reliability of Engineering Systems and Structures - Proceedings of the 9th Int. Conference on Structural Safety and Reliability, Millpress, Rotterdam, pp 1701–1707

[42] Möller B, Graf W, Hoffmann A, Steinigen F (2005) Numerical simulation of rc-structures with textile reinforcement. Computers & Structures 83:1659–1688

[43] Müller H, Möller B (1985) Lineare und physikalisch nichtlineare Statik von Faltwerken. Schriftenreihe Bauforschung – Baupraxis, Heft 155, Verlag Bauinformation, Berlin

[44] Näther W, Körner R (2002) On the variance of random fuzzy variables. In: Bertoluzza C, Gil MA, Ralescu DA (eds) Statistical Modelling, Analysis and Management of Fuzzy Data, Physica-Verlag, Heidelberg, pp 25–42

[45] Nguyen SH (2003) Modellierung unscharfer zeitabhängiger Einflüsse als Fuzzy-Prozess - Anwendung auf die Schädigung von Tragwerken und auf die Simulation von Erdbeben. Dissertation, TU Dresden, Veröffentlichungen des Lehrstuhls für Statik, Heft 6, Dresden

[46] Niederreiter H (1992) Random Number Generation and Quasi-Monte Carlo Methods. Society for Industrial and Applied Mathematics, Philadelphia, Pennsylvania

[47] Nigussie TG (2005) Investigation on Sediment Transport Characteristics and Impacts of Human Activities on Morphological Processes of Ethiopian Rivers: Case Study of Kulfo River, Southern Ethiopia. Dissertation, TU Dresden, Dresdner Wasserbauliche Mitteilungen, Heft 30, Dresden

[48] Oeser M (2004) Numerische Simulation des nichtlinearen Verhaltens flexibler mehrschichtiger Verkehrswegebefestigungen. Dissertation, TU Dresden, Veröffentlichungen des Lehrstuhls für Statik, Heft 7, Dresden

[49] Ohno-Machado L (1996) Medical Applications of Artificial Neural Networks: Connectionist Model of Survival. Dissertation, Stanford University, Standford

[50] Park SK, Miller KW (1988) Random number generators: good ones are hard to find. Communications of the ACM 31(10):1192–1201

[51] Petryna Y (2004) Schädigung, Versagen und Zuverlässigkeit von Tragwerken des konstruktiven Ingenieurbaus. Schriftenreihe des Instituts für Konstruktiven Ingenieurbau, Heft 2, Ruhr-Universität Bochum, Bochum

[52] Pönitz S, Schaller MB (2002) Ergebnisbericht über die Inklinometer- und Extensometermessungen im Messzeitraum Dezember 1998 bis November 2002 vom 26.11.2002 von der Gesellschaft für Geomechanik und Baumesstechnik mbH Espenhain. Gesellschaft für Geomechanik und Baumesstechnik mbH, Espenhain

[53] Puri ML, Ralescu D (1986) Fuzzy random variables. Journal of Mathematical Analysis and Applications (114):409–422

[54] Refenes AP (1995) Neural Networks in the Capital Markets, 1st edn. John Wiley & Sons Ltd., Chinchester New York Weinheim

[55] Rehkugler H, Kerling M (1995) Einsatz neuronaler netze für analyse- und prognose-zwecke. In: Matschke MJ, Sieben G, Schildbach T (eds) Betriebswirtschaftliche Forschung und Praxis. Heft 3, Verlag Neue Wirtschafts-Briefe, Herne Berlin, pp 306–324

[56] Rissanen J (1978) Modelling by shortest data description. Automatica (14):465–471

[57] Rubinstein RY (1981) Simulation and the Monte Carlo Method. John Wiley & Sons Ltd., New York

[58] Rumelhart DE, Hinton GE, Williams RJ (1986) Learning internal representations by error propagation. In: Rumelhart DE, McClelland JL (eds) Parallel Distributed Processing: Explorations in the Microstructure of Cognition, MIT Press, Cambridge, MA, vol 1, pp 318–362

[59] Schlittgen R (2001) Angewandte Zeitreihenanalyse. Oldenbourg, München Wien

[60] Schlittgen R, Streitberg B (2001) Zeitreihenanalyse, 9th edn. Oldenbourg, München Wien

[61] Schwarz G (1978) Estimating the dimension of a model. Annals of Statistics (6):461–464

[62] Sickert JU, Beer M, Graf W, Möller B (2003) Fuzzy probabilistic structural analysis considering fuzzy random functions. In: 9th International Conference on Applications of Statistics and Probability in Civil Engineering, Millpress, Rotterdam, pp 379–386

[63] Sickert JU, Graf W, Reuter U (2005) Application of fuzzy randomness to time-dependent reliability. In: Augusti G, Schueller GI, Ciampoli M (eds) Safety and Reliability of Engineering Systems and Structures - Proceedings of the 9th Int. Conference on Structural Safety and Reliability, Millpress, Rotterdam, pp 1709–1716

[64] Spaethe G (1992) Die Sicherheit tragender Baukonstruktionen, 2nd edn. Springer, Wien New York

[65] Stralkowsky CM, Wu SM, DeVor RE (1974) Charts for the interpretation and estimation of the second order moving average and mixed first order autoregressive-moving average models. Technometrics (16):275–285

[66] Viertl R, Hareter D (2004) Generalized bayes' theorem for non-precise a-priori distribution. Metrika (59):263–273

[67] Werkmeister S, Steven B, Alabaster D, Arnold G, Oeser M (2005) 3d finite element analysis of accelerated pavement test results from new zealand's captif facility. In: Horvli I (ed) Proceedings of the 7th International Conference on the Bearing Capacity of Roads, Railways and Airfields, BCA'05, Trondheim, pp 363–402

[68] Wilson G (1967) Factorization of the covariance generating function of a pure moving average process. SIAM Journal of Numerical Analysis (6):1–7

[69] World Meteorological Organization (ed) (2005) Protecting the ozone layer – A priority for WMO. World Meteorological Organization, Geneva

[70] Zell A (2000) Simulation Neuronaler Netze, 3rd edn. Oldenbourg, München Wien

[71] Zimmermann HJ (1992) Fuzzy set theory and its applications. Kluwer Academic Publishers, Boston London

Index